Geophysical Monograph Series

Including

IUGG Volumes

Maurice Ewing Volumes

Mineral Physics Volumes

GEOPHYSICAL MONOGRAPH SERIES

Geophysical Monograph Volumes

41 **Gondwana Six: Stratigraphy, Sedimentology, and Paleontology** *Garry D. McKenzie (Ed.)*
42 **Flow and Transport Through Unsaturated Fractured Rock** *Daniel D. Evans and Thomas J. Nicholson (Eds.)*
43 **Seamounts, Islands, and Atolls** *Barbara H. Keating, Patricia Fryer, Rodey Batiza, and George W. Boehlert (Eds.)*
44 **Modeling Magnetospheric Plasma** *T. E. Moore and J. H. Waite, Jr. (Eds.)*
45 **Perovskite: A Structure of Great Interest to Geophysics and Materials Science** *Alexandra Navrotsky and Donald J. Weidner (Eds.)*
46 **Structure and Dynamics of Earth's Deep Interior (IUGG Volume 1)** *D. E. Smylie and Raymond Hide (Eds.)*
47 **Hydrological Regimes and Their Subsurface Thermal Effects (IUGG Volume 2)** *Alan E. Beck, Grant Garven, and Lajos Stegena (Eds.)*
48 **Origin and Evolution of Sedimentary Basins and Their Energy and Mineral Resources (IUGG Volume 3)** *Raymond A. Price (Ed.)*
49 **Slow Deformation and Transmission of Stress in the Earth (IUGG Volume 4)** *Steven C. Cohen and Petr Vaníček (Eds.)*
50 **Deep Structure and Past Kinematics of Accreted Terranes (IUGG Volume 5)** *John W. Hillhouse (Ed.)*
51 **Properties and Processes of Earth's Lower Crust (IUGG Volume 6)** *Robert F. Mereu, Stephan Mueller, and David M. Fountain (Eds.)*
52 **Understanding Climate Change (IUGG Volume 7)** *Andre L. Berger, Robert E. Dickinson, and J. Kidson (Eds.)*
53 **Plasma Waves and Istabilities at Comets and in Magnetospheres** *Bruce T. Tsurutani and Hiroshi Oya (Eds.)*
54 **Solar System Plasma Physics** *J. H. Waite, Jr., J. L. Burch, and R. L. Moore (Eds.)*
55 **Aspects of Climate Variability in the Pacific and Western Americas** *David H. Peterson (Ed.)*
56 **The Brittle-Ductile Transition in Rocks** *A. G. Duba, W. B. Durham, J. W. Handin, and H. F. Wang (Eds.)*
57 **Evolution of Mid Ocean Ridges (IUGG Volume 8)** *John M. Sinton (Ed.)*
58 **Physics of Magnetic Flux Ropes** *C. T. Russell, E. R. Priest, and L. C. Lee (Eds.)*
59 **Variations in Earth Rotation (IUGG Volume 9)** *Dennis D. McCarthy and Williams E. Carter (Eds.)*
60 **Quo Vadimus *Geophysics for the Next Generation* (IUGG Volume 10)** *George D. Garland and John R. Apel (Eds.)*
61 **Cometary Plasma Processes** *Alan D. Johnstone (Ed.)*
62 **Modeling Magnetospheric Plasma Processes** *Gordon K. Wilson (Ed.)*
63 **Marine Particles Analysis and Characterization** *David C. Hurd and Derek W. Spencer (Eds.)*
64 **Magnetospheric Substorms** *Joseph R. Kan, Thomas A. Potemra, Susumu Kokubun, and Takesi Iijima (Eds.)*
65 **Explosion Source Phenomenology** *Steven R. Taylor, Howard J. Patton, and Paul G. Richards (Eds.)*
66 **Venus and Mars: Atmospheres, Ionospheres, and Solar Wind Interactions** *Janet G. Luhmann, Mariella Tatrallyay, and Robert O. Pepin (Eds.)*
67 **High-Pressure Research: Application to Earth and Planetary Sciences (Mineral Physics Volume 3)** *Yasuhiko Syono and Murli H. Manghnani (Eds.)*
68 **Microwave Remote Sensing of Sea Ice** *Frank Carsey, Roger Barry, Josefino Comiso, D. Andrew Rothrock, Robert Shuchman, W. Terry Tucker, Wilford Weeks, and Dale Winebrenner*
69 **Sea Level Changes: Determination and Effects (IUGG Volume II)** *P. L. Woodworth, D. T. Pugh, J. G. DeRonde, R. G. Warrick, and J. Hannah*
70 **Synthesis of Results from Scientific Drilling in the Indian Ocean** *Robert A. Duncan, David K. Rea, Robert B. Kidd, Ulrich von Rad, and Jeffrey K. Weisset (Eds.)*
71 **Mantle Flow and Melt Generation at Mid-Ocean Ridges** *Jason Phipps Morgan, Donna K. Blackman, and John M. Sinton (Eds.)*
72 **Dynamics of Earth's Deep Interior and Earth Rotation** *J.-L. Le Mouël, D. E. Smylie, and T. Herring (Eds.)*

Maurice Ewing Volumes

1 **Island Arcs, Deep Sea Trenches, and Back-Arc Basins** *Manik Talwani and Walter C. Pitman III (Eds.)*
2 **Deep Drilling Results in the Atlantic Ocean: Ocean Crust** *Manik Talwani, Christopher G. Harrison, and Dennis E. Hayes (Eds.)*
3 **Deep Drilling Results in the Atlantic Ocean: Continental Margins and Paleoenvironment** *Manik Talwani, William Hay, and William B. F. Ryan (Eds.)*
4 **Earthquake Prediction—An International Review** *David W. Simpson and Paul G. Richards (Eds.)*
5 **Climate Processes and Climate Sensitivity** *James E. Hansen and Taro Takahashi (Eds.)*
6 **Earthquake Source Mechanics** *Shamita Das, John Boatwright, and Christopher H. Scholz (Eds.)*

IUGG Volumes

1 **Structure and Dynamics of Earth's Deep Interior** *D. E. Smylie and Raymond Hide (Eds.)*
2 **Hydrological Regimes and Their Subsurface Thermal Effects** *Alan E. Beck, Grant Garven, and Lajos Stegena (Eds.)*
3 **Origin and Evolution of Sedimentary Basins and Their Energy and Mineral Resources** *Raymond A. Price (Ed.)*
4 **Slow Deformation and Transmission of Stress in the Earth** *Steven C. Cohen and Petr Vaníček (Eds.)*
5 **Deep Structure and Past Kinematics of Accreted Terrances** *John W. Hillhouse (Ed.)*
6 **Properties and Processes of Earth's Lower Crust** *Robert F. Mereu, Stephan Mueller, and David M. Fountain (Eds.)*
7 **Understanding Climate Change** *Andre L. Berger, Robert E. Dickinson, and J. Kidson (Eds.)*
8 **Evolution of Mid Ocean Ridges** *John M. Sinton (Ed.)*
9 **Variations in Earth Rotation** *Dennis D. McCarthy and William E. Carter (Eds.)*
10 **Quo Vadimus Geophysics for the Next Generation** *George D. Garland and John R. Apel (Eds.)*
11 **Sea Level Changes: Determinations and Effects** *P. L. Woodworth, D. T. Pugh, J. G. DeRonde, R. G. Warrick, and J. Hannah (Eds.)*
12 **Dynamics of Earth's Deep Interior and Earth Rotation** *J.-L. Le Mouël, D. E. Smylie, and T. Herring (Eds.)*

Mineral Physics Volumes

1 **Point Defects in Minerals** *Robert N. Schock (Ed.)*
2 **High Pressure Research in Mineral Physics** *Murli H. Manghnani and Yasuhiko Syona (Eds.)*
3 **High Pressure Research: Application to Earth and Planetary Sciences** *Yasuhiko Syono and Murli H. Manghnani (Eds.)*

Geophysical Monograph 73
IUGG Volume 13

Environmental Effects on Spacecraft Positioning and Trajectories

A. Vallance Jones
Editor

American Geophysical Union
International Union of Geodesy and Geophysics

Published under the aegis of the AGU Books Board.

Library of Congress Cataloging-in-Publication Data

Environmental effects on spacecraft positioning and trajectories / A. Vallance Jones, editor.
p. cm. — (Geophysical monograph ; 73) (IUGG ; v. 13)
Based on papers presented at a Union Symposium held at the Twentieth General Assembly of the International union of Geodesy and Geophysics, which took place in Vienna in August 1991.
ISBN 0-87590-464-5
1. Space trajectories—Congresses. 2. Space vehicles—Tracking--Congresses. 3. Space vehicles—Aerodynamics—Congresses.
I. Jones, A. Vallance (Alister Vallance), 1924-
II. International Union of Geodesy and Geophysics. III. Series.
IV. Series: IUGG (Series) ; v. 13.
TL 1075.E55 1993
629.4'1—dc20 93-7793
CIP

ISSN: 0065-8448
ISBN 0-87590-464-5

Printed in the United States of America.

CONTENTS

PREFACE

This book combines a presentation of the remarkable progress that has been made in techniques for determining the exact positions of satellites, with an account of the basic science of the atmospheric environment affecting both satellite trajectories and position measurements. Recent outstanding advances make it possible to measure spacecraft and ground point positions with accuracies of the order of a few centimeters or better. Such measurements are of growing importance in navigational applications as well as in many related problems such as gravitational field determinations, sea surface topography, earth crustal movements and studies of earth rotation.

Conditions in both the ionized and the neutral components of the earth's atmosphere affect the accuracy with which the position of spacecraft can be measured by radio and optical techniques. A related problem is the very accurate prediction of spacecraft trajectories so that positions can be predicted or interpolated where direct measurements are not available. Longer-term predictions of the evolution of spacecraft orbits are often required. In making such predictions it is essential to take into account fluctuations in drag forces due to geomagnetically driven changes in atmospheric density and winds. Other environmental effects which must be considered are gravity waves, radiation pressure and thermal and electric drag.

While some modern position measuring techniques succeed in cancelling out almost all the error introduced by the atmosphere, an understanding of the nature and magnitude of environmental effects is important. With other techniques corrections for errors must be made, and consequently information about atmospheric and ionospheric conditions is important and must be provided by supplementary measurements or else derived from empirical or basic models. Such models are also very important for longer-term predictions of satellite trajectories. For these reasons, this book pays considerable attention to the present status of models of the ionosphere and the neutral atmosphere. The theory of drag forces on satellites is another important topic considered.

The book is based on papers presented at a Union Symposium held at the Twentieth General Assembly of the International Union of Geodesy and Geophysics, which took place in Vienna in August 1991. A major component of the book is a set of invited review papers by leading specialists in different aspects of the subject. All these review papers were adapted by their authors for publication in this book. In addition, manuscripts were prepared for inclusion by almost all the authors of contributed papers. These were presented in poster format at the symposium. In the book, the contributed papers are included as far as possible in a logical sequence among the review articles.

The first article, by Thomas P. Yunck, describes the effects of the ionosphere and neutral atmosphere on the radio signals employed in some modern positioning systems, as well as the effect of the neutral atmosphere on optical signals employed in laser sounding systems. Techniques to correct or mitigate the effects of the atmosphere and ionosphere are reviewed, and examples of present and expected future achievements are presented. The second article, by Shaun Quegan, is devoted to the status of attempts to model and predict conditions in the ionosphere. Both empirical and physically based models are discussed. Models of the latter type attempt to trace the interaction of the solar wind with the magnetosphere via electric fields and particle precipitation, and to predict large- and small-scale perturbations of the ionosphere and thermosphere and their effect on radio positioning signals. The article by R. Leitinger discusses the difficult problem of correcting for horizontal gradients in ionization. The following article, by Yinn-Nien Huang, describes a method of obtaining charts of radio propagation delays from combined observations of polar and geostationary satellites. Next, Tetsuro Kondo and Michito Imae describe a technique whereby total electron content measurements, derived from mapping the signals from GPS satellites, are used to obtain the total electron content needed to correct VLBI measurements. P. Escudier, N. Picot, and O. Z. Zanife discuss a method of making a global model of total electron content using DORIS bifrequency Doppler positioning measurements. This model will be used to correct POSEIDON single-frequency radar altimeter data. The next article, by G. K. Hartmann, deals with tropospheric effects on modern radio astronomical observations and satellite geodesy using radio techniques for which tropospheric constituents, particularly water vapor, constitute the limiting source of error. Current methods of correcting for this error are discussed. T. L. Killeen, A. G.

Burns, R. M. Johnson and F. A. Marcos provide an authoritative treatment of the present status of different types of models of upper atmospheric densities, tides and winds. The article by J. C. Ries, C. K. Shum and B. D. Tapley provides the main discussion of the effect of atmospheric and other surface forces on satellite orbits. These are primarily atmospheric drag and solar radiation pressure, although for the most accurate measurements, earth radiation pressure and the effect of the re-emission of heat by the satellite can be significant. Satellite drag effects in particular cases are discussed in next three articles. The first, by F. Nouël, J. P. Berthias, A. Guitart, A. Piuzzi and C. Valorge, deals with drag effects on the SPOT2 satellite as measured by the DORIS system; the second, by François Barlier and François Mignard, discusses various drag forces effective at very high altitudes; and the third, by Christiane Berger and François Barlier, deals with the influence of thermospheric drag, winds and gravity waves on a satellite designed to study small irregularities in the earth's gravitational field. The article by M. N. Vlasov provides one example of an attempt to set up a numerical time-dependent upper atmosphere density-temperature-composition model that could be used to predict the evolution of satellite orbits. The final article, by Rita C. Sagalyn and Sidney A. Bowhill, discusses progress in the short- and medium-term prediction of geomagnetic storms which are ultimately responsible for the effects described in many of the preceding papers.

It should be emphasized that the volume consists of manuscripts written independently by 14 different authors and consequently there is some overlap in the material presented, as well as some differences in the details of the equations and the symbols used. The variety in the approaches adopted where there is overlap in the review papers may be helpful in providing a deeper insight into the topics discussed.

The subjects for the invited review papers were largely chosen by the convenors to provide broad coverage of the original symposium topic. The range of contributed papers provides some further indication of the topics that are under active study at present, although not all the possibly related areas of study are covered to the same degree.

The original symposium was organized by a committee consisting of myself as convenor and William Melbourne, Jo Ann Joselyn, François Barlier, Karen Labitzke and Michael Parke as co-convenors representing the four component associations of IUGG concerned with geodesy, aeronomy and geophysics, meteorology and atmospheric physics, and the physical sciences of the oceans.

All the papers appearing in this book were reviewed by at least two referees who in many cases provided extremely useful constructive criticism and suggestions. I should like to acknowledge the very valuable assistance in this respect provided by V. Ashkenazi, P. Bauer, G. Beutler, F. K. Brunner, A. G. Burns, N. Crooker, B. A. Emery, Clyde Goad, J. Gosling, A. E. Hedin, J. A. Klobuchar, G. E. Lanyi, R. Leitinger, A. Milani, P. Moore, T. Moore, R. S. Nerem, Anna Nobili, J. J. Olivero, G. Petit, Ch. Reigber, J. C. Ries, R. G. Roble, R. J. Sica, J. J. Sojka, G. Thuillier, K. F. Wakker, E. R. Westwater, and P. Willis.

Special thanks are due to François Barlier and Jo Ann Joselyn, who acted as co-editors and provided invaluable advice in the process of identifying reviewers qualified to referee some of the more specialized topics covered in this book. I should also like to acknowledge the essential help of my co-convenors for providing invaluable suggestions for the invited speakers and the eventual authors of the review papers in this book.

A. Vallance Jones
Editor

FOREWORD

The scientific work of the International Union of Geodesy and Geophysics (IUGG) is primarily carried out through its seven associations: IAG (briefly, Geodesy), IASPEI (Seismology), IAVCEI (Volcanology), IAGA (Geomagnetism), IAMAP (Meteorology), IAPSO (Oceanography), and IAHS (Hydrology). The work of these associations is documented in various ways.

Environmental Effects on Space Positioning and Trajectories is one of a group of volumes published jointly by IUGG and AGU that are based on work presented at the Inter-Association Symposia as part of the IUGG General Assembly held in Vienna, Austria, in August 1991. Each symposium was organized by several of IUGG's member associations and comprised topics of interdisciplinary relevance. The subject areas of the symposia were chosen such that they would be of wide interest. Also, the speakers were selected accordingly, and in many cases, invited papers of review character were solicited. The series of symposia were designed to give a picture of contemporary geophysical activity, results, and problems to scientists having a general interest in geodesy and geophysics.

In view of the importance of these interdisciplinary symposia, IUGG is grateful to AGU for having put its unique resources in geophysical publishing expertise and experience at the disposal of IUGG. This ensures accurate editorial work, including the use of peer reviewing. So the reader can expect to find expertly published scientific material of general interest and general relevance.

Helmut Moritz
President, IUGG

Coping with the Atmosphere and Ionosphere in Precise Satellite and Ground Positioning

THOMAS P. YUNCK

Jet Propulsion Laboratory, California Institute of Technology, Pasadena, California

The earth's ionosphere and neutral atmosphere act in several ways to hinder the recovery of satellite orbits and the determination of ground point positions with space techniques. The ionosphere alters the propagation of signals employed by VLBI, GPS, and other radio (but not optical) systems. To first order, the ionosphere affects the phase and group delay observables with equal magnitude but opposite sign; the resulting delays are inversely proportional to the frequency squared. These properties can be exploited to achieve a high degree of delay calibration. The remaining higher order effects can be calibrated by modeling the electron distribution and magnetic field. The neutral atmosphere affects signal propagation somewhat differently. At radio frequencies the delay is largely frequency-independent and offers no simple intrinsic means of calibration. At optical frequencies, however, there is a degree of frequency dependence that may one day be exploited for delay calibration. At altitudes below about 1000 km, the atmosphere has a more direct impact on Earth satellites, creating drag and lift forces that alter satellite motion and complicate orbit determination and orbit prediction. Several newly developed techniques go a long way toward eliminating errors resulting from both the propagation effects and the dynamic effects of the atmosphere. With these techniques the three major space geodetic techniques—VLBI, SLR, and GPS—can achieve millimeter precision in measurements over hundreds of kilometers and centimeter precision across the globe. Variations on these techniques promise centimeter-level orbit accuracy for even the most dynamically unpredictable satellites.

INTRODUCTION

The neutral atmosphere and its ionization products have long been impediments in space geodesy. Here we review the mechanisms by which the atmosphere and ionosphere corrupt precise orbit determination and ground positioning, and the diversity of techniques now in use to minimize the resulting errors. The paper is organized into three sections: 1) ionospheric effects on RF signal propagation and radio positioning; 2) effects of the neutral atmosphere on geodetic positioning; and 3) the dynamic effect of the atmosphere—drag and lift—on low Earth orbiters and the consequences for precise orbit determination.

The three major space geodetic technologies are very-long-baseline interferometry (VLBI), satellite laser ranging (SLR), and the Global Positioning System (GPS) used in various differential forms. In some cases the specific effects of the atmosphere or ionosphere depend strongly on the technique used—the ionosphere, for example, has little effect on optical signals but can seriously degrade radio positioning. In other cases the general principles apply to all systems with only minor variations. Most examples here are taken from GPS, which is used in the full range of applications and is subject to all of the important media effects.

Figure 1 depicts the GPS constellation, which, when completed in 1993, will consist of 24 GPS satellites in 12-hour (20,200-km altitude) circular orbits [Milliken and Zoller, 1978; Spilker, 1978]. Each satellite broadcasts navigation signals on two L-band frequencies: 1.57542 GHz (L1) and 1.2276 GHz (L2). The beamwidths of the GPS signals extend roughly 3,000 km beyond the limb of the earth as viewed from the GPS satellites. Each carrier is modulated with a pseudorandom ranging code that enables the receiver to determine precisely and unambiguously the arrival time of each code bit. Since the transmission time (according to the satellite clock) is known, this measurement determines the range plus the time offset between the satellite and receiver clocks, a quantity known as the pseudorange. In addition, the receiver can continuously measure and count the phase of the two L-band carriers to provide Doppler and integrated Doppler data. Typical 1-sec measurement precisions are 20-100 cm for pseudorange and 0.5-5.0 mm for phase, though bear in mind that phase data contain unknown integer-cycle biases while pseudorange measurements are absolute.

Figure 2 illustrates the basic system configuration for precise GPS-based geodesy and satellite orbit determination [Yunck et al., 1985, 1990; Thornton et al., 1985]. In this scheme, the positions of all points of interest, such as the user satellite or geodetic ground points, are determined with respect to a small network of

Environmental Effects on Spacecraft Positioning and Trajectories
Geophysical Monograph 73, IUGG Volume 13

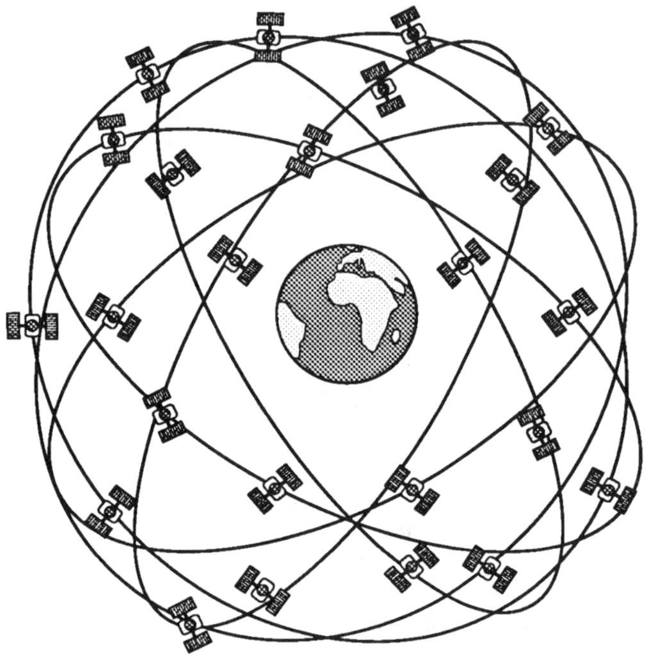

Fig. 1. The full GPS constellation, to be completed in 1993, will consist of 24 satellites in 12-hour circular orbits inclined at 55 degrees. Each satellite broadcasts dual L-band signals within a cone that extends approximately 3000 km beyond the earth limb.

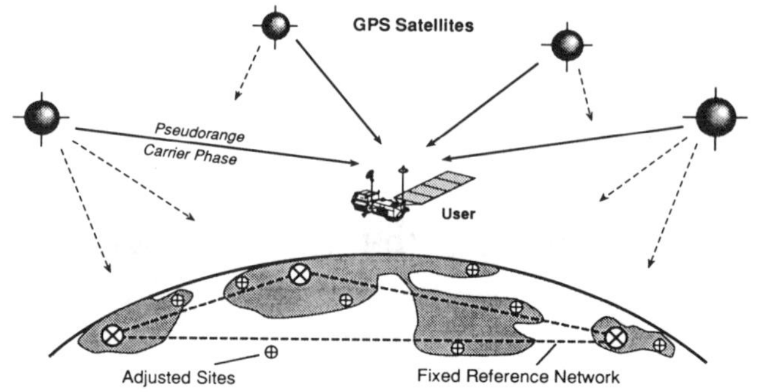

Fig. 2. Precise ground positioning and orbit determination with GPS depends on a network of ground receivers, usually with a few at precisely known reference points. The most powerful technique solves at once for GPS orbits, non-reference receiver positions, and miscellaneous parameters.

known reference points called fiducial sites. In the general approach, receivers placed at the fiducial and other sites continuously collect phase and pseudorange data from all GPS satellites in view. The data from all sites are then brought together and, in the most rigorous variation, a simultaneous least squares solution is carried out. Adjusted parameters typically include the non-fiducial site positions, GPS orbits, user satellite orbits (if any), receiver and satellite clock offsets, carrier phase biases, and miscellaneous quantities such as atmospheric delays and satellite force parameters. The data from all ground sites are subject to significant atmospheric and ionospheric effects. While data from the low orbiting user generally escapes atmospheric effects, if the satellite is below about 2000 km the ionospheric effect can be sizable, and below about 1000 km the satellite is subject to significant aerodynamic drag.

The Ionosphere

While the ionosphere itself is vastly complex and its behavior understood imperfectly, its effects on electromagnetic signals are well characterized. To start from the beginning, the ionizing radiation from the sun liberates some electrons from atmospheric molecules, giving rise to equal numbers of positive and negative ions. As these attractive charges tend rapidly to recombine, the density of free ions is roughly ten times greater during the hours of direct sunlight than at night, when new ions are not being generated. An electromagnetic signal—an oscillating electromagnetic field—encountering these ions exerts on them an oscillating force with a magnitude proportional to the charge. The positive ions, being 2,000 to 60,000 times heavier than the electrons, are little affected and can hereafter be ignored. The electrons, however, respond with significant motion; that motion gives rise to further electromagnetic effects which result in an advance of signal phase. Figure 3 is an estimated electron density profile derived from the Bent ionospheric model [Llewellyn and Bent, 1973] with inputs representative of Boulder, Colorado, for March 1980. It shows the characteristic density peak at about 350 km altitude. Figure 4 shows the daily variation of zenith electron content observed during April 1979, in Hamilton, Mass. The seasonal transition makes the ionosphere unusually erratic in April, as can be seen by the variability of the daytime peak. (Both figures are provided courtesy of J. Klobuchar, Phillips Laboratory.)

Below about 2,000 km altitude, where virtually the entire ionosphere resides, the earth's magnetic field enters the picture in a significant way. Electrons set in motion by an electromagnetic signal are constrained to follow curved paths around the magnetic field lines. Being of the same charge, all electrons curve in the same direction, introducing an asymmetry, or handedness. The ionospheric signal delay now depends on two additional factors: the signal polarization and the angle between the direction of propagation and the magnetic field lines. A general signal can be expressed as the sum of left and right circularly polarized signals, which, because of the handedness of electron motion, exhibit different phase velocities through the combined ionosphere and magnetic field. As a result, the plane of polarization of a linearly polarized signal will rotate, a phenomenon known as Faraday rotation.

In all, the ionosphere has at least seven distinct effects on RF signals. First there is the phase advance, or increased phase velocity. Second, because the ionosphere consists of physical matter, the group velocity is retarded; that is, information carried by the signal travels more slowly. Third, the signal is refracted, bending the propagation path. Fourth, because the ionospheric electron content is constantly changing, the signal frequency is Doppler shifted. Fifth, as we've noted, the plane of polarization is rotated in the presence of the magnetic field. Sixth, because the propagation velocity through the ionosphere depends on signal frequency, the frequency components of a broadband signal are dispersed.

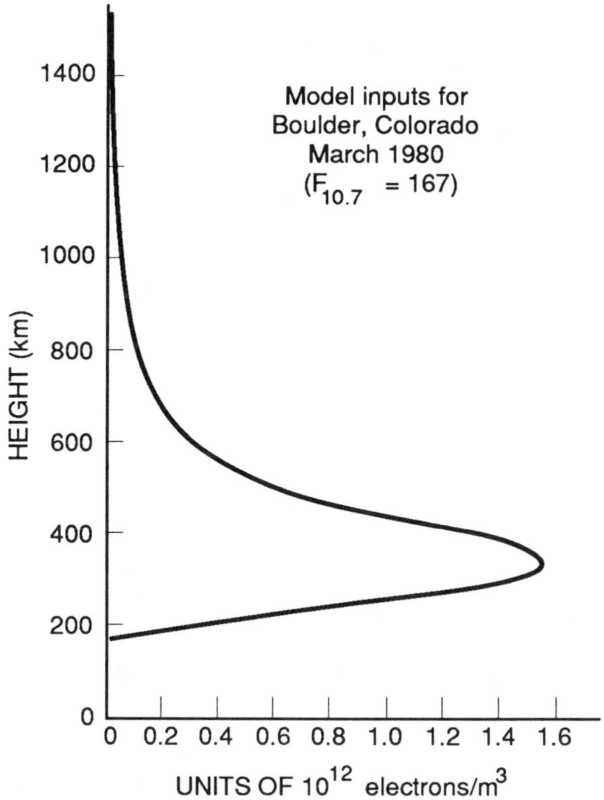

Fig. 3. Estimated vertical profile of ionospheric electron density, derived from the Bent ionospheric model, with inputs for Boulder, CO, March 1980. Adapted from Klobuchar [1985b].

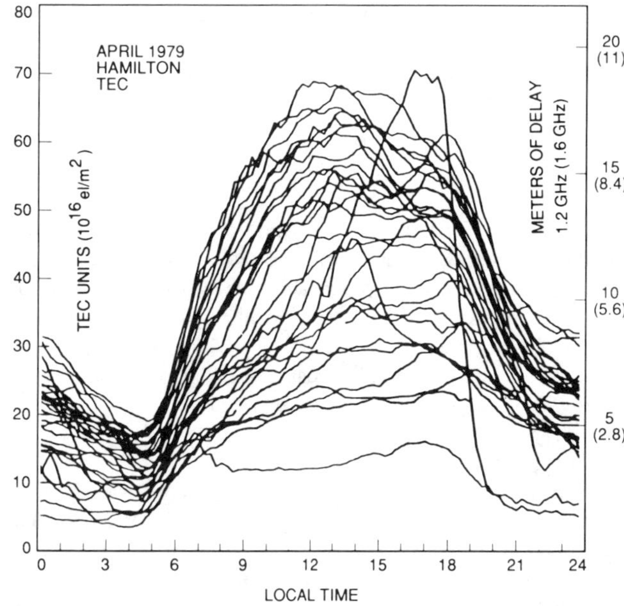

Fig. 4. Daily zenith total electron content, and the associated zenith delay in meters, observed over 30 consecutive days in Hamilton, Mass., during April of 1979. (Courtesy of J. Klobuchar, Phillips Laboratory.)

Seventh, as a result of both refraction and diffraction (by ionospheric structures), the signal will travel along multiple paths which cross and recombine, causing amplitude and phase scintillation. For in-depth treatment of these and the topics that follow, see, for example, Bassiri and Hajj [1992], Brunner and Gu [1991], Rishbeth and Garriot [1969], Davis [1966], Papas [1965], and Jursa [1985].

Phase and Group Delay

The ionospheric contribution to RF phase can be expressed in a power series expansion:

$$\phi_{ion} = \frac{p}{f} + \frac{q}{f^2} + \frac{r}{f^3} + \cdots \tag{1}$$

where

$$p = k_1 \int N\, d\ell = k_1 \text{TEC} \tag{2}$$

$$q = \pm k_2 \int N \mathbf{B} \cdot \mathbf{d}\ell \tag{3}$$

f is the frequency, N is the electron density, TEC is the integrated or total electron content along the raypath (also called the slant electron content, or SEC), **B** is the magnetic field vector, ℓ is a unit vector along the direction of propagation, and r is a more complicated expression involving N and **B**. The magnetic field comes into play only in the 2nd and higher order terms; when **B** is zero, those terms vanish. The + and - in (3) apply to the left and right circularly polarized components of the signal (also called the ordinary and extraordinary waves), respectively. It is easily shown that the simple first order term is overwhelmingly the dominant term. In many applications employing frequencies above 1 GHz, the higher order terms can be ignored. Although the first order term can be substantial (more than 50 m at GPS L-band frequencies), its simple form allows direct means of calibration.

For now, let us ignore the higher order terms in (1) and write

$$\phi_{ion} \cong \frac{k \cdot \text{TEC}}{f} \tag{4}$$

There are two different measurements that can be used to estimate signal propagation delay. One is the phase delay, which is derived from the phase measurement at a single frequency and is defined as transmitted phase minus received phase divided by frequency. Measurement precision is generally proportional to signal frequency. The ionospheric effect on phase delay is given by

$$\tau_{\phi_{ion}} \equiv -\frac{\phi_{ion}}{f} \cong -\frac{k \cdot \text{TEC}}{f^2} \tag{5}$$

The second type of delay measurement is the group delay, defined as the derivative with respect to frequency of transmitted phase minus received phase. Measurement of this derivative requires observations over a non-zero bandwidth, with measurement precision being proportional to the bandwidth. GPS pseudorange is a group delay measurement. The ionospheric effect on group delay is given by

$$\tau_{g_{ion}} \equiv -\frac{d\phi_{ion}}{df} \cong \frac{k \cdot TEC}{f^2} \qquad (6)$$

Notice that the ionospheric phase and group delays are of equal magnitude but opposite sign. Only the group delay gives the true (and necessarily positive) propagation delay of a wave packet centered at frequency f. The apparent negative delay in (5) is a consequence of ionospheric phase advance—the curious phenomenon in which phase velocity increases, while photons themselves (and any information they carry) are retarded.

Calibrating the Ionosphere

Left uncalibrated, the ionosphere can introduce large errors in positioning and orbit determination with radio techniques. The delay effect on raw GPS L-band observables taken from the ground can be as small as 1 m at the zenith during the nighttime and can surpass 50 m at low elevations during the day. As we have seen, this delay is inversely proportional to frequency squared and thus is substantially reduced at higher frequencies. In X-band (8 GHz) VLBI measurements it can still reach 1 m or more during the day. Although the mapping from raw measurement error to final position error is complex and depends on the tracking geometry and solution strategy employed, to first order the resulting error in position will be comparable to the measurement error. For precise orbit determination (which we define as a position error of less than 10 cm) of low Earth satellites, and for precise geodesy (baseline measurement error of less than 1 part in 10^8, or 1 cm over 1000 km) calibration of ionospheric delay is essential.

Ionospheric models. One form of calibration that can be useful to single-frequency users is the introduction of an ionospheric delay model. The GPS satellites broadcast a crude delay model consisting of eight parameters to describe the ionospheric delay for the entire globe [Klobuchar, 1987]. Those parameters, regularly updated, are sent to users as part of the 50 bit/sec data message broadcast continuously on both L-band frequencies. Given the surpassing complexity of the ionosphere, one may doubt that a serviceable global model can be distilled to eight parameters. A number of experimental studies have shown that the GPS model consistently exceeds its design goal of 50% accuracy, in some regions reaching 70% [Feess and Stephens, 1987; Coco et al., 1990]. On the basis of percent accuracy, the model provides superior performance during the daytime when the gross effect is large and poorer performance (sometimes exceeding 100% error) during the night. In absolute terms, the nighttime error is still relatively small.

The surprising accuracy of the simple GPS model can, unfortunately, be only marginally improved upon with more strenuous modeling efforts. Because of the inherently chaotic nature of the ionosphere, which shows typical day-to-day variations of 20-25% and monthly average variations of ~10% [Klobuchar, 1985a, b], models comprising hundreds of parameters cannot consistently surpass 75% accuracy. Ultimately, model-based ionospheric calibrations are useful only for moderate accuracy applications; they are not now and probably will never be suitable for precise work as we have defined it here.

Single-frequency group+phase delay. Perhaps the least appreciated technique for ionospheric calibration available to some single-frequency users of GPS, VLBI, and other radio tracking systems is the formation of a new observable by taking the simple average of the group and phase delay observables. Noting the respective ionospheric effects described by Eqs. 5 and 6, we can write simplified expressions for the total phase and group delay observables as

$$\tau_\phi \cong \tau - \frac{k \cdot TEC}{f^2} + \text{bias} + \varepsilon_\phi \qquad (7)$$

and

$$\tau_{grp} \cong \tau + \frac{k \cdot TEC}{f^2} + \varepsilon_{grp} \qquad (8)$$

where τ represents the geometric and all other non-ionospheric delays encountered by the signal and ε represents the measurement error. (To first order, the ionospheric delay can be regarded as the only dispersive or frequency-dependent delay in the system, and thus τ is the same in both equations.) Taking the average of (7) and (8), we have

$$\frac{\tau_\phi + \tau_{grp}}{2} \cong \tau + \text{bias}' + \frac{\varepsilon_\phi + \varepsilon_{grp}}{2} \qquad (9)$$

Observe that (9) represents an ionosphere-free (ignoring higher order terms) measurement having the form of the phase delay observable—that is, the total delay plus a bias. The bias is known to within the error of the single frequency group delay measurement. An analogous technique, formulated in terms of ionospheric delay change, was first investigated and validated by MacDoran [1970] using time-differenced range and integrated Doppler data from Mariners VI and VII. The idea has received little attention since.

It must be noted that bandwidth limitations keep the group delay observable typically 20-100 times less precise than phase delay. The random error in the combined observable is therefore dominated by the group delay error, which is reduced by half. In a sense, the combined observable represents the worst of both worlds: It has the reduced information content (the bias) of the phase delay and the approximate noise level of the group delay. But in return it has removed a potentially dominant systematic error. It happens that this can be a very favorable trade in applications where relatively long data arcs (a few hours or more) and thus a high degree of data smoothing are the rule—such as precise geodesy and orbit determination. As we shall see in a later example, the single-frequency GPS user can achieve surprising accuracy with this approach.

Dual frequency calibration. For the most precise work dual frequency calibration is required. This technique can be applied independently to phase-only data, thus preserving much of the precision of the carrier measurement. It can also be applied separately to group delay data or, in a more efficient variation, to both data types at once [Wu et al., 1991]. Suppose we have delay measurements (either group or phase) τ_1 and τ_2 at two frequencies f_1 and f_2 given (in simplified form) by

$$\tau_i \cong \tau + \frac{p}{f_i^2} \tag{10}$$

where, again, only the dispersive ionospheric term is separated from the other delays. We can now form the ionosphere-free observable τ_c as a simple linear combination of τ_1 and τ_2

$$\tau_c \equiv \left(\frac{f_2^2}{f_2^2 - f_1^2}\right)\tau_2 - \left(\frac{f_1^2}{f_2^2 - f_1^2}\right)\tau_1 \cong \tau \tag{11}$$

Substituting (10) into (11) we can easily confirm that the first-order ionospheric term has been removed. If the biased phase observable is used, a transformed bias appears in the calibrated observable. Although this technique does not generally push the noise level of the phase measurement close to that of the group delay measurement, it does amplify the noise, often by a significant amount. To compute the noise increase, let us rewrite (11) as

$$\tau_c = c_1 \tau_2 - c_2 \tau_1 \tag{12}$$

If we assume the random measurement error on the two frequencies is independent with a common standard deviation σ, then the error σ_c of the combined observable is

$$\sigma_c = \mathrm{sqrt}\,(c_1^2 + c_2^2)\,\sigma \tag{13}$$

Taking the expressions for c_1 and c_2 from (11) we find that the noise amplification at GPS frequencies is approximately a factor of three. Although in principle a more advanced filtering strategy might be brought to bear to reduce this multiplier, taking into account the particular spectral characteristics of the ionospheric delay, little work has been done in this area. Here VLBI has a distinct advantage: the wide frequency separation of the standard S- and X-band systems yields a noise multiplier of only 1.08.

Comparing the single-frequency and dual-frequency calibration techniques we see that the precision advantage of the carrier phase observable is in fact substantially less than it first appears. In practical GPS applications, when the effects of atmosphere, multipath, and other factors are taken into account, pseudorange precision is typically 50 times worse than phase precision. The single-frequency averaging technique cuts the pseudorange noise level in half, while the dual-frequency combination multiplies the phase noise by 3. Thus there is a net swing of a factor of six, and the relative disadvantage of the single-frequency approach is seen to be about a factor of eight, not 50. In many applications, the effect of random phase error is below that of other error sources in the total error budget, and the net penalty for the single-frequency approach is therefore still less. An orbit determination example comparing these techniques is given in the final section.

Faraday rotation and secondary GPS calibration. Tracking systems operating at lower frequencies are subject to significant higher order effects. The U.S. Tranet Doppler system, for example, which operates at 150 and 400 MHz, can see higher order effects of 1 m or more. While this presents a problem for precision tracking, the effect can offer another means of ionospheric calibration. The second order term on the right side of (1) is the Faraday rotation of a linearly polarized signal. For frequencies in the range of hundreds of MHz this rotation angle can be a full cycle or more. If the initial plane of polarization of a linearly polarized signal is known, as it is for certain geosynchronous satellites, then the net rotation due to the ionosphere and magnetic field can be observed at the receiver. If a simple model of the magnetic field is provided, Eq. 3 can be inverted to estimate the total electron content along the signal path (or at least over that portion of the path where the **B** field is significant). Simple mapping functions can then be used to estimate the TEC in other directions. In this way, the 2nd order effect on one signal may be used to calibrate the first order effect on another. This technique has been used for several decades to calibrate the ionosphere for NASA's deep space tracking network.

More recently, NASA has introduced dual-frequency GPS receivers into the network for this purpose. These have the twin advantages of making more accurate TEC measurements and observing multiple satellites well distributed in the sky. The data acquired at one site over a several hour period can be represented two-dimensionally; a 2-D polynomial can then be fitted to the data for interpolation to other points of interest. NASA has now adopted this technique for operational ionospheric calibration [Royden et al., 1984; Lanyi and Roth, 1988] to support the tracking of deep space probes. Because of the interpolation required, however, accuracies degrade to tens of centimeters at the GPS L1 frequency, and about half that at typical S-band spacecraft frequencies.

Higher order calibrations. In modern high frequency measurement systems the higher order effects are small, typically a few millimeters for GPS and considerably smaller for VLBI. Under suitable conditions, however, such as a high daytime TEC and a signal propagating nearly parallel to the magnetic field lines (e.g., near the poles), the 2nd and 3rd order effects on GPS can reach several centimeters [Bassiri and Hajj, 1992; Brunner and Gu, 1991]. For orbit determination this may never be a problem, but in the era of millimeter GPS geodesy, serious thought must be given to calibrating the higher order effects.

This can be done quite easily by a technique somewhat complementary to the Faraday rotation technique for the 1st order calibration. In this case we can use the dual-frequency TEC calibration to provide the absolute scale of the ionospheric effect; then, using models of the magnetic field and the ionospheric electron distribution, we can evaluate (3), and the corresponding 3rd order equation, to estimate the 2nd and 3rd order terms. Brunner and Gu [1991] describe a practical calibration technique for GPS measurements that should keep the total ionospheric error below 2 mm, an estimate also supported by the work of Bassiri and Hajj [1992].

In future systems, frequencies can be added for direct higher order calibrations. A third frequency will permit full calibration of the 2nd order effect, and a 4th frequency can remove the 3rd order effect. In fact, GPS broadcasts on a third (L3) frequency as well. Unfortunately, L3 is activated only intermittently, on the order of minutes per day. And in any case it is poorly positioned, roughly midway between L1 and L2. The resulting large noise multiplier would not justify its use to remove so small an effect [Bassiri, 1990].

The Atmosphere: Effects on Positioning

In this section we consider the delay effects of the neutral atmosphere both on the raw measurements used in precise positioning and on the resulting position solutions. In one respect this is a simpler issue than the ionosphere. At radio frequencies the atmosphere is non-dispersive: There is no frequency dependence in the delay and the group and phase delays are the same. On the other hand, since it was those very distinctions that we exploited to calibrate the ionosphere, the atmospheric calibration task becomes quite different and often less effective.

For our purposes, the neutral atmosphere can be resolved into two distinct constituents, known informally as the *wet* and *dry* components. The wet component consists of suspended water vapor and liquid water, while the dry (or hydrostatic) component, consists of everything else. The total zenith delay of the two components is typically 2.0-2.4 m, of which typically 90% is due to the dry portion. Despite its dominance of the total delay, the dry component is far less problematical for precise positioning: It is highly uniform, stable, and predictable over large regions and, with the simple expedient of surface pressure measurements, can be modeled (at zenith) with an accuracy of 1 cm or better. If the atmosphere consisted of only the dry component, its delay contribution would be of no more concern than the higher order ionospheric delay at L-band since its (usually) agreeable uniformity permits a simple and accurate estimation of the residual model error directly from the positioning measurements.

The focus of attention in the atmosphere, therefore, and indeed the limiting error in the most precise forms of radio frequency space geodesy, is the wet delay. While it is frequently well behaved, particularly in dry climates, the wet component can be highly variable in both time and space. We can think of the wet component as consisting of cells of water vapor drifting through a region, giving rise to temporal and spatial delay variations. It is not unusual to see the zenith delay change by as much as 5 cm in an hour, although 1 or 2 cm is more typical. The total wet zenith delay may range from nearly zero in very dry conditions to more than 50 cm. The delay is amplified by the greater slant range through the atmosphere at low elevation angles, increasing by about a factor of ten at 5 degrees.

Calibrating the Wet Delay

At radio frequencies there are three calibration strategies in widespread use today: 1) application of seasonal models, often supplemented with surface temperature and humidity data; 2) estimation of wet delay based on brightness temperature measurements made with water vapor radiometers; and 3) estimation of total zenith delay directly from the positioning data as part of the position solution.

Seasonal models alone are typically accurate to 4 or 5 cm at zenith, with the error scaling approximately as the inverse sine of the elevation angle down to about 15 degrees (at which point a more exact mapping function should be invoked). This accuracy of course depends a good deal on season and location. In less variable regions, such as the Mojave Desert in California during the long dry season, 3 cm may be more typical. While it is customary to include surface temperature and humidity data to tune the model, there is growing evidence that little is actually gained from this, presumably because local surface data reflect poorly the wet structure at higher altitudes and at some distance from the observing site. Real improvements in the seasonal model correction require more comprehensive synoptic measurements, which are seldom available.

Better calibrations are obtained with water vapor radiometers (WVRs) [Janssen, 1985], which actually measure atmospheric brightness temperature at various water vapor emission lines between about 20 and 32 GHz. WVR's can be pointed alternately at the signal sources being observed or put through a range of tipping curves that sample the full sky (though in some cases they are fixed at the zenith). The usually broad WVR beamwidths preclude observations at low elevation angles and prevent a fine focus on the sources, which can limit the achievable accuracy. The brightness temperatures, moreover, must be converted to wet delay by a complex inversion algorithm which is itself subject to limiting assumptions. In addition, the measurement requires an internal absolute calibration, which can drift off by the equivalent of a centimeter or more of delay. Various other sources of WVR error are documented in Elgered et al. [1991]. In practice, WVR performance is seen to vary a good deal between units, between designs, and between experiments. Properly functioning WVRs are typically accurate to 1 cm or better along the line of sight; they can, however, drift off by several centimeters. It is therefore common practice when using WVR data in geodetic applications to solve simultaneously for a constant zenith delay bias.

Extensive experimentation with GPS and VLBI data has shown that solving for a constant zenith delay bias over several hours, or a piecewise constant bias over longer periods, can produce an excellent atmospheric delay correction, whether WVR data are available or not. In fact it is seen that when this strategy is employed, the inclusion of WVR corrections generally leads to only small improvement, or even none at all, as measured by daily baseline repeatability and agreement with other techniques. The resulting measurement accuracy typically corresponds to a 1-2 cm zenith delay error.

A more sophisticated approach now gaining favor is to solve for the zenith delay as a continuously varying stochastic parameter. It came a something of a surprise to early investigators that space geodetic data possessed sufficient strength to permit such comprehensive monitoring of atmospheric delay; but five years of experimentation and refinement have shown that this approach can be applied with success in both GPS and VLBI, permitting sub-centimeter tracking of rapid changes in the zenith delay. A popular approach in GPS analysis is to model the zenith delay as a random walk, with the delay solution adjusted every few minutes under a constraint that permits a delay change of about 1 cm in 1 hr, or 5 cm in 24 hrs [Lichten and Border, 1987; Lichten, 1990]. This is executed as part of a Kalman filter solution. Other stochastic models, such as a first order Gauss-Markov process, are also used with success in both GPS and VLBI [Herring et al., 1990].

Figure 5 gives two examples of the stochastic solution for zenith delay. The first plot compares a GPS-based solution directly with a WVR-based measurement, while the second compares GPS- and VLBI-based solutions. The total (wet+dry) delay is shown in

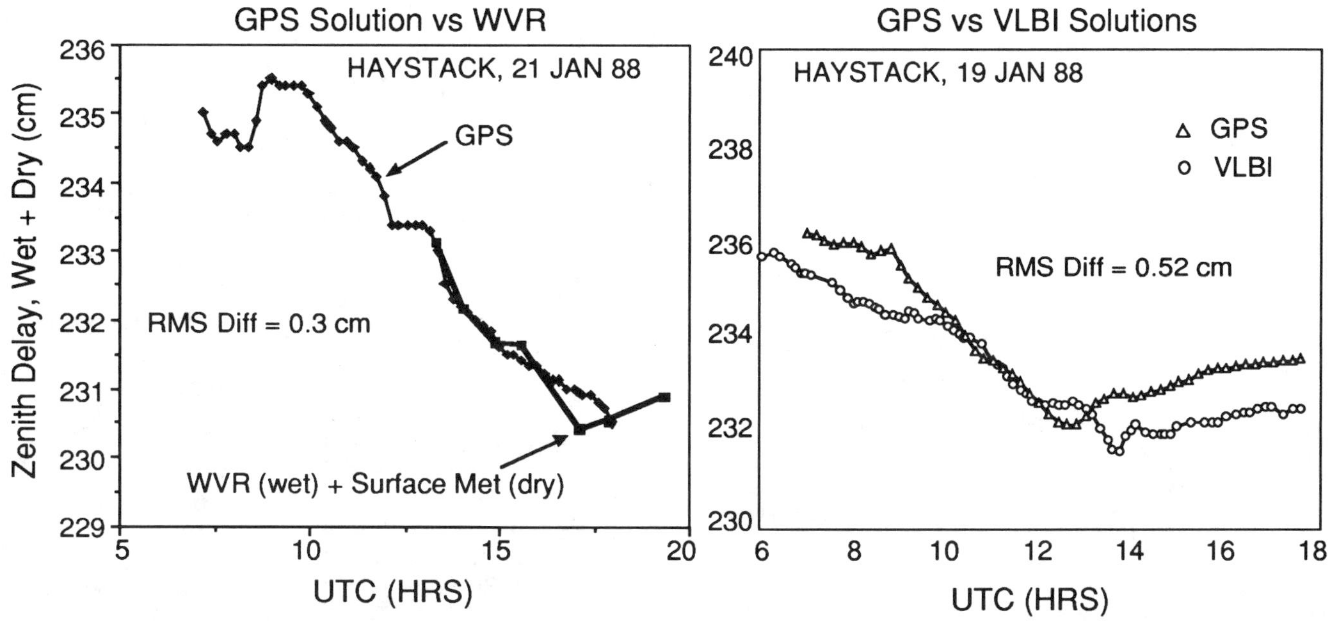

Fig. 5. Comparison of GPS-derived solutions for atmospheric zenith delay with water vapor radiometer measurements (left) and very long baseline interferometry solutions.

all cases. Both plots cover periods of rapid change in the zenith delay. In the first plot, the RMS departure of the GPS and WVR solutions is only 3 mm over 5 hrs. It should be noted that no bias has been removed between the two solutions, so this suggests (though does not prove) absolute accuracy at that level. The GPS and VLBI solutions show an RMS agreement of about 5 mm over 11 hrs. In similar comparisons, GPS, VLBI, and WVRs have shown agreements ranging from 3 to 12 mm, with most below 10 mm. Somewhat surprisingly, therefore, we find that GPS, designed as a military positioning system, provides one of the best means available for measuring the atmospheric wet delay. NASA is now moving to adapt its GPS-based ionospheric calibration system to provide atmospheric calibrations for deep space tracking as well.

Stochastic zenith delay estimation is essentially empirical. A physical approach, based on the theory that wet delay variation is characterized by Kolmogorov turbulence transported by winds [Treuhaft and Lanyi, 1987], has recently been proposed [Treuhaft and Lowe, 1991; Treuhaft, 1992]. Known as the observable covariance matrix approach, this technique employs an estimator that incorporates a covariance matrix for delay fluctuations derived from a physical model. Analysis suggests that if the covariance matrix is accurate, baseline errors could be reduced by a factor of two or three in comparison with solutions employing stochastic zenith delay estimation. Expected uncertainties in the model parameters, however, which include a structure constant, height, and wind velocity, might take away much of this advantage [Treuhaft, 1992]. This technique has not yet been tested experimentally.

Atmospheric Error and Vertical Position

Let us now see what effect an atmospheric zenith delay error has on position solutions, particularly the zenith component. It happens that there is a strong correlation between the error in the zenith atmospheric delay and the vertical component of a point position or baseline solution. To see why this is so, we note that the signal delay due to the atmosphere is approximately inversely proportional to the sine of the elevation angle, θ_{el}, down to about 15 degrees, and is given by

$$\tau_{atm} \cong \frac{\tau_z}{\sin \theta_{el}} \tag{14}$$

where t_z is the atmospheric delay at zenith. The change in signal propagation time $\Delta\tau_v$ due to a change ΔV in the vertical position of the receiver is given approximately by

$$\Delta\tau_v \cong \Delta V \sin \theta_{el} \tag{15}$$

While these equations are functionally quite different, they can look very similar over a wide range of realistic elevation angles. Figure 6 gives a plot of the two functions for t_z = 1 cm and ΔV = -2 cm, over elevation angles from 10 to 80 degrees [from Treuhaft, 1992]. Note that a bias, representing a clock offset adjustment equal to 3 cm of delay, has been added to the $\Delta\tau_v$ curve. Over the range of 30 to 80 degrees elevation the curves are strikingly similar, and only when observations are made below 20 degrees do they clearly separate. This implies that (1) a fixed error in the atmospheric zenith delay will cause an error in the vertical position solution which will increase as lower elevation data are included; and (2) an attempt to solve simultaneously for the zenith delay and the position will benefit from the addition of the distinguishing low elevation data. In other words, low elevation data are harmful if a fixed zenith delay error is present and helpful if

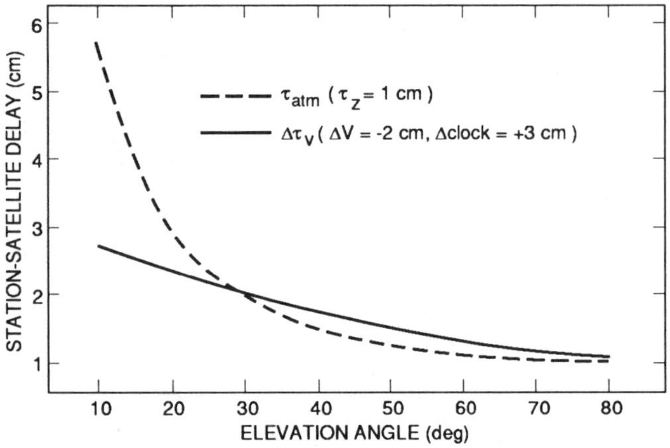

Fig. 6. Comparison of the observed delay change resulting from a 1 cm change in zenith atmospheric delay (dashed line) and from a -2 cm change in vertical position. A 3 cm clock bias has been added to the solid curve. Adapted from Treuhaft [1992].

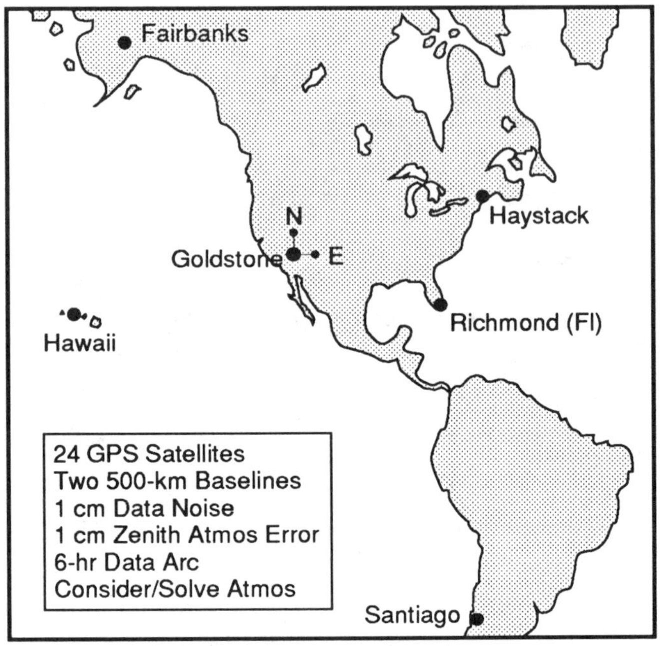

Fig. 7. Map of the eight observing sites used in atmospheric error covariance analysis.

the delay is adjusted in the solution, provided the mapping function remains valid.

The large vertical effect of an atmospheric delay error is not simply a consequence of the geometric weakness in the vertical exhibited by all space geodetic techniques. Because GPS, VLBI, and SLR can in principle make observations around the full 360 degrees of azimuth, but over less than a hemisphere above the horizon, the natural strength of all techniques is generally a factor of two to three greater in the horizontal than the vertical. This fact is familiar to space geodesists and is reflected in both the formal errors and repeatabilities of baseline results. But what might be called the atmosphere-induced weakness in the vertical can far surpass the inherent geometric weakness.

Covariance studies. We can evaluate this effect with a simple covariance analysis. Figure 7 shows a typical network that might be used in a regional GPS geodetic campaign in North America. In this network we've introduced two fictitious sites that form 500-km baselines with the NASA site at Goldstone, California: one to the north and one to the east. Three North American fiducial sites (Goldstone, Haystack, and Richmond) were held fixed to establish the reference frame within which the GPS orbits and receiver positions were determined. The covariance analysis, performed with JPL's OASIS software [Wu and Thornton, 1985], evaluates both the formal error due to random measurement error (exactly as it's computed in a real solution) and the additional systematic errors arising from model errors in unadjusted parameters. Among the latter, we shall focus on the effect of a bias in the zenith delay model. The assumptions and solution strategy are summarized in Table 1.

Two different solution strategies are evaluated. In the first, the atmospheric delay model is held fixed and the effect of a 1 cm zenith delay error, such as would be typical for a good water vapor radiometer pointed at the zenith, is considered. (For the larger seasonal model errors the results can simply be scaled in proportion.) In the second we examine the effect of adjusting the zenith delay in the solution according to a random walk model. In both cases the elevation cutoff angle is varied from zero to 45 degrees.

Figure 8 gives the results of the fixed-model strategy for the horizontal and vertical baseline components, averaged for the two baselines. The two-dimensional horizontal error is shown, computed as the root-sum-square (RSS) of the North and East component errors. The typical one-dimensional horizontal error would thus be lower by the square root of 2. Note the large vertical scale difference (a factor of ten) between the two plots. Both plots show the same characteristic pattern of errors with changing elevation angle. As the elevation cutoff is lowered, the geometric observing strength is increased; the computed formal error from random data noise goes down and the solution would appear to improve. At the same time, however, the slant range through the atmosphere grows

TABLE 1. Model for Atmospheric Covariance Analysis

System Characteristics	
Number of Ground Sites:	8 (with 3 N. American fiducials)
Number of GPS Satellites:	24
Data Types:	L1 and L2 unbiased carrier phase (ambiguities resolved)
Data Interval:	5 Minutes
Data Noise:	1 cm
Adjusted Parameters and A Priori Errors	
GPS Satellite States:	2 m; 0.2 mm/sec, each component
GPS and Receiver Clock Biases:	3 msec (modeled as white noise)
Zenith Atmosph. Delay (Case 1):	10 cm (modeled as random walk)
Fixed Errors Evaluated	
Zenith Atmosph. Delay (Case 2):	1 cm

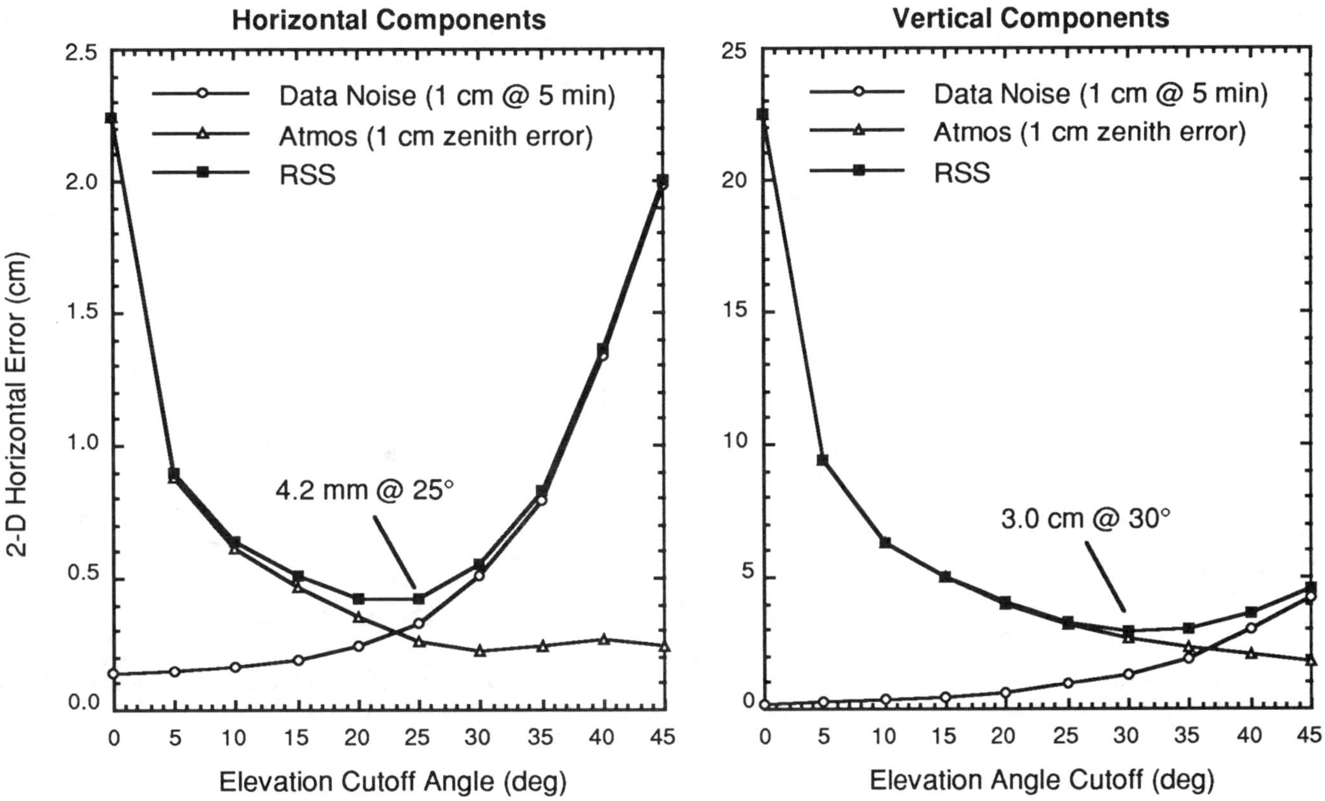

Fig. 8. Covariance analysis of horizontal (2-D RSS) and vertical baseline errors assuming a 1 cm error in zenith atmospheric delay. Results from two 500-km baselines are averaged.

larger—markedly so below 15 degrees; the zenith delay error is correspondingly amplified, as is the resulting error in the baseline solution. Consequently, there is an optimum elevation angle at which the error sum is a minimum. The optimum angle will of course depend on the magnitudes of the data noise and zenith delay error.

What is most striking, however, is the huge difference between vertical and horizontal errors. At the optimum points the vertical error is about seven times the horizontal error—well above the factor of two or three expected from the structural weakness in the vertical. The reason for this can be seen more clearly in Fig. 9, which plots the ratio between the vertical and horizontal errors. The data noise ratio, which reflects the relative observing strength, has an average value of about 2.5. The atmospheric error ratio, however, is much higher, with an average value of more than 10. To first order, both ratios are roughly constant with elevation angle. Dealing effectively with atmospheric delay error is therefore crucial to gaining control of the vertical position error.

The strategy of solving for the atmospheric delay is intended to ameliorate this problem by exploiting the strong information on atmospheric delay at low elevations. Figure 10 shows the covariance results obtained when the zenith delay is modeled as a random walk, with a new adjustment made every five minutes. In this case there is only one curve—data noise—for the horizontal and vertical components; because atmospheric delay is now an adjusted parameter its effect shows up as part of the formal data noise error. (There is an important caveat to this which we shall deal with shortly.) We therefore see only the characteristic data noise pattern: the solution becomes uniformly stronger as the elevation cutoff is reduced.

Note, however, that although the atmospheric error is now gone, the data noise error at a given elevation cutoff is considerably greater than in the fixed-model solutions. At 30 degrees, for example, the vertical data noise error is about 6 cm in Fig. 10 and less than 2 cm in Fig. 8. This is simply because a good deal of data strength is required for the continual readjustment of the atmospheric delay solution. Overall, however, if we can reliably go to low elevation angles, the solutions are improved in comparison with the fixed model solution. We can again gain some insight by examining the ratio of vertical to horizontal error, shown in Fig. 11. The ratio is no longer generally constant, but shows a systematic trend. Above about 25 degrees the ratio tends to flatten out at a value of six or seven, or not much below the pure atmospheric error ratio of about ten in the previous case. This reflects the weak separation of vertical position and zenith atmospheric signatures at high elevation angles (Fig. 6), which offers small assistance in reducing the atmosphere-induced error. Below 25 degrees the error ratio declines steadily, reaching the inherent structural limit of ~2.5 at zero degrees.

A caution is in order here. From Figs. 10 and 11 we might be

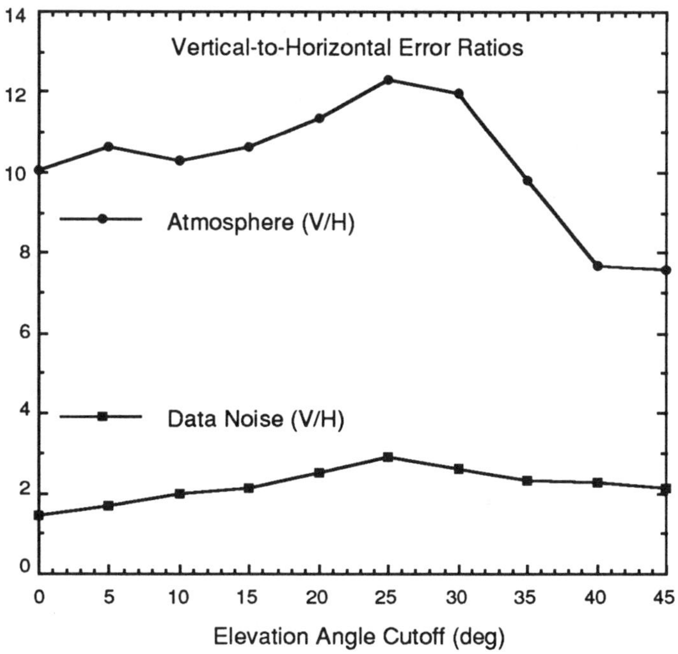

Fig. 9. Ratios of vertical to horizontal (2-D RSS) baseline errors, separated into the effects of a fixed 1 cm atmospheric zenith delay error and random measurement error.

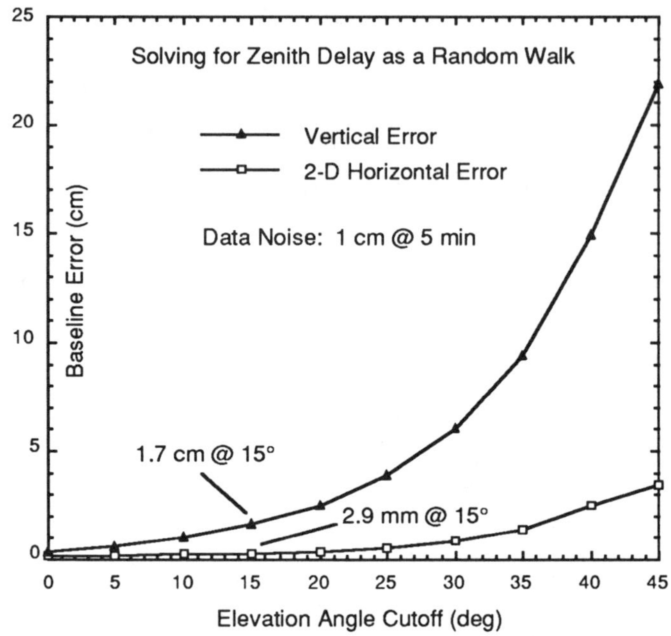

Fig. 10. Covariance analysis of horizontal (2-D RSS) and vertical baseline errors when the zenith atmospheric delay is adjusted as a random walk. Because the atmospheric mapping function loses fidelity below about 15°, results at those elevations are not reliable.

tempted to conclude that observations should be pushed to the lowest achievable elevation angles and all data used in the solution. There are two problems with this. First, the solution for the atmospheric delay is based on an assumption of a smooth and predictable variation with elevation angle; observations at low elevations are related to the zenith by a pre-determined theoretical mapping function, of which there are a number to choose from [Chao, 1972; Lanyi, 1984; Davis et al., 1985]. Given the irregularity of the atmosphere and the long slant ranges (and thus remote atmospheric sampling) at low elevations, the mapping function cannot be trusted below some elevation limit. When data are employed from elevation angles where the mapping function is no longer valid, the solution will generally get worse. Second, unless a highly directional antenna is used (as is often the case in VLBI), data collected at low elevation angles may be corrupted by multipath—reflected signals from the ground and objects near the ground. This again causes the data to be unrepresentative of the local atmospheric delay and the solution is degraded.

Extensive experimentation with both GPS and VLBI has shown that the optimum elevation cutoff is generally in the vicinity of 10-15 degrees. Because of multipath, it is seldom beneficial to go below about 12 degrees with standard GPS antennas. With directional VLBI antennas, however, the limits tend to be atmospheric; when nature cooperates and the atmosphere is uniform (and the mapping function valid at low elevations), it can be profitable to go as low as 5 degrees, while at other times it can be harmful to go below 15 degrees [Davis et al., 1985]. Unfortunately, it can be difficult to tell which days are good and which are bad, and the penalty for a wrong guess can be substantial. For operational purposes a conservative VLBI cutoff angle of 10-15 degrees may therefore be advisable.

Optical positioning. At optical frequencies the gross atmospheric dry delay is nearly the same as at radio frequencies. Sensitivity to the wet component, however, is less by about a factor of 100 [Lambeck, 1988], and thus a major issue at radio frequencies—the irregularity and variability of the wet delay—is not a concern at optical frequencies. Standard atmospheric delay models are therefore all that are typically used, for example, in Lageos geodetic solutions. There is, however, a slight frequency dependence of the atmospheric delay at optical frequencies. A delay model reflecting that dependence, such as that of Marini and Murray [1973], is therefore advantageous.

The general sparsity of laser data tends to inhibit the effective solution for atmospheric delays for individual observing sessions. To achieve the requisite data strength for state-of-the-art geodetic solutions with satellite laser data, data arcs as long as 3 months are often used [Smith et al., 1990]. Errors in the atmospheric delay models and other system models are generally expected to average down to small values over such intervals. As more laser geodetic satellites are launched in the 1990s, the data strength for geodetic solutions will improve and direct solution for atmospheric delays may become attractive.

One area of current research is to exploit the frequency dependence of the atmospheric delay at optical frequencies. This dispersion is relatively weak, with a delay change on the order of 10% (~20 cm) for a frequency change of a factor of two. The instrumental precision required to exploit this difference at the millimeter level is therefore extremely high, corresponding to a timing

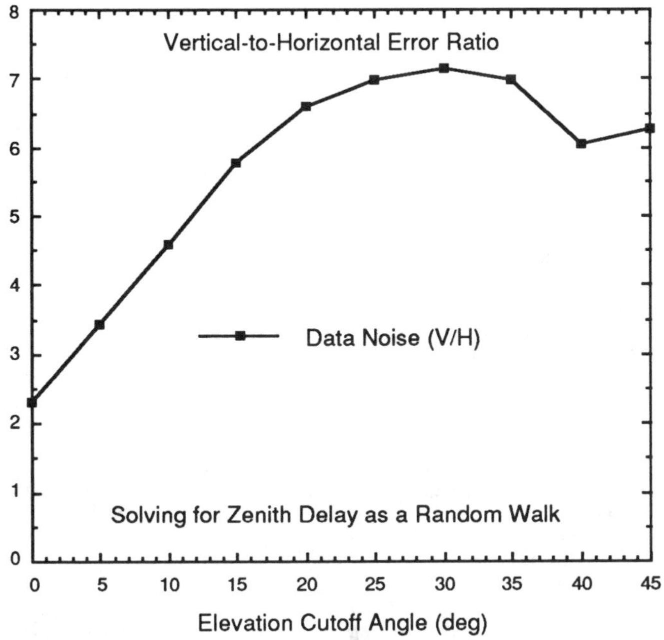

Fig. 11. Ratio of vertical to horizontal (2-D RSS) baseline error when the zenith atmospheric delay is adjusted as a random walk.

resolution of about 1 ps [Degnan, 1985]. Two- and three-color laser systems have been developed which may eventually reduce atmospheric error, in a manner analogous to dual-frequency ionosphere calibration, to 5 mm or better [Abshire and Gardner, 1985]. At the moment it is fair to say that in their most powerful configurations all three techniques—SLR, VLBI, and GPS—provide comparable accuracy for a wide range of geodetic measurements.

Low earth orbiters. We have focused here on the atmospheric delay as it affects ground (or near-ground) positioning, where measurements are made directly through the atmosphere. Lasers and GPS are of course both used to track various types of Earth satellites. Since laser satellite tracking also traverses the atmosphere directly from the ground to the orbiter (or, in the case of the proposed Geoscience Laser Ranging System, vice-versa) the atmospheric effects are much the same as in ground positioning. An orbiter carrying GPS, on the other hand, looks upward and there is no direct atmospheric error. Very accurate orbit solutions, however, require a reference network to track the GPS satellites simultaneously from the ground, introducing atmospheric errors. For sub-decimeter applications, of which the first example is TOPEX/Poseidon, advanced atmospheric calibration is needed. Error studies for TOPEX and other platforms indicate that the stochastic adjustment strategy will be the most effective, limiting the indirect atmospheric effects on the orbit solution to less than 1 cm.

Aerodynamic Drag and Orbit Determination

We now turn to precise orbit determination of low Earth satellites and a very different atmospheric problem: aerodynamic drag. For satellites below about 1000 km altitude, aerodynamic drag (and its companion, lift) becomes an important source of orbit error. Indeed, it is asserted by Ridgway et al. [1990] that "As gravity models become increasingly accurate, atmospheric drag is emerging as the dominant error in existing altimeter-satellite orbits." To understand why, we must know something about the classical technique of dynamic orbit determination, which, of necessity, has been used for all low Earth satellites, at least until recently.

Dynamic Orbit Determination

The technique begins with a set of tracking measurements (range or Doppler, for example) along with mathematical models of the forces acting on the satellite and of the satellite's physical properties. Those models in turn provide a model of satellite acceleration over time, from which, by double integration, one can derive a nominal trajectory. In principle, all that's required to obtain the orbit solution is to determine the two constants of integration—the initial position and velocity, also known as the epoch state. That is done by a least squares procedure (or its equivalent) that finds the epoch state for which the resulting trajectory best fits the tracking data. To improve the fit, one can simultaneously adjust various of the model parameters (drag, solar pressure, or gravity coefficients, for example). Ultimately, however, orbit accuracy will be limited by how faithfully the force models, fixed or adjusted, describe the real forces acting on the satellite. When the accuracy goal is centimeters over continuous fitting arcs of many days, the force models must be extraordinarily precise. At lower altitudes, by far the largest of the forces (apart from gravity itself) acting on the satellite is aerodynamic drag; it is also among the most unpredictable.

Drag, Lift, and Atmospheric Density

Atmospheric density (and the consequent aerodynamic force) decreases approximately exponentially with increasing altitude. A crude "onion-skin" atmospheric density model is sometimes useful in simple dynamic studies. Such models have a succession of concentric density layers, each represented by an exponential decay with altitude. In one example [Wu et al., 1986], the density $\rho(h)$ at altitude h is given by

$$\rho(h) = \rho_i \exp[-k_i (h-h_i)] \quad (16)$$

where the index i indicates the layer within which h falls. The parameters k_i are selected so that the densities agree at layer boundaries. Below 1000 km or so, the density gradient is quite steep. For example, in the late 1980s, when the planned altitude for the Earth Observing System (EOS) was lowered from 824 to 705 km, the expected drag effect increased by about an order of magnitude.

While this model is serviceable for simple studies, the atmospheric reality is vastly more complex. Atmospheric density at any point and time depends on such factors as the atmospheric constituents, temperature, the diurnal atmospheric bulge, the current and mean solar flux, the declination of the sun, the geomagnetic activity index, and the latitude of the point of interest. Since the dependencies are not in general well understood, accurate knowledge of these quantities does not necessarily lead to accurate knowledge of atmospheric density. Because the consequences of geomagnetic activity are especially hard to model or predict, orbit error due to drag mismodeling tends to be particularly large during geomagnetic storms [Ridgway et al., 1990]. Moreover, current atmospheric density models suffer from a lack of data above 500 km altitude and thus are less reliable at the higher altitudes where all precise altimetric platforms are or will be located.

Even if the atmospheric density were accurately known, the modeling problem would not be simple, since that quantity must then be converted to aerodynamic acceleration. Given an atmospheric density ρ, the aerodynamic acceleration vector **a** is given approximately by

$$\mathbf{a} = \frac{1}{2}\rho\frac{A}{m}V^2(C_D\,\mathbf{v} - C_L\,\mathbf{L}) \tag{17}$$

where A is the spacecraft cross-sectional surface area in the velocity direction, m is its mass, V is the magnitude of velocity, C_D is the drag coefficient, C_L is the lift coefficient, **v** is a unit vector in the velocity direction, and **L** is a unit vector in the direction of lift. For a complex spacecraft with varying physical characteristics (e.g., rotating body or solar arrays), just computing an accurate cross-section can be a challenge. The real difficulties, however, lie in the coefficients for drag and lift. (Drag has typically 5-10 times the magnitude of lift—see, for example, Stark [1986].) These depend on such factors as the shape and composition of the spacecraft elements, the angle of attack of each element, the ratio of diffuse to specular molecular reflection, the ratio of both incident and reflected molecular speed to spacecraft speed, the mean molecular mass, and the spacecraft surface temperature. Many of these quantities are difficult to know accurately and, in any case, the computation required to apply them faithfully, were they known, would be prohibitive.

In precise orbit determination, a three-step approach is widely used to deal with aerodynamics:

1) Model the shape and surface properties of the spacecraft as accurately as possible for the computing power available. For TOPEX/Poseidon, precomputed attitude-dependent cross sections are used.

2) Compute the atmospheric density with the best available data and algorithms, again within the limits imposed by computing power.

These steps are carried out prior to the orbit solution itself. Bearing in mind, however, that "because of...substantial fluctuations, models of air density...based on readily available parameters such as the geomagnetic indices, are of limited use in estimating the drag forces with any degree of accuracy" [Lambeck, 1988], the crucial third step is:

3) Estimate adjustments to the drag (and possibly lift) coefficient at suitable intervals as part of the dynamic orbit solution.

The Seasat mission of 1978 offers a good example of this strategy in operation. Seasat had an ungainly shape and flew at an altitude of 800 km, where drag begins to be a serious problem. It was tracked with one-way Doppler data from the U.S. Tranet network and the orbit determination was performed at the Goddard Space Flight Center [Haines et al., 1992; Ridgway et al., 1990; Marsh et al. 1980]. Figure 12 shows post-fit residuals for the along-track component over a 7-day fitting arc, which coincided with a significant geomagnetic storm. The fit was performed with two different atmospheric density models, DTM [Barlier et al., 1978] and MSIS [Hedin, 1987]. In the solution shown, a single drag coefficient was adjusted for the 7-day arc. Note that the two models disagree with one another by about as much as they disagree with the observations—about 30 m. When the solution strategy was modified to adjust the drag coefficient every day (not shown), the along-track residuals came down to about 3 m. An alternative strategy, in which the drag models were tuned with the initial post-fit residuals, produced 7-day residuals of 3-5 m. Errors in altitude tend to be considerably lower, with the best Seasat results estimated at 30-35 cm, of which perhaps 20 cm is due to drag mismodeling. While a notable improvement, this suggests the difficulty of modeling drag with centimeter accuracy for a complex satellite at low altitude.

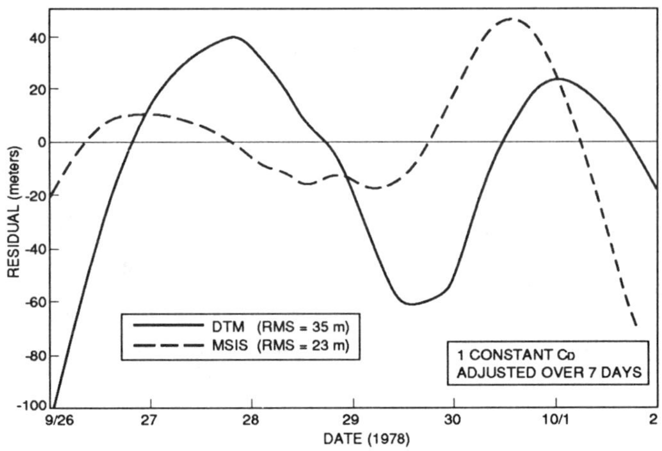

Fig. 12. Post-fit along-track residuals from Seasat dynamic orbit solutions made with two different atmospheric density models during a period of high geomagnetic activity. A single drag coefficient was adjusted for the 7-day arc. Optimized strategies reduce the residuals to about 3 m. (Plot courtesy S. Klosko, Goddard Space Flight Center.)

Drag Avoidance and Drag Correction

There are other ways to deal with drag, at least for specially designed missions. Figure 13 shows three different Earth satellites for which sub-decimeter orbit accuracy is required. In the upper left is Lageos, an example of the "cannonball" solution in which a small dense sphere is lofted to a high altitude, effectively avoiding the problem altogether. Drag, solar radiation pressure, and other forces acting on Lageos can be modeled to within a few centimeters for arcs of more than 30 days. For more conventional platforms the problem is not so easily disposed of, and for EOS in particular, a huge platform at 705 km, the prospects for centimeter drag modeling are bleak.

Two other approaches seek to measure drag and other nonconservative forces directly, to permit real time or after-the-fact compensation. In the simpler approach, an accelerometer is flown (either three-axis, or one-axis aligned along track) and its output recorded for later calibration of the non-gravitational accelerations. This strategy will be employed on the Aristoteles geopotential mapping mission, which will carry multiple accelerometers to measure gravity gradients, though these will double as drag calibrators. A more ambitious and costly approach is to

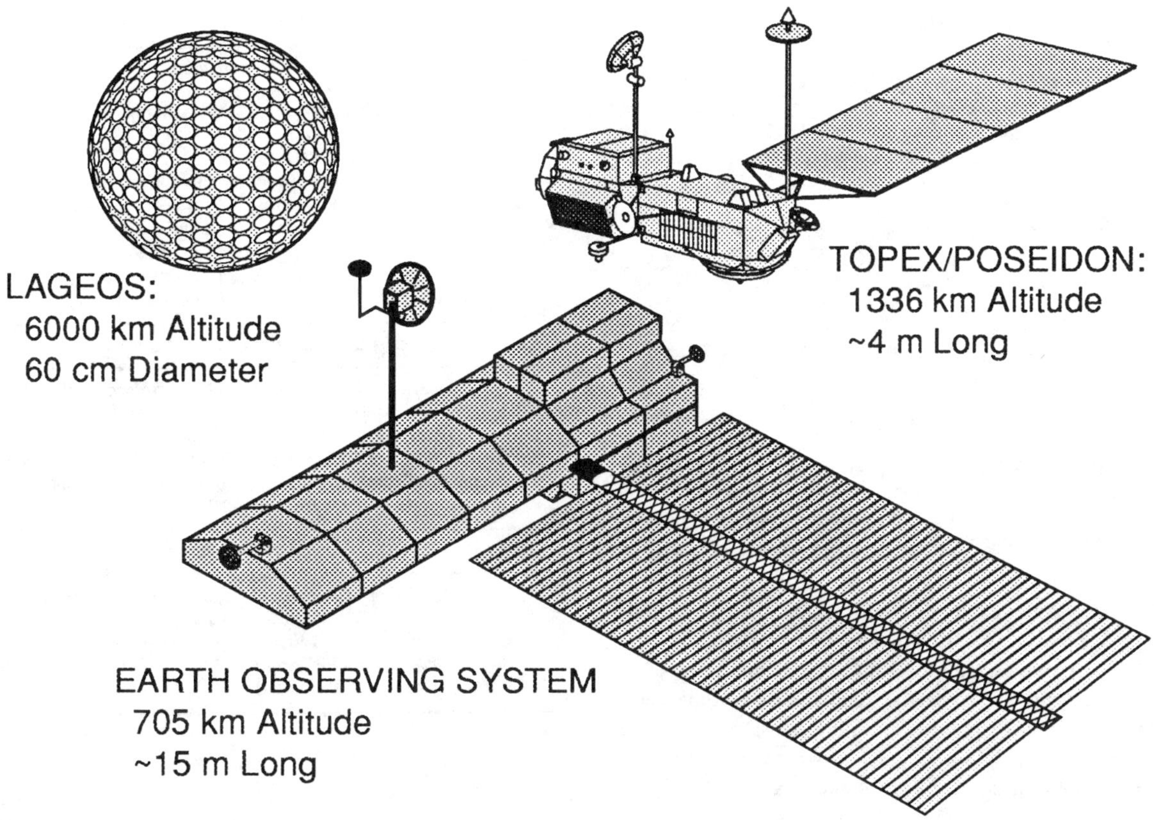

Fig. 13. Three current or planned Earth satellites seeking sub-decimeter orbit accuracies. Dynamic modeling accuracy degrades for larger vehicles and lower altitudes.

provide active velocity compensation by continuous low level thrusting. A proof mass is held suspended in an evacuated chamber, isolated from all non-gravitational forces. As drag decelerates the spacecraft, the mass will approach the chamber wall; this motion is sensed and a velocity correction applied, keeping a constant separation between the mass and the wall. This way the spacecraft is forced to follow the ideal "drag-free" trajectory of the proof mass. Two proposed spaceborne physics experiments, Gravity Probe B and the Satellite Test of the Equivalency Principle (STEP), plan to use this technique. Since a great deal of thrusting fuel is needed (e.g., in the form of evaporating liquid helium) such missions can typically last less than one year.

GPS-Based Kinematic Tracking

Many remote sensing satellites must fly at relatively low altitudes and few can easily accommodate active drag compensation. EOS is one example; others include the Space Station, proposed to fly at about 400 km altitude, and the Space Shuttle, which frequently flies even lower. At those altitudes, drag model errors can be enormous. It has recently been discovered that such seemingly lost causes (from a precise tracking standpoint) can in fact be tracked continuously in three dimensions with an accuracy of a few centimeters. This is made possible by the comprehensive observability offered by GPS, and it opens up new possibilities for precise earth observation from low orbiting platforms of opportunity. The technique is called kinematic tracking, and it differs from the popular techniques of kinematic and pseudo-kinematic surveying in two important ways: it does not require an accurate a priori position fix and it does not require (or even attempt) the resolution of phase ambiguities.

To see how this works, refer to the diagram in Fig. 14. The dashed line represents the true trajectory of a low orbiter, subject to an assortment of varying forces that may include drag, gravity, solar radiation pressure, thrusting, outgassing, and thermal radiation. The smooth solid curve represents the dynamic orbit solution obtained with a particular set of force models, some of which may have been adjusted in the solution. Because these models are imperfect, the orbit solution fails to track the detailed path variations. As a result, the orbit solution will give rise to prominent post-fit residuals, as in the Seasat solutions of Fig. 12.

Note that with tracking systems other than GPS, coverage will be incomplete. That is, there will seldom if ever be measurements in all three dimensions at once. Typically there will be data from only one site, and occasionally (or, in the case of lasers, frequently) from none at all. Solution strategies like the classical dynamic technique, which employ an orbit model to fill in the missing information, are therefore required.

Now suppose the orbiter in Fig. 14 carries a GPS receiver which acquires data continuously from at least four satellites. If a classical dynamic solution is carried out (even with very crude models)

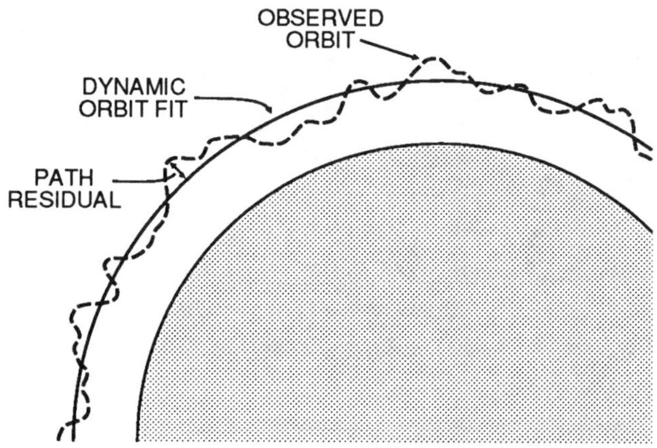

Fig. 14. In kinematic tracking, an underlying dynamic solution (solid curve) is corrected by, in effect, adding back the 3-D post fit residual at every time point to yield an observed trajectory determined solely by the measurements. Dynamic modeling error is eliminated.

it will yield post-fit residuals in all three spatial dimensions at all times. This means that by simple geometry we can reconstruct the departure of the observed trajectory from the dynamic orbit solution at every point. In other words, the satellite state is completely observable and need not be inferred from imperfect models. We can therefore devise a solution strategy in which an initial dynamic solution supplies absolute three-dimensional residuals from which the complete residual trajectory is then constructed. The observed path is recovered and dynamic model errors drop out entirely. The resulting orbit accuracy is largely independent of altitude or the presence of unpredictable dynamics, including maneuvers. The principal remaining error sources are measurement noise (the effect of which is modulated by changing geometry) and reference frame errors. The ionospheric and atmospheric correction techniques discussed here continue to apply.

In practice the kinematic solution can be obtained in a single filter step that estimates stochastic corrections to the classical dynamic solution. By varying the constraints on the stochastic corrections, we can tune the solution either to depend entirely on geometry and thus be purely kinematic, or to combine optimally the dynamic and geometric information [Yunck et al., 1990; Wu et al. 1991].

Carrier-Only Kinematic Tracking

Notice we have not specified which GPS data types—carrier phase (integrated Doppler) or pseudorange—are to be used in kinematic tracking. In principle the technique will work with either or both. The only qualification is that if the highest accuracy is desired, carrier phase data must be included. Although the solution will converge more quickly to its limiting accuracy if carrier phase and pseudorange are combined (because pseudorange helps to constrain the phase biases and clock offsets), the same limiting accuracy can be approached with phase data alone.

It may seem improbable that we can achieve high accuracy kinematically with a phase observable that is not itself absolute—that contains an arbitrary bias. But we need only observe that the initial dynamic solution provides estimates for those biases, rendering the residuals (and thus the final trajectory construction) absolute. It should be stressed that this is not the same as phase ambiguity resolution, which is neither required nor sought. It should also be noted that for high accuracy this must be carried out in the context of differential tracking (Fig. 2) in which ground sites, GPS orbits, atmospheric and other parameters are adjusted in the conventional way.

TABLE 2. Model for Shuttle Orbit Determination Analysis

System Characteristics	
Orbit (circular):	300 km, polar
Number of Ground Sites:	6 (with 3 DSN fiducial sites)
Number of GPS Satellites:	24
Flight Antenna Field of View:	Hemispherical
Flight Receiver Tracking Capacity:	8 satellites (L1 and L2)
Data Types:	L1 and L2 pseudorange L1 and L2 carrier phase
Data Interval:	5 Minutes
Smoothed Data Noise:	10-cm pseudorange
Adjusted Parameters and A Priori Errors	
Shuttle Epoch State:	1 km; 1 m/sec, each component
GPS Satellite States:	2 m; 0.2 mm/sec, each comp.
Carrier Phase Biases:	10 km
GPS and Receiver Clock Biases:	3 msec (modeled as white noise)
Non-DSN Ground Locations:	10 cm each component
Zenith Atmospheric Delay Error:	10 cm (modeled as random walk)
Fixed Errors Evaluated	
DSN Reference Positions:	3 cm each component
GM of Earth Uncertainty:	1 part in 10^8
Earth Gravity Model Error:	50% GEM10-GEML2 (20x20)
Atmospheric Drag :	10%
Solar Radiation Pressure :	10%

Covariance Analysis for Precise Tracking

The carrier-only kinematic technique permits us to achieve accurate tracking of low-altitude, dynamically unpredictable satellites with simple dual-frequency codeless receivers or even with a single-frequency unit. Recall our single-frequency technique for removing the ionospheric delay from GPS data—averaging phase and pseudorange. That technique yields a biased observable, analogous to (though less precise than) carrier phase. With the carrier-only kinematic strategy we can apply this observable to obtain orbit solutions free of drag and other force model errors. Such solutions will be less precise than those obtained with the dual-frequency phase observable, but surprisingly precise nevertheless.

A covariance study of GPS-based satellite orbit determination is presented here, combining several of the calibration techniques highlighted in this paper: single- and dual-frequency ionosphere calibration; stochastic atmospheric delay estimation; and kinematic correction of drag and other dynamic model errors. The key assumptions are shown in Table 2. We should note that the data sample interval is 5 minutes, and that the random data noise on those samples is assumed to have been smoothed (e.g., against a polynomial) over that interval. The resulting data noise levels are typical of modern high performance receivers.

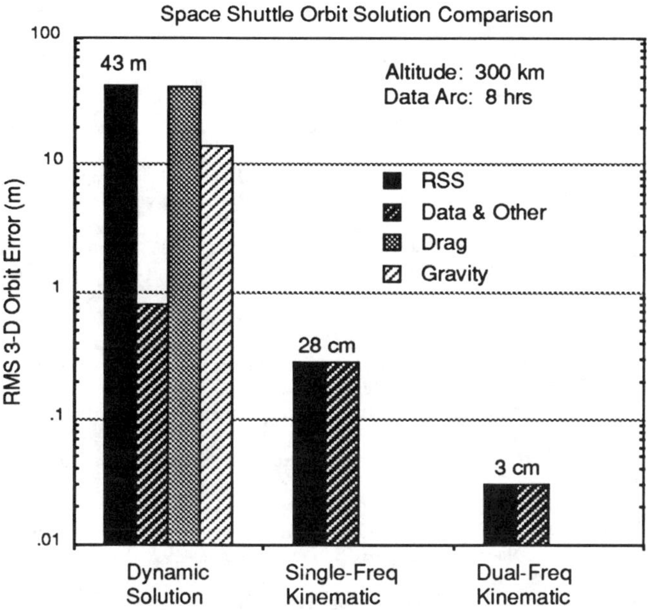

Fig. 15. Covariance analysis for Space Shuttle orbit determination at 300 km altitude, comparing three different solution strategies. RSS represents the total root sum square of the three error components.

For this analysis we have chosen a demanding test: the Space Shuttle flying in a polar orbit at an altitude of 300 km, where ionospheric delay, drag, and gravity effects are at high levels. The 11 ground receivers are of course subject to the maximum atmospheric delays, which are estimated stochastically. Figure 15 presents the results, in the form of the estimated 3-D orbit error computed as the RMS over the orbit arc, for three different solution strategies: dual-frequency dynamic, single-frequency kinematic, and dual-frequency kinematic. In the first case, the dual-frequency correction is of little value since the errors from drag and gravity mismodeling are so great. The RMS position error is tens of meters. Although one can seek to improve this by tuning the drag and lift coefficients and the gravity model, it would be extremely difficult to bring the dynamic error below a few meters.

With the kinematic techniques, the overwhelming dynamic model errors vanish and orbit accuracy improves by two to three orders of magnitude, falling to about 2 cm per component with dual-frequency phase. The order-of-magnitude difference between the two kinematic cases is explained entirely by the higher data noise on the single-frequency observable. (That case, incidentally, assumed single-frequency receivers at all ground sites as well.) The GPS kinematic tracking techniques are to be demonstrated for the first time in 1992 on the Extreme Ultra-Violet Explorer (single-frequency) and TOPEX/Poseidon (dual-frequency) satellites. Because of the limited six-satellite TOPEX receiver tracking capacity and the nominal six-site ground network, the orbit error for TOPEX is expected to be about double that shown in Fig. 15.

Conclusions

The ionosphere and electrically neutral atmosphere interfere in a variety of ways with precise positioning both on the ground and in Earth orbit. Over the years the efforts of many investigators have produced an arsenal of techniques to minimize the resulting errors. The three major space geodetic techniques—VLBI, GPS, and SLR—now routinely achieve baseline measurement accuracies of a few parts in 10^9, or millimeters over 1000 km. In addition, SLR reaches centimeter accuracy in tracking Lageos and other high-orbiting laser geodetic satellites. By straightforward extension of ground-based GPS techniques together with kinematic correction for drag and other dynamic modeling errors, we can expect to see centimeter tracking accuracy extended to dynamically unpredictable satellites at the lowest orbit altitudes. While the atmosphere and ionosphere will continue to make their presence felt, they will diminish as a threat as space geodesy comes of age in the 1990s.

Acknowledgement. I wish to thank Sien Wu, Steve Lichten, Bruce Haines, and George Hajj of JPL; Sassan Bassiri of Sharif University of Technology, Tehran, Iran; John Klobuchar of Phillips Laboratory; and Steve Klosko of ST Systems Corporation for helpful discussions. The research described in this paper was carried out in part by the Jet Propulsion Laboratory, California Institute of Technology, under contract with the National Aeronautics and Space Administration.

References

Abshire, J. B. and S. Gardner, Atmospheric refractivity corrections in satellite laser ranging, *IEEE Trans. Geosci. & Remote Sensing, 23*, 414-425, 1985.

Barlier, F., C. Berger, J. L. Falin, G. Kockarts and G. Thuillier, A thermospheric model based on satellite drag data, *Annales de Geophysique, 34*, 9-24, 1978.

Bassiri, S., Three-frequency ranging systems and their applications to ionospheric delay calibration, JPL *Telecom. and Data Acquisition Prog. Rept. 42-103*, 14-20, Jul-Sep 1990.

Bassiri, S. and G. A. Hajj, Higher-order ionospheric effects on the accuracy of the Global Positioning System observables and means of modeling them, JPL *Telecom. and Data Acquisition Prog. Rept.*, 1992 (in press).

Bishop, G. J. and J. A. Klobuchar, Ranging errors due to disturbances in the polar ionosphere, *Proc. Inst. of Navigation GPS-90*, 175-179, Colorado Springs, CO, Sept 1990.

Brunner, F. K. and M. Gu, An improved model for the dual frequency ionospheric correction of GPS observations, *Manuscripta Geodaetica, 16*, 205-214, 1991.

Chao, C. C., A model for tropospheric calibration from daily surface and radiosonde balloon measurements, *Tech, Memo. 391-350*, Jet Propulsion Lab., Pasadena, CA, 1972.

Christodoulidis, D. C., D. E. Smith, R. Kolenkiewicz, P. J. Dunn, S. M. Klosko, M. H. Torrence, S. K. Fricke and S. Blackwell, Observing tectonic plate motions and deformations from satellite laser ranging, *JGR, 90*, 9249-9264, 1985.

Coco, D. S., C. Coker and J. R. Clynch, Mitigation of ionospheric effects for single frequency GPS users, *Proc. Inst. of Navigation GPS-90*, 169-168, Colorado Springs, CO, Sept 1990.

Davis, J. L., T. A. Herring, I. I. Shapiro, A. E. E. Rogers and G. Elgered, Geodesy by radio interferometry: Effects of atmospheric modeling errors on estimates of baseline length, *Radio Science, 20*, 1593-1607, 1985.

Davis, J. L., T. A. Herring and I. I. Shapiro, Effects of atmospheric modeling errors on determination of baseline vectors from very long baseline interferometry, *JGR, 96*, 643-650, 1991.

Davis, K., *Ionospheric Radio Propagation*, Dover, New York, 1966.

Degnan, J. J., Satellite laser ranging: current status and future prospects, *IEEE Trans. Geosci. & Remote Sensing, 23*, 398-413, 1985.

Dixon, T. H. and S. Kornreich Wolf, Some tests of wet tropospheric calibration for the CASA UNO Global Positioning System experiment, *Geophys. Res. Lett., 17*, 203-206, 1990.

Elgered, G., J. L. Davis, T. A. Herring and I. I. Shapiro, Geodesy by radio interferometry: Water vapor radiometry for estimation of the wet delay, *JGR, 96*, 6541-6555, 1991.

Feess, W. A. and S. G. Stephens, Evaluation of GPS ionospheric time-delay model, *IEEE Trans. Aerosp. & Elect. Sys., 23*, 332-338, 1987.

Gu, M. and F. K. Brunner, Theory of the two frequency dispersive range correction, *Manuscripta Geodaetica, 15*, 357-361, 1990.

Haines, B. J., G. H. Born, R. G. Williamson and C. J. Koblinsky, Application of the GEM-T2 gravity field to altimetric satellite orbit computation, submitted to *JGR Oceans*, Feb 1992.

Hedin, A. E., MSIS-86 thermospheric model, *JGR, 92*, 4649-4662, 1987.

Herring, T. A., J. L. Davis and I. I. Shapiro, Geodesy by radio interferometry: The application of Kalman filtering to the analysis of very long baseline interferometry data, JGR, 95, 12561-12581, 1990.

Janssen, M. A., A new instrument for the determination of radio path delay due to atmospheric water vapor, *IEEE Trans. Geosci. & Remote Sensing, 23*, 485-490, 1985.

Jacchia, L. G., Revised static models of the thermosphere and exosphere with empirical temperature profiles, *Smith Astrophys. Obs. Special Report no. 332*, 1971.

Jursa, A. S., Ed., *Handbook of Geophysics and the Space Environment*, Air Force Geophysics Laboratory, 1985.

Klobuchar, J. A., Ionospheric time delay effects on earth-space propagation, in A. S. Jursa, Ed., *Handbook of Geophysics and the Space Environment*, Air Force Geophysics Laboratory, pp. 10-84 to 10-88, (1985a).

Klobuchar, J. A., Ionospheric total electron content (TEC), in A. S. Jursa, Ed., *Handbook of Geophysics and the Space Environment*, Air Force Geophysics Laboratory, pp. 10-89 to 10-96, (1985b).

Klobuchar, J. A., Ionospheric time-delay algorithm for single-frequency GPS users, *IEEE Trans. Aerosp. & Elect. Sys., 23*, 325-331, 1987.

Klobuchar, J. A. and P. H. Doherty, The statistics of ionospheric time delay for GPS ranging on L1, *Proc. Inst. of Navigation GPS-90*, 161-168, Colorado Springs, CO, Sept 1990.

Klobuchar, J. A., Ionospheric effects on GPS, *GPS World*, 48-51, Apr 1991.

Lambeck, K., *Geophysical Geodesy*, Clarendon Press, Oxford, 1988.

Lanyi, G. E., Tropospheric calibration in radio interferometry, *Proc. Int'l. Symp. on Space Techniques for Geodynamics*, IAG/COSPAR, Sopron, Hungary, 184-195, 1984.

Lanyi, G. E. and T. Roth, A comparison of mapped and measured total ionospheric electron content using global positioning system and beacon satellite observations, *Radio Sci., 23*, 483-492, 1988.

Lichten, S. M., Precise estimation of tropospheric path delays with GPS techniques, *TDA Progress Report, 42-100*, 1-12, 15 Feb 1990.

Lichten, S. M., Estimation and filtering for high-precision GPS positioning applications, *Manuscripta Geodaetica, 15*, 159-176, 1990a.

Lichten, S. M. and W. I. Bertiger, Demonstration of sub-meter GPS orbit determination and 1.5 parts in 10^8 three-dimensional baseline accuracy, *Bull. Geod., 63*, 167-189, 1989.

Lichten, S. M. and J. S. Border, Strategies for high-precision Global Positioning System orbit determination, *JGR, 92*, 12751-12762, 1987.

Llewellyn, S. K. and R. B. Bent, Documentation and description of the Bent ionospheric model, AFCRL-TR-73-0657, AD 772733, 1973.

MacDoran, P. F., A first-principles derivation of the differenced range versus integrated Doppler (DRVID) charged-particle calibration method, *JPL Space Programs Summary 37-62*, Vol II, March 1970.

Marini, J. W. and C. W. Murray, Jr., Correction of laser range tracking data for atmospheric refraction at elevation angles above 10 degrees, Goddard Space Flight Center, Greenbelt, MD, *NASA Tech. Rep. X-591-73-351*, 1973.

Marsh, J. G. and R. G. Williamson, Precision orbit analyses in support of the Seasat altimeter experiment, *J. Astronaut. Sci., 28*, 345-369, 1980.

Milliken, R. J. and C. J. Zoller, Principles of operation of NAVSTAR and system characteristics, *Navigation, 2*, 95-106, 1978.

Papas, C. H., *Theory of Electromagnetic Wave Propagation*, McGraw-Hill, New York, 1965.

Ridgway, J. R., S. Klosko and C. J. Koblinsky, Analysis of SEASAT residuals to extract atmospheric drag information (abstract), *EOS Trans. AGU, 71*, 1277, 1990.

Rishbeth, H. and O. K. Garriott, *Introduction to Ionospheric Physics*, Academic Press, New York, 1969.

Royden, H. N., R. B. Miller and L. A. Buennagel, Comparison of NAVSTAR satellite L band ionospheric calibrations with Faraday rotation measurements, *Radio Sci., 19*, 798-804, 1984.

Scharroo, R., K. F. Wakker, B. A. C. Ambrosius and R. Noomen, On the along-track acceleration of the LAGEOS satellite, *JGR, 96*, 729-740, 1991.

Smith, D. E., R. Kolenkiewicz, P. J. Dunn, J. W. Robbins, M. H. Torrence, S. M. Klosko, R. G. Williamson, E. C. Pavlis, N. B. Douglas and S. K. Fricke, Tectonic motion and deformation from satellite laser ranging to LAGEOS, *JGR, 95*, 22013-22041, 1990.

Spilker, J. J., GPS signal structure and performance characteristics, *Navigation, 25*, 29-54, 1978.

Stark, J., Aerodynamic modelling of spacecraft for precise orbit determination, *Proc. 2nd Int'l. Symp. on Spacecraft Flight Dynamics*, Darmstadt, FRG, 20-23 Oct 1986 (*ESA SP-255*, Dec. 1986).

Tapley, B. D., B. E. Schutz and R. J. Eanes, Station coordinates, baselines, and earth rotation for LAGEOS laser ranging: 1976-1984, *JGR, 90*, 9235-9248, 1985.

Thornton, C. L., J. L. Fanselow and N. A. Renzetti, GPS-based geodetic measurements utilizing the global positioning system satellites, *Space Techniques and Geodesy*, Academic Press, London, 1985.

Tralli, D. M., T. H. Dixon and S. A. Stephens, The effect of wet tropospheric path delays on estimation of geodetic baselines in the Gulf of California using the Global Positioning System, *JGR, 93*, 6545-6557, 1988.

Treuhaft, R. N. and G. E. Lanyi, The effect of the dynamic wet troposphere on radio interferometric measurements, *Radio Sci., 22*, 251-265, 1987.

Treuhaft, R. N. and S. Lowe, A measurement of planetary relativistic deflection, *Astron. J., 102*, 1879-1888, 1991.

Treuhaft, R. N., Tropospheric and charged particle propagation errors in very long baseline interferometry, *Proc. IAG Symp. on Refraction of Transatmospheric Signals in Geodesy*, The Hague, The Netherlands, 19-22 May 1992.

Wu, S. C. and Thornton, C. L., OASIS—A new GPS covariance and simulation software system, *Proc. 1st Int. Symp. on Precise Positioning with GPS*, National Geodetic Information Center, Rockville, MD, May 1985.

Wu, S. C., W. I. Bertiger, J. S. Border, S. M. Lichten, R. F. Sunseri, B. G. Williams, P. J. Wolff, J. T. Wu, *OASIS Mathematical Description, V. 1.0*, JPL D-3139 (internal document), 1986.

Wu, S. C., T. P. Yunck and C. L. Thornton, Reduced-dynamic technique for precise orbit determination of low earth satellites, *J. Guidance, Control, and Dynamics, 12*, 24-30, Jan-Feb 1991.

Wu, S. C., J. T. Wu and W. G. Melbourne, Equivalent GPS measurements for efficient estimation process, *AAS paper 91-538*, AAS/AIAA Astrodynamics Specialist Conference, Durango, CO, 19-22 Aug 1991.

Yunck, T. P., W. G. Melbourne and C. L. Thornton, GPS-based satellite tracking system for precise positioning, *IEEE Trans. Geosci. & Remote Sensing, 23*, 450-457, 1985.

Yunck, T. P., S. C. Wu, J. T. Wu and C. L. Thornton, Precise tracking of remote sensing satellites with the Global Positioning System, *IEEE Trans. Geosci. & Remote Sensing, 28*, 108-116, Jan 1990.

Yunck, T. P., S. C. Wu, S. M. Lichten, W. I. Bertiger, U. J. Lindqwister and G. Blewitt, Toward centimeter orbit determination and millimeter geodesy with GPS, *Proc. 5th Int. Geodetic Symp. on Satellite Positioning*, Las Cruces, NM, 272-281, Mar 1989.

Thomas P. Yunck, Jet Propulsion Laboratory, California Institute of Technology, Pasadena, California.

Modelling and Prediction of Ionospheric Disturbances Affecting the Accuracy of Position Measurements (including Scintillations)

SHAUN QUEGAN

Department of Applied and Computational Mathematics, University of Sheffield, Sheffield S10 2TN, UK.

Variations in the electron density at ionospheric heights can have significant impact on systems using trans-ionospheric propagation paths. Effects include fading for space-based radars, loss of phase coherence in synthetic aperture formation and errors in transforming signal delay into distance for altimeters. As our physical understanding grows, empirical models of the global ionosphere and the occurrence of irregularities and scintillations are beginning to be complemented by physically-based models for the interaction of the solar wind with the earth's magnetosphere/ionosphere/thermosphere system. These models embody mechanisms, such as density or temperature gradients, large electric fields and particle precipitation, which can give rise to large and small scale variations in the ionospheric plasma density, as well as mechanisms for distributing irregularities in space. Parallel advances in understanding the physics of plasma instabilities, and development of propagation tools such as phase-screen models, open up the prospect of using global-scale models embodying the small-scale physics of irregularities to assess the effects of ionospheric variability on positional measurement and more general systems which rely on ionospheric propagation paths.

INTRODUCTION

Space-based microwave systems used for remote sensing and positioning use the ionosphere as a propagation path. At these frequencies, the presence of free electrons causes slight deviations of the refractive index from unity. Over the long propagation paths involved, these deviations are large enough to introduce significant errors into measurements of distance based on signal delay. When the ionosphere is irregular, slight variations in phase induced during the passage through the ionosphere can also give rise to large phase and amplitude fluctuations observed at the ground. These are known as scintillations. Phase scintillation is particularly troublesome for systems, such as synthetic aperture radar (SAR), which rely on phase coherence; for a SAR, even if scintillation is not severe enough to prevent image formation, uncorrected phase errors can cause the beam direction to be tilted, leading to positional errors.

Positional accuracy requirements for some applications are very stringent, as a few examples illustrate. The TRANSIT system is a satellite-based geodetic positioning system designed to give single point positional accuracy to approximately 1 metre; relative positioning of two or more points can be obtained to within 20 - 30 cm [Ashkenazi and Sykes, 1980]. The Global Positioning System (GPS) is primarily designed to provide high accuracy instantaneous three-dimensional positioning. Though this instantaneous positioning is less accurate than TRANSIT can provide (albeit by observations over periods of the order of three days), GPS is intended to provide relative positions on the order

Environmental Effects on Spacecraft Positioning and Trajectories
Geophysical Monograph 73, IUGG Volume 13
Published in 1993 by the International Union of Geodesy and Geophysics and the American Geophysical Union.

of 1 cm, by using interferometric techniques [Hothem and Fronczek, 1983]. Very long base interferometry (VLBI) uses correlation of extra-galactic radio signals received at very widely spaced ground stations to investigate crustal dynamics to centimetric accuracy [Ashkenazi and McLintock, 1982]. Satellite altimetry makes precise measurements of the time delay suffered by a microwave pulse on the path to and from the satellite's nadir point. Accuracy of the order of 10 cm enabled the Seasat altimeter to provide striking images of how ocean currents and sea-bed topography leave distinct signatures in mean sea-level and sea-level variability [Marsh and Martin, 1982; see also Lorell et al., 1982]. Such information will be further enhanced by the altimeter carried on the ERS-1 satellite.

Quantification (and, in some cases, correction) of ionospheric effects on such systems relies on models of the ionospheric electron density structure. This is dependent on a wide range of parameters, such as time (local, magnetic, universal and day of the year), solar and magnetic activity, electric fields, etc. Figure 1 gives some idea of the problem. It shows idealised electron density measurements by a satellite orbiting in the noon-midnight meridian. There are a number of large-scale structures, such as the mid-latitude trough, and there are several regions showing fine structure, such as the cusp. Empirical models provide some basis for describing these structures, but they are being increasingly complemented by physically-based models, as our understanding of ionospheric processes grows. Physical models are particularly valuable in the high-latitude regions, where ionospheric behaviour is at its most dynamic. This is because comprehensive databases are not available for these regions, but also because the averaging involved in formulating statistical models causes smoothing out of the large gradients and sharp boundaries which are characteristic of the ionosphere at these latitudes.

Existing models essentially fall into two types, corresponding to different spatial scales. Macro-scale models are concerned with the large-scale structures observed in the ionosphere, such as the mid-latitude trough, the equatorial anomaly, the auroral zone, etc. Micro-scale models deal with the irregularities which give rise to scintillation, with spatial scales from a few metres up to tens or hundreds of kilometres. The two scales are related, since the physical conditions giving rise to irregularities are dependent on the macro-scale ionospheric structure and on the processes which give rise to this structure.

In this paper we describe the basic physics of signal delay and scintillation caused by propagation through

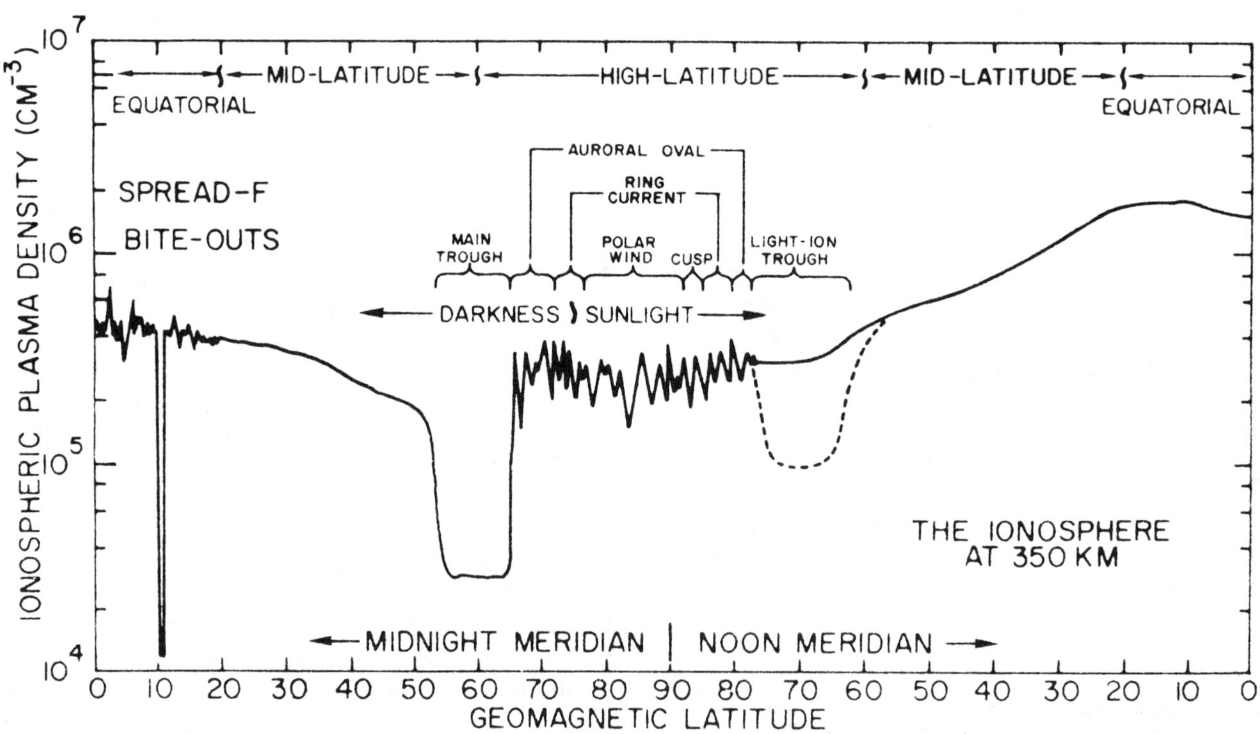

Fig. 1. Idealised plot of electron densities observed by a satellite orbiting in the noon-midnight meridian (from Szuszczewicz et al., 1983).

the ionosphere. The types of ionospheric model needed to quantify these effects and their present state of development are reviewed. Because this involves (at least) two major research areas with a vast literature, the treatment is not meant to be comprehensive, and is biased by the author's own work and interests. For the same reason, the choice of references is selective; key references for further reading are provided. We summarise by identifying some of the areas where advances in modelling and prediction of ionospheric disturbances and their consequences are likely.

Basic Concepts And Examples

The starting point for analysis of ionospheric effects on position measurements is the ionospheric refractive index μ. For space systems operating at VHF frequencies or above, μ can be well-approximated by the expression

$$\mu^2 = 1 - \frac{Ne^2}{4\pi^2\epsilon_0 m_e f^2} \tag{1}$$

where e is the electron charge, m_e is the electron mass, ϵ_0 is the permittivity of free space, N is the electron density, and f is the (cyclic) frequency of the propagating wave. (For a full description of μ see, for example, Ratcliffe (1972) or Egeland et al. (1973)). For frequencies exceeding 30 MHz, we can further approximate μ by

$$\mu = 1 \mp 40.3\frac{N}{f^2} \tag{2}$$

where N is now expressed in electrons m^{-3}. In this expression the negative sign corresponds to the phase refractive index, while the plus sign gives the group refractive index. The latter is what is needed to describe signal delay in a propagating wave. The wave is slowed by a factor μ compared to free space propagation as it traverses the ionosphere, which causes a change in apparent (one-way) path length

$$\begin{aligned}\Delta l &= \int_S (\mu - 1) ds \\ &= \frac{40.3}{f^2}\int_S N ds \\ &= \frac{40.3}{f^2} \times \text{TEC}\end{aligned} \tag{3}$$

Here we are treating the radiation as a ray travelling along the path S between the transmitter and receiver. The integrated electron density along this path is the total electron content (TEC).

The transmitter and receiver may be located on the satellite, as would be the case for an altimeter or synthetic aperture radar (SAR), or they may be separated, one on the ground and one on the satellite, as would occur in, for example, the Global Positioning System. For the systems discussed in this paper the path S may be considered as effectively a straight line, i.e., the frequencies are high enough to prevent significant bending of the wave vector from its initial direction. This would no longer be a viable assumption at lower frequencies. (Indeed, at frequencies less than some critical value in the HF band, refractive tilting of the wave-vector would be sufficient to prevent any penetration of the ionosphere; this principle is exploited by ionosondes and over-the-horizon radar).

The ionosphere therefore induces a range error which is proportional to TEC and inversely proportional to the square of the frequency. TEC shows considerable spatial and temporal variation. Figure 2 shows monthly median TEC contours inferred from ATS-3 satellite beacon measurements at a variety of stations in the northern hemisphere. The range of values exhibited is from 3×10^{16} electrons m^{-2} to 2.8×10^{17} electrons m^{-2}, and there are clear diurnal and spatial variations. These are median values, and individual measurements of the order of 10^{18} electrons m^{-2} have been observed. The range delay associated with a TEC of 5×10^{17} electrons m^{-2} for a variety of systems using the ionosphere as a propagation path is shown in Table 1.

For positioning measurements relying on a precise knowledge of range delay, the ionospheric electron content therefore imposes a positive bias on the measurement. This bias, which varies with time and position, can be removed if two frequencies are available. It is easy to show that if the apparent path length at frequency f_i is l_i (i = 1, 2), then the true path length is

$$l = \frac{f_1^2 l_1 - f_2^2 l_2}{f_1^2 - f_2^2} \tag{4}$$

(Here we are ignoring second-order effects, which become significant at lower frequencies, and tropospheric delay). Not all systems use two frequencies. For example, the altimeter on the ERS-1 satellite is a single frequency instrument operating at 13.8 GHz. In this case, correction for the ionosphere must rely on models for TEC (possibly in a time-average sense if multiple observations are available). Such models are not necessarily easy to derive, because of their dependence on system geometry, horizontal gradients in electron density, and the earth's magnetic field geometry.

The magnetic field plays an important role because of the fact that, above approximately 180 km, the dynamics of the plasma can, to a good approximation, be decoupled into motion perpendicular and parallel to

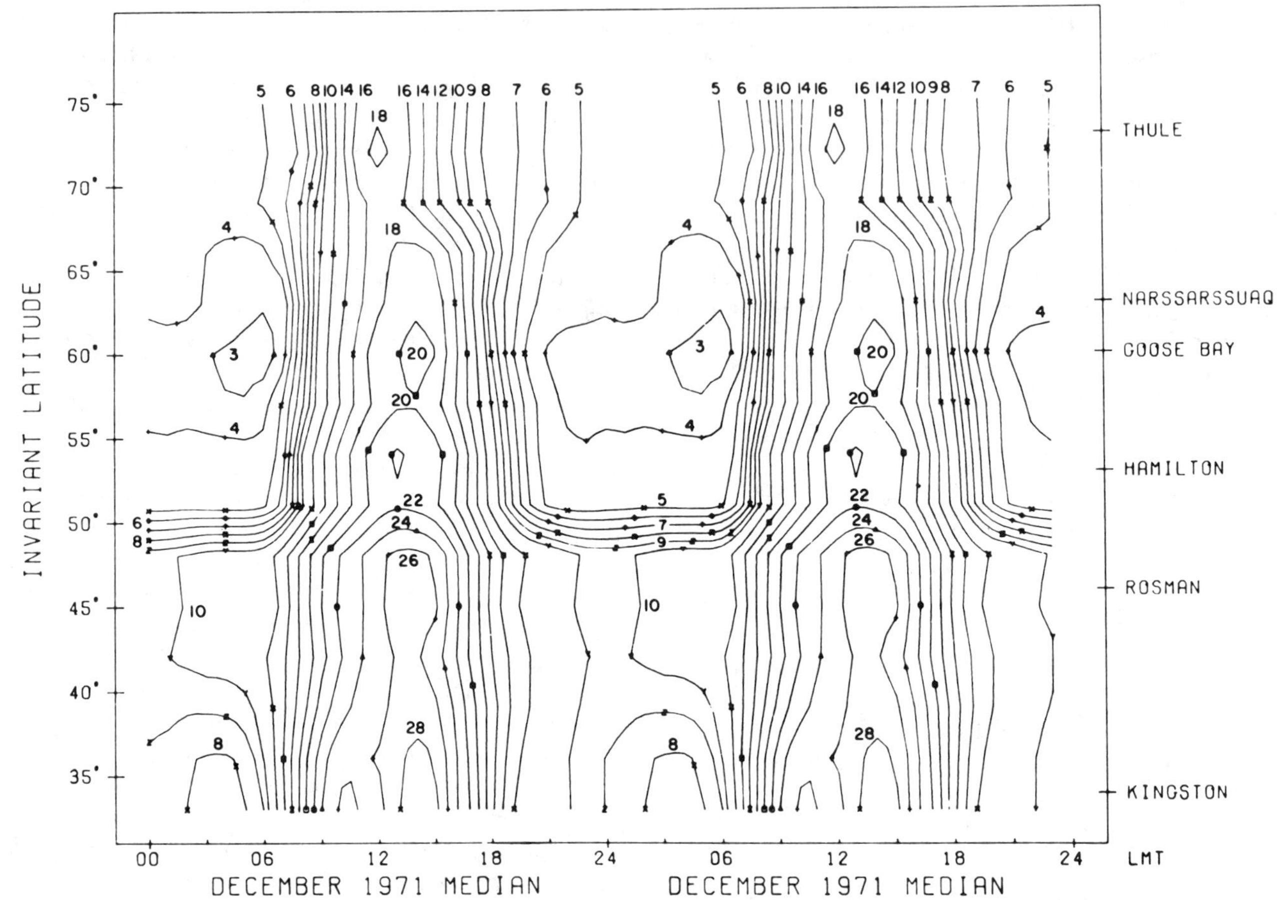

Fig. 2. Monthly median ATS 3 TEC contours for December 1971 as a function of invariant latitude and local time. The contour levels are in units of 10^{16} electrons m^{-2} (from Mendillo and Klobuchar, 1975).

the magnetic field [Schunk, 1975]. A typical plasma distribution along the field is shown in Figure 3. At the lowest altitudes molecular species predominate in a chemically controlled regime. With increasing height, the dominant ion becomes O^+ and diffusion becomes increasingly important. Electron density maximises at the F2-peak, which occurs typically at an altitude of 300-400 km, and roughly marks the transition from chemical to diffusive control [Ratcliffe, 1972]; this maximum density is denoted NmF2. At greater heights H^+ becomes the dominant ion, the transition height being dependent on local time, solar cycle, and season. (Under some circumstances, He^+ can become dominant in a layer between the O^+ and H^+ regimes [Quegan et al., 1984; Bailey and Sellek, 1990], but in most regions and at most times this does not occur).

TABLE 1. Range Delays for a TEC of 5×10^{17} el. m^{-2}

System	Frequency (GHz)	Δl (m)
TRANSIT	0.15	900
GPS	1.2	14
VLBI	8.4	0.3
Seasat altimeter	14.6	0.1

Along a single field line, the relative contribution to the TEC from the different regimes cannot be described in an elementary way, but a rough rule of thumb is that approximately 2/3 of the TEC occurs above the F2-peak and 1/3 below. This means that horizontal gradients in N give rise to variations in TEC on different field-lines. However, the ray-path from a satellite to the ground may cut through many field lines, giving problems in converting observed or modelled ionospheric densities into propagation effects for any particular geometry. For the purposes of range delay calcu-

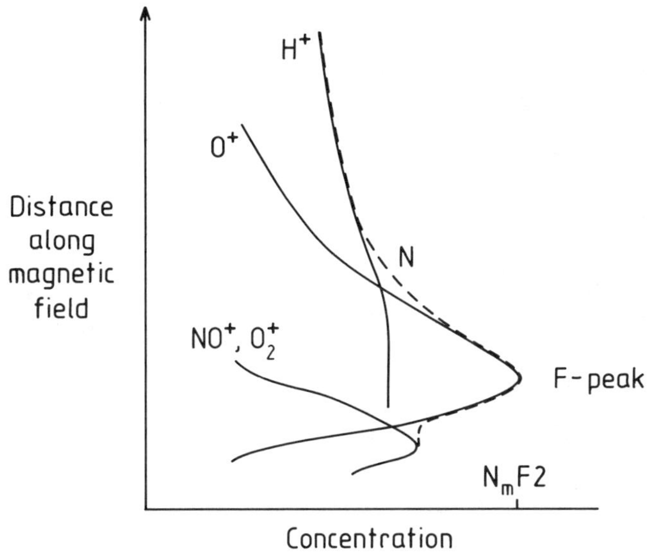

Fig. 3. Typical structure of the dominant ion distributions in the magnetic field-aligned direction for the dayside ionosphere. The electron density at each point is the sum of the ion densities.

lations it is often convenient to regard the ionosphere as being a slab of thickness L with constant electron density NmF2, so that

$$L \times \text{NmF2} = \text{TEC} \tag{5}$$

This gives rise to slab thicknesses of the order of 50-200 km.

Delay caused by propagation through a uniform ionospheric slab gives rise to range positioning errors. A very different form of positioning error can be produced by an irregular ionosphere. In this case, the phase fronts of a plane wave will no longer be planar after traversing the ionosphere (see Figure 4). For weak scattering, this phase perturbation will be the only significant effect; amplitude effects developed in the slab are negligible, so that the ionosphere can be treated as a phase-changing screen. The perturbed wave can be represented as a collection of plane waves, each with the same wavelength but with its own wave vector; this is the wave spectrum of the perturbed signal. However, this immediately leads us to expect interference effects, as Figure 5 illustrates. Here two plane waves of the same wavelength λ but with different wave-vector directions are propagating towards the ground. In one wave the wave fronts are parallel to the ground; in the other, the wave fronts (marked as dashed lines) make an angle θ with the ground. If the waves are in-phase at the point P, then the difference between the phases of the waves at O will be $kz(1 - \cos\theta)$ where $k = 2\pi/\lambda$ is the wave-number and z is the distance from P to the observation point O. Hence the waves will be in phase at O if $z(1 - \cos\theta) = n\lambda$ for an integer n, and will be out of phase if $z(1 - \cos\theta) = (n + 1/2)\lambda$. Moving horizontally away from O, the phase difference will be $k[z(1 - \cos\theta) - x\sin\theta]$, giving rise to a sinusoidal spatial pattern of constructive and destructive interference. This spatial pattern is what we mean by scintillation. The derivation of the wave spectrum and its propagation properties are described in a classic paper by Ratcliffe (1956).

This elementary description can be placed in a more comprehensive context by using linear filtering concepts. The propagation of a wave from the base of the ionospheric slab to the ground can be described by the convolution

$$g(\mathbf{x}) = f(\mathbf{x}) ** h(\mathbf{x}) = \int_{-\infty}^{\infty}\int_{-\infty}^{\infty} f(\mathbf{x} - \mathbf{y})h(\mathbf{y})d\mathbf{y} \tag{6}$$

Here $\mathbf{x}$ is a two-dimensional spatial variable, $g(\mathbf{x})$ is the observed signal at the ground, $f(\mathbf{x})$ is the signal on exit from the ionospheric slab, and $h(\mathbf{x})$ is the Fresnel filter:

$$h(\mathbf{x}) = \frac{1}{\lambda z}\exp(-j\pi|\mathbf{x}|^2/\lambda z) \tag{7}$$

where z is the distance from the base of the slab to the ground. Since $f(\mathbf{x})$ is a random phase-perturbed wave, it can be written

$$f(\mathbf{x}) = e^{j\phi(\mathbf{x})} \tag{8}$$

where $\phi(\mathbf{x})$ is a random phase function.

One of the most useful ways of describing the variation of random functions is by the spatial autocorrelation function or power spectrum. In the following, the notation $\Phi_\alpha(\mathbf{u})$, where $\mathbf{u}$ is the two-dimensional spatial wave number, denotes the spatial power spectrum of a general random function $\alpha(\mathbf{x})$. Equation (6) then leads to the relation

$$\Phi_g(\mathbf{u}) = \Phi_f(\mathbf{u})\ |H(\mathbf{u})|^2 \tag{9}$$

where

$$h(\mathbf{x}) \longleftrightarrow H(\mathbf{u}) \tag{10}$$

is a two-dimensional Fourier transform. In fact it is easy to see that $|H(\mathbf{u})|^2 = 1$, so that equation (9) appears to be vacuous. However, f and g are complex signals, and observationally we are interested in their amplitude and phase. Under weak scattering conditions, the variance of $\phi(\mathbf{x})$ is small, and it is then possible to show [Yeh and Liu, 1982] that

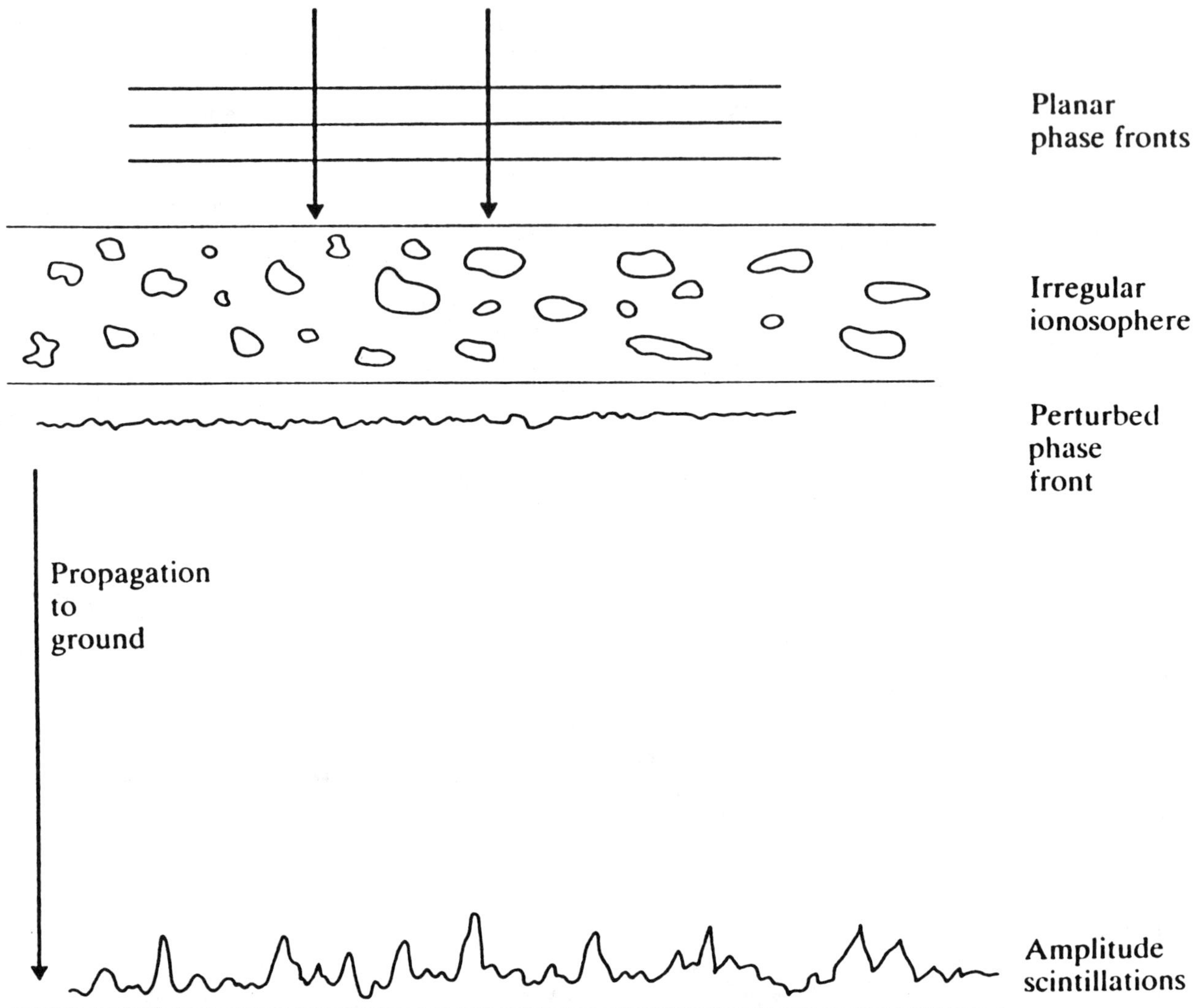

Fig. 4. The origin of scintillations; a plane wave normally incident on an irregular ionosphere suffers phase perturbations; interference effects then develop as the disturbed wave propagates to the ground.

$$\chi(\mathbf{x}) \approx \phi(\mathbf{x}) ** h_1(\mathbf{x}) \tag{11}$$

and

$$S_1(\mathbf{x}) \approx \phi(\mathbf{x}) ** h_2(\mathbf{x}) \tag{12}$$

Here χ and S_1 are the log amplitude and phase of the observed wave, and h_1 and h_2 are filter functions:

$$h_1(\mathbf{x}) = \text{Re } h(\mathbf{x}) = \frac{1}{\lambda z} \cos\left(\frac{\pi}{\lambda z}|\mathbf{x}|^2\right) \tag{13}$$

and

$$h_2(\mathbf{x}) = -\text{Im } h(\mathbf{x}) = \frac{1}{\lambda z} \sin\left(\frac{\pi}{\lambda z}|\mathbf{x}|^2\right) \tag{14}$$

Then

$$\Phi_\chi(\mathbf{u}) = \Phi_\phi(\mathbf{u}) \sin^2\left(\frac{\lambda z|\mathbf{u}|^2}{4\pi}\right) \tag{15}$$

and

$$\Phi_{S_1}(\mathbf{u}) = \Phi_\phi(\mathbf{u}) \cos^2\left(\frac{\lambda z|\mathbf{u}|^2}{4\pi}\right) \tag{16}$$

Hence to find the spatial spectrum of log amplitude and phase at the ground, all we need to know is the power spectrum of phase on exit from the ionospheric slab. The filter functions in Eqs. (13)-(16) do the rest for us. But the power spectrum of phase can be readily related to the spatial variation of TEC in the slab and to the three-dimensional electron density fluctuations ΔN in the ionospheric slab by the relations

$$\begin{aligned} \Phi_\phi(\mathbf{u}) &= A_1 \Phi_{\text{TEC}}(\mathbf{u}) \\ &= A_2 \Phi_{\Delta N}(\mathbf{u}, 0) \end{aligned} \tag{17}$$

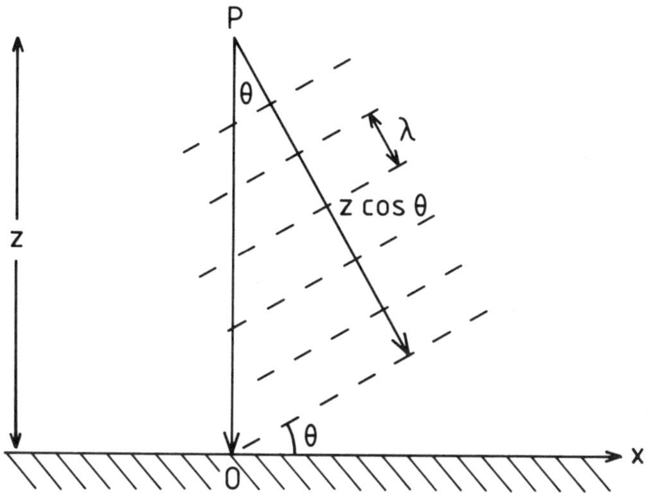

Fig. 5. Interference effects arising between two components of the wave spectrum whose wave vectors have different directions.

where A_1 and A_2 are constants (see Yeh and Liu (1982) for their values) and the 0 in the argument of $\Phi_{\Delta N}$ indicates evaluation at the top of the ionospheric slab.

Observations from a variety of sources indicate that electron density fluctuations exhibit a power-law spectrum, i.e., $\Phi_{\Delta N}(\mathbf{k})$ varies as $|\mathbf{k}|^{-p}$, where $\mathbf{k}$ is a three-dimensional wave number (since electron density fluctuations occur in three dimensions) and p is known as the spectral index. From Eq. (17) this implies that the two-dimensional spatial power spectrum of phase also has a power law with spectral index p. (It should be noted that care is needed in reading the literature since not all authors use the same nomenclature. Yeh and Liu (1982) define the spectral index p as above. Other workers prefer to define the spectral index by the directly observable index of the temporal power spectrum of phase, often with the same notation, p. As discussed below, these two spectral indices differ by 1).

In Figure 6 we attempt to illustrate and demystify the rather abstract treatment above. The top curve in Figure 6 is the filter function of Eq. (15); the second curve is a power-law spectrum of phase at the base of the ionosphere; and the third, their product, is the spectrum of log amplitude observed at the ground. Figure 7 shows the analogous results for the phase spectrum observed at the ground. We also show in Figure 8 the results of a slightly more general treatment using the Rytov solution for a wave propagating through an irregular slab of thickness $L = 100$ km. This takes into account the effects of scattering on the amplitude of the wave as it traverses the ionosphere. This treatment again gives

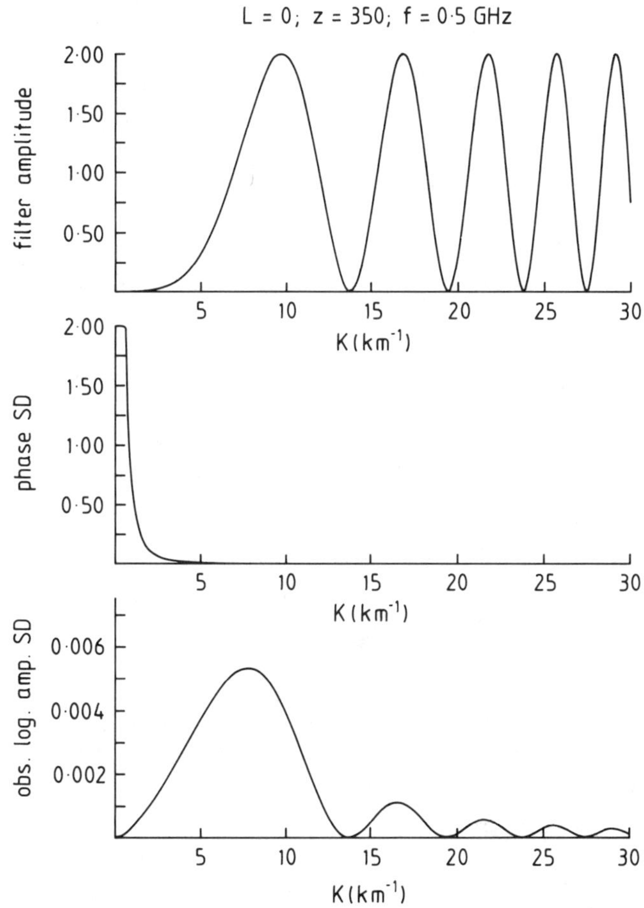

Fig. 6. The generation of amplitude scintillations by Fresnel filtering. The top panel shows the filter function. The middle panel shows the phase spectrum of the wave emerging from the base of the ionospheric slab, corresponding to a power law spectrum for ionospheric density fluctuations with a spectral index of 4.5. The bottom panel gives the resulting spatial spectrum of log amplitude observed at the ground. The calculations are for a thin phase screen ($L = 0$), at an altitude $z = 350$ km for a wave of frequency 500 MHz.

rise to a filter function, which is shown in the top panel of Figure 8. An obvious inference from these plots is that Fresnel filtering suppresses the contribution of the low frequencies in the spatial phase variation at the base of the ionospheric slab to the amplitude variation observed at the ground. Using Eq. (15), we see that the filtering term is a maximum for spatial wavenumbers satisfying

$$|\mathbf{u}| = \pi\sqrt{\frac{2}{\lambda z}} \tag{18}$$

Since wavelength is given by $2\pi/|\mathbf{u}|$, this corresponds

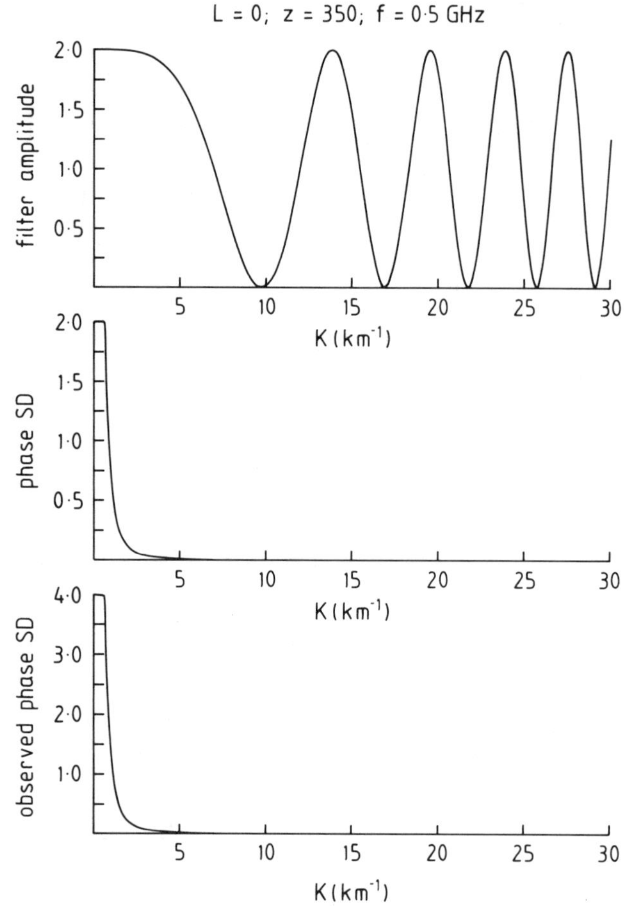

Fig. 7. As for Fig. 6, but for phase scintillations.

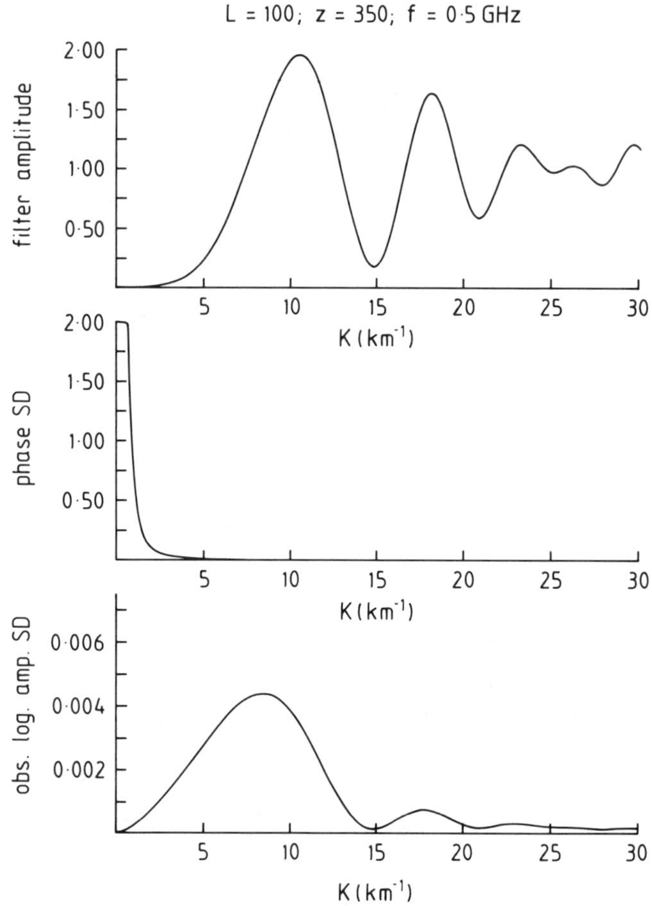

Fig. 8. As for Fig. 6, but using the Rytov solution, which allows amplitude fluctuations inside the ionospheric slab. This modifies the filter function, as shown in the top panel. The slab is of thickness $L = 100$ km, and its upper boundary is at an altitude $z = 350$ km.

to a wavelength of $\sqrt{2\lambda z}$. We can interpret this as implying that ionospheric irregularities with a scale size of the order of $\sqrt{2\lambda z}$ contribute most power to the observed pattern. By contrast, Figure 7 shows that phase fluctuations are most affected by large-scale irregularities in the ionosphere.

These calculations tell us that a plane wave normally incident on a stationary irregular ionosphere will give rise to phase and amplitude scintillations. For very many systems (such as those using a single receiver) these manifest themselves not as spatial but as temporal variations. This occurs because of relative motion of the ionosphere, the transmitter and/or the receiver. Yeh and Liu (1982) carry out an analysis of this type of situation. For interpretation of observational data, the most significant results of this analysis are that a power-law form for $\Phi_{\Delta N}$ gives rise to a power-law with index $p-1$ for the temporal power spectrum of phase, while the temporal power spectrum of log amplitude should be flat at the lower frequencies with a "roll-off" to a power-law with index $p-1$ at high frequencies. The analysis can also be extended to take account of the relative geometry of the propagation vector, the magnetic field, and the ionospheric layer, and to allow for a finite beam width.

Examples of scintillation are given in Figures 9 and 10. Figure 9 shows scintillation in the intensity of the signal from the TRIAD satellite, observed at Poker Flat, Alaska. Figure 10 shows the temporal power spectrum of amplitude and phase on a 40 MHz link observed at Boulder, Colorado. At high frequencies, the spectra have a linear trend in the logarithmic plot used; this indicates that the spectra obey a power law, with index ν given by the slope of the plot, i.e., $\nu = 3.3$. The amplitude spectrum rolls off at low frequencies. As discussed above, this behaviour is expected on the basis

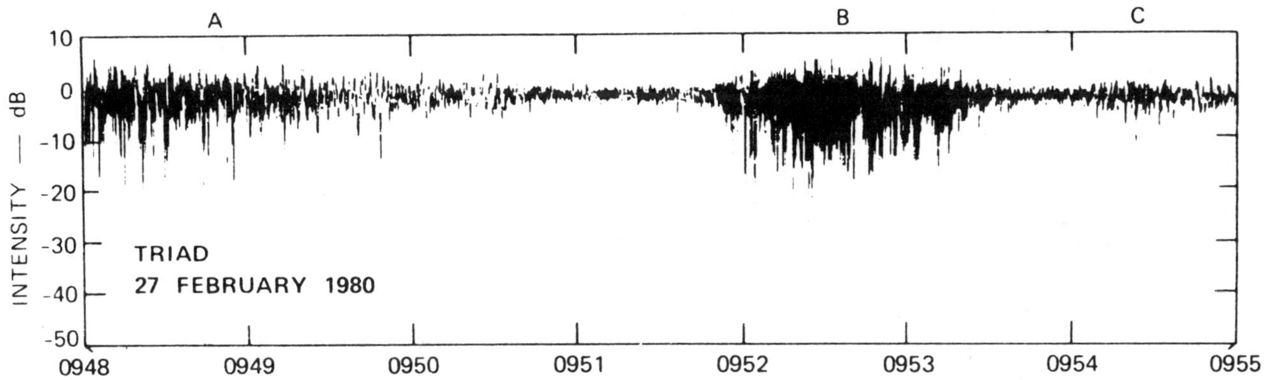

Fig. 9. Observations of amplitude scintillation generated by the signal from the TRIAD satellite (from Rino and Vickrey, 1982).

of weak scattering theory, and these spectra are consistent with a power law spectrum for the electron density fluctuations with spectral index 4.3.

When spaced receivers or transmitters are used, the spatial variations can have direct consequences. An example of such a system is a space-borne SAR. A SAR obtains high spatial resolution by saving the returns from many emitted pulses, correcting for the changes in phase associated with the propagation path, and then adding coherently. In this way a large antenna is synthesised, with the remarkable property that its spatial resolution at any range R is independent of R and of the operating wavelength, and is given to a good approximation by half the length of the real antenna [Elachi, 1988]. The key to this performance is accurate correction for the propagation-induced changes in phase occurring while the beam is being synthesised. Spatial variability in the phase degrades performance. Figure 11 shows how a highly irregular ionosphere can affect the SAR beam shape. At shorter wavelengths (corresponding to shorter synthetic antennas) the beam is hardly affected, but as wavelength increases the effects become progressively worse. The main effects can be seen as shifting in the position of the peak (corresponding to a positioning error), increase in the side-lobes (giving loss of contrast) and increase in the width of the main lobe (giving degraded resolution). At L-band ($\lambda = 24$ cm), effects are so severe as to completely destroy the structure of the antenna gain pattern, leading effectively to complete loss of the imaging capability. Note that the positional shifts are given in metres, and are in the direction of satellite motion.

Model requirements

To meet the needs of the variety of systems using the ionosphere as a propagation path, three types of ionospheric model are needed. We can think of these as:

(1) DC: describing the mean ambient ionosphere as a function of time, space, and geophysical conditions, and needed to assess the likely effects of ionospheric delay on single ray paths;
(2) Low Frequency: describing the large scale variability in time and space about the DC level, so that the variation in predictions of ionospheric delay can be quantified;
(3) High Frequency: describing the structure of irregularities as a function of time and space.

We also need models to relate the macroscopic processes generating the large scale variability to the microscopic processes producing irregularities. Finally, to relate these to system effects we need propagation models, but these are not discussed here.

One approach to modelling is purely empirical, by the establishment of large comprehensive databases and averaging. Though such models are of considerable value, a possibly more fruitful approach to dynamic processes (which are the norm in many regions of the ionosphere) is through modelling the physical processes underlying the observations. We here provide an overview of the approaches taken by such physical models, and of their uses.

DC Models

The Navier-Stokes equations for a dilute plasma provide the basis for modelling ionospheric behaviour. They are a coupled set of partial differential equations, of which the first three are the continuity, momentum, and energy equations. Our knowledge of the physics of the ionosphere is reflected in these equations and in the parameters they require as inputs. Over the last 15 years schemes to solve these equations for ionospheric condi-

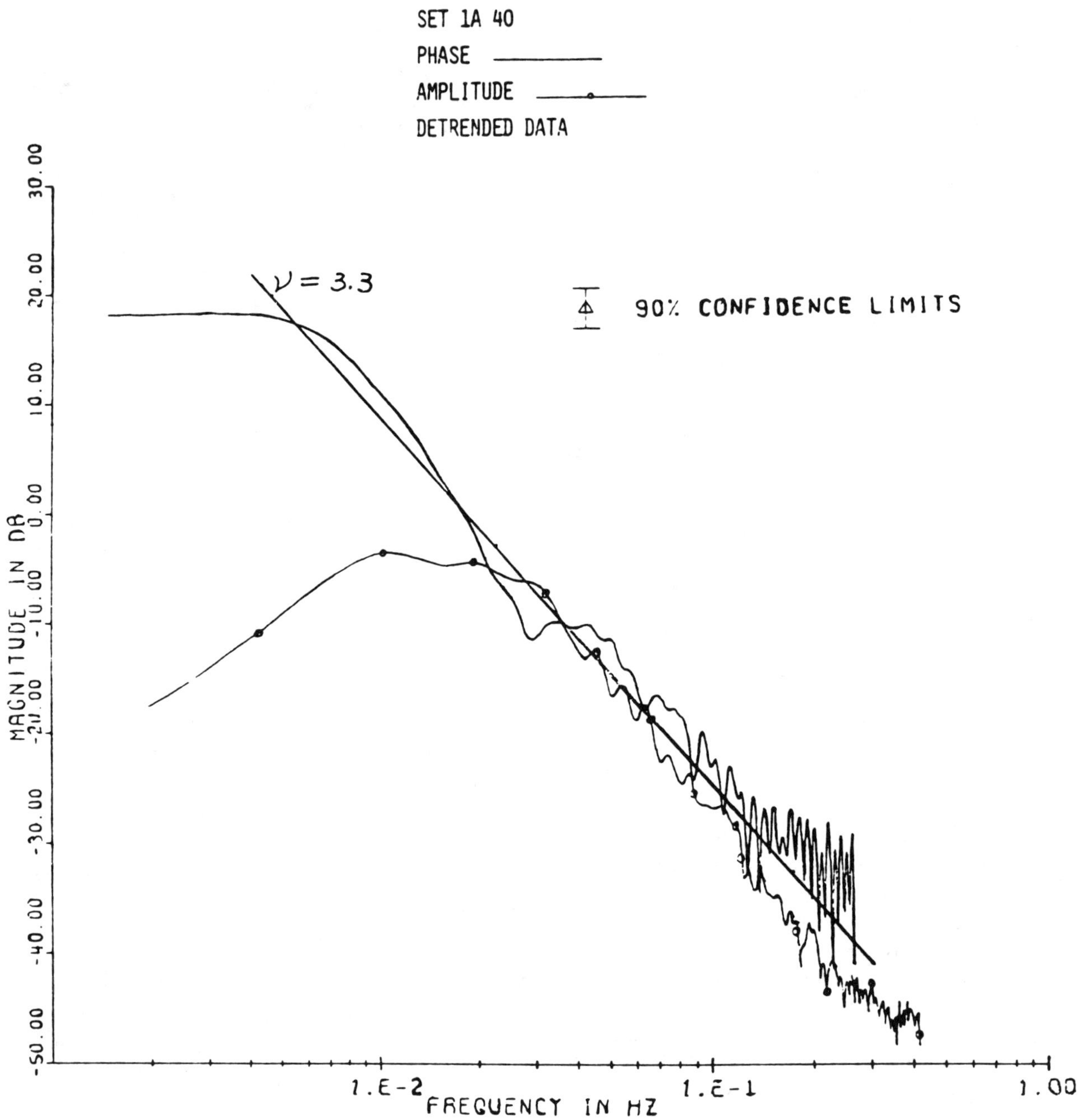

Fig. 10. Amplitude and phase power spectra observed on a 40 MHz channel at Boulder, Colorado (from Myers et al., 1979).

tions have been developed and applied by a number of workers [Quegan et al., 1982; Sojka and Schunk, 1985, and refs.]. As viable and efficient methods of solving the partial differential equations have developed, applying them to modelling or simulation of ionospheric behaviour has become more and more dependent on reliable models of the processes driving these equations.

As an example, the continuity equation for species i (ion or electron) can be written as:

$$\frac{\partial N_i}{\partial t} = P_i - \beta_i N_i - \nabla.(N_i v_i) \qquad (19)$$

where N_i is the number density, P_i is the production rate, β_i is the loss coefficient, and v_i is the velocity. In

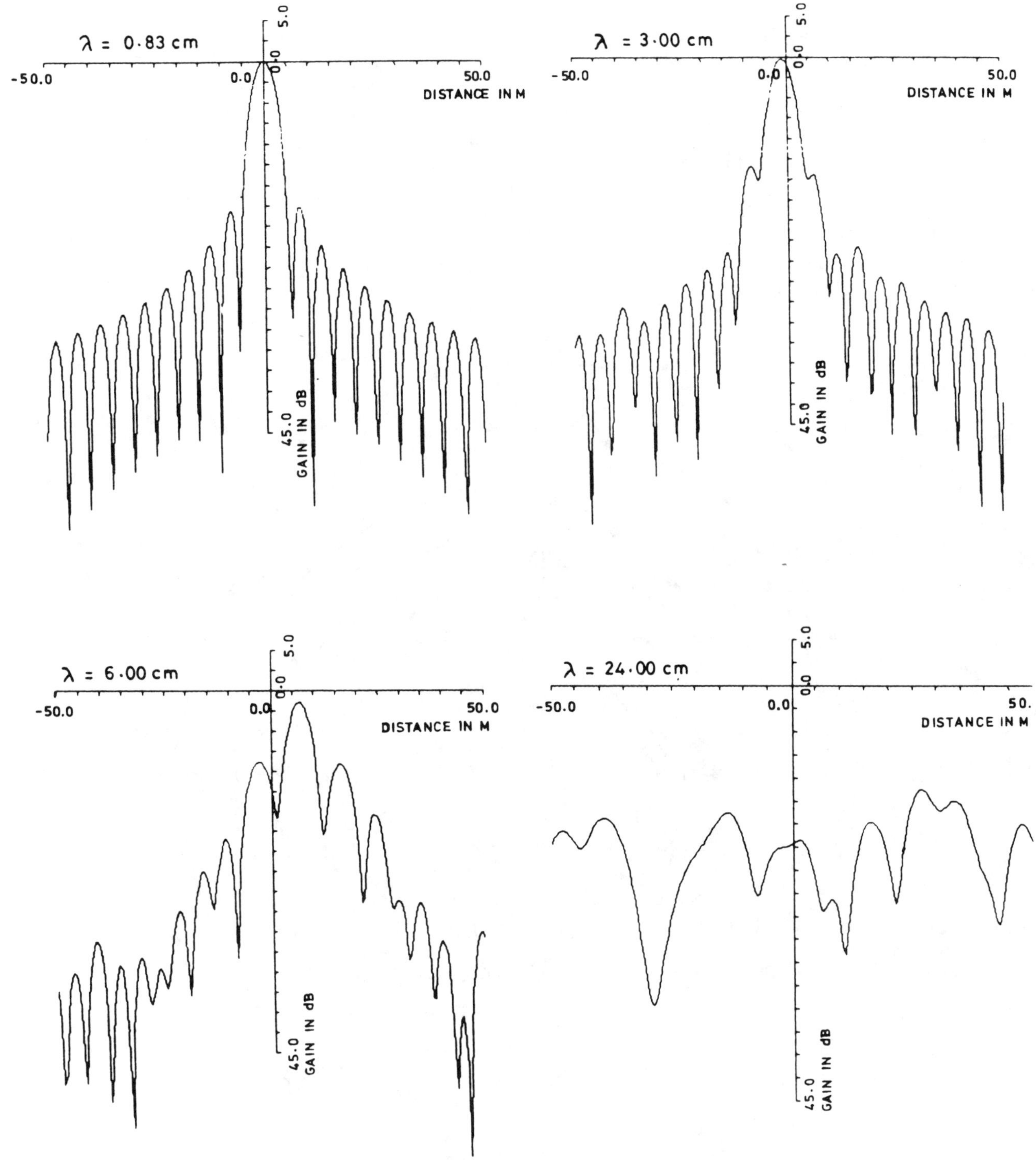

Fig. 11. Simulated synthetic aperture radar gain patterns for a disturbed ionosphere. The ionosphere is the same in each case, but the wavelength of the system (λ) is increasing, as indicated on the separate plots (from Quegan and Lamont, 1986).

order to drive this equation, the production and loss rates must be provided. This needs underlying models for the neutral atmosphere and atmospheric photochemistry, particle precipitation and its interaction with the atmosphere, and the three-dimensional temperature distribution for the species. While photochemical models are well-developed, models for precipitation are much less reliable because precipitation is highly varying in space and time. It is also intimately related to the electric field structure (see below).

The divergence term needs the velocity of the species. This is provided by the next equation up in the set, the momentum equation. (This is always the case for the Navier-Stokes system. It is not a closed system of equations, since each successive equation effectively involves a velocity moment provided by the next equation up in the system. Closure is achieved by making extra assumptions, such as negligible stress tensor). To solve the momentum equation, we then find that a model for the ambient electric and magnetic fields must be provided externally. The electric field model is a key input, and we show in Figure 12 a representation of the equipotentials of such a model for the northern high-latitude ionosphere in a fixed sun-earth reference

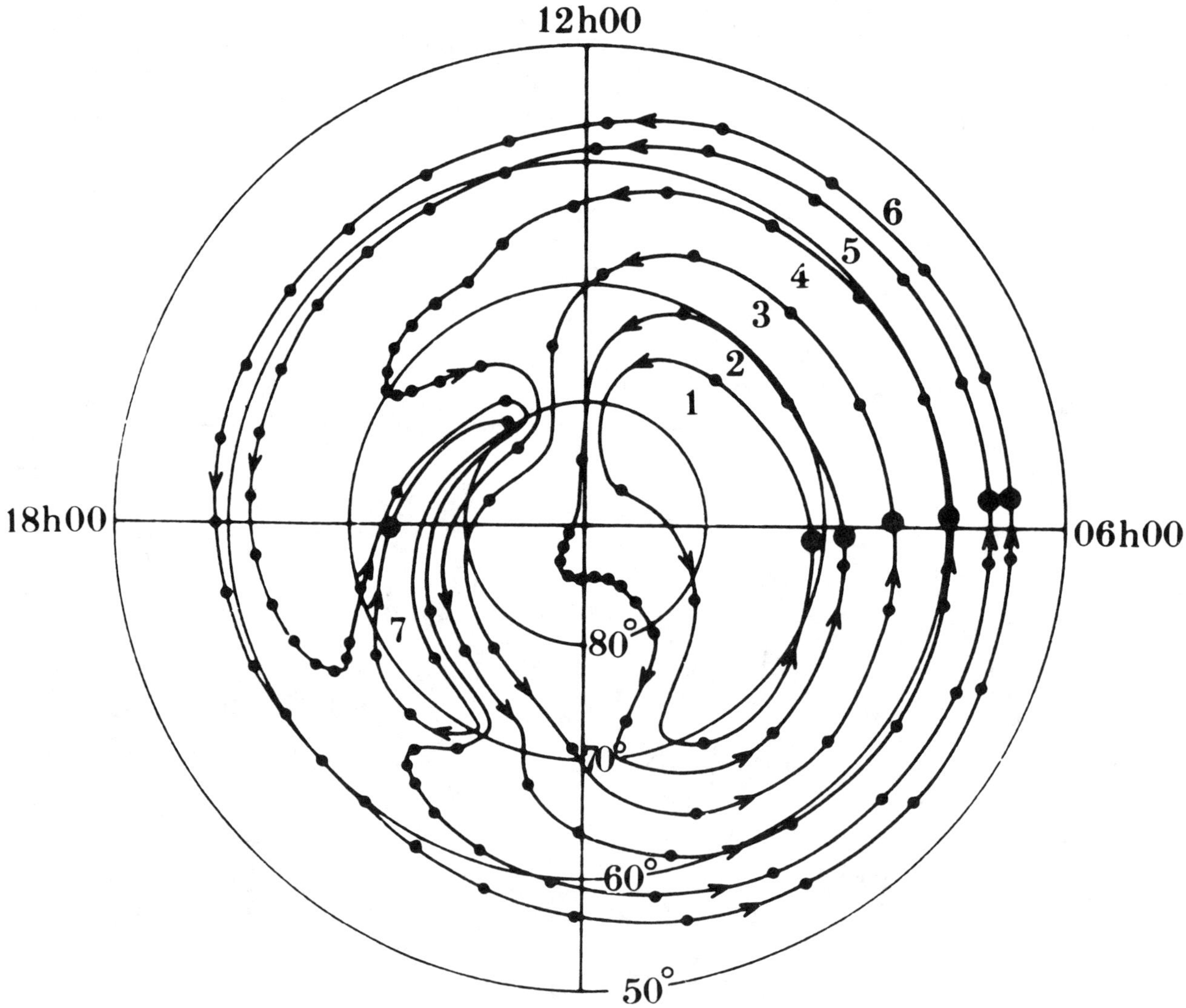

Fig. 12. Examples of convection paths of plasma at 300 km in the northern hemisphere under the influence of an imposed magnetospheric electric field, depicted in the fixed sun-earth reference frame. The large dots indicate the starting points used in the calculation of the paths, and successive dots indicate a travel time of one hour. The numbers on the paths correspond to the listed potentials of the paths.

frame. Since plasma travels along equipotentials, this Figure also shows how the plasma is convecting; dots on the flow lines correspond to 1 hour of travel time. The polar cap boundary in this Figure is a circle of radius 15° whose centre is on the noon-midnight meridian and displaced 5° towards midnight. This boundary marks a convection reversal in the corotating frame of reference. On theoretical and observational grounds it is expected to be embedded in a region of particle precipitation, but there is still uncertainty on how precipitation and electric field boundaries are related, particularly under time-dependent conditions.

Finally, Eq. (19) needs boundary conditions. There are good physical grounds for assuming chemical equilibrium at low altitudes, so that $N_i = P_i/\beta_i$ provides a lower boundary value for N_i. However, the upper boundary requires some model for flows into and out of the ionosphere (unless a full solution along closed magnetic field lines connecting the northern and southern hemispheres is carried out). Again, this is an externally imposed condition, based on some model.

A minimal list of the inputs needed to drive a solution of only the continuity and momentum equations for the global ionosphere is:

- a solar spectrum and photochemistry model;
- a model for the global electric field;
- a magnetic field model;
- a model for particle precipitation and its interaction with the atmosphere;
- a model for electron temperatures;
- a model for the composition, temperature, and dynamics of the neutral atmosphere;
- an upper boundary flow condition, if full inter-

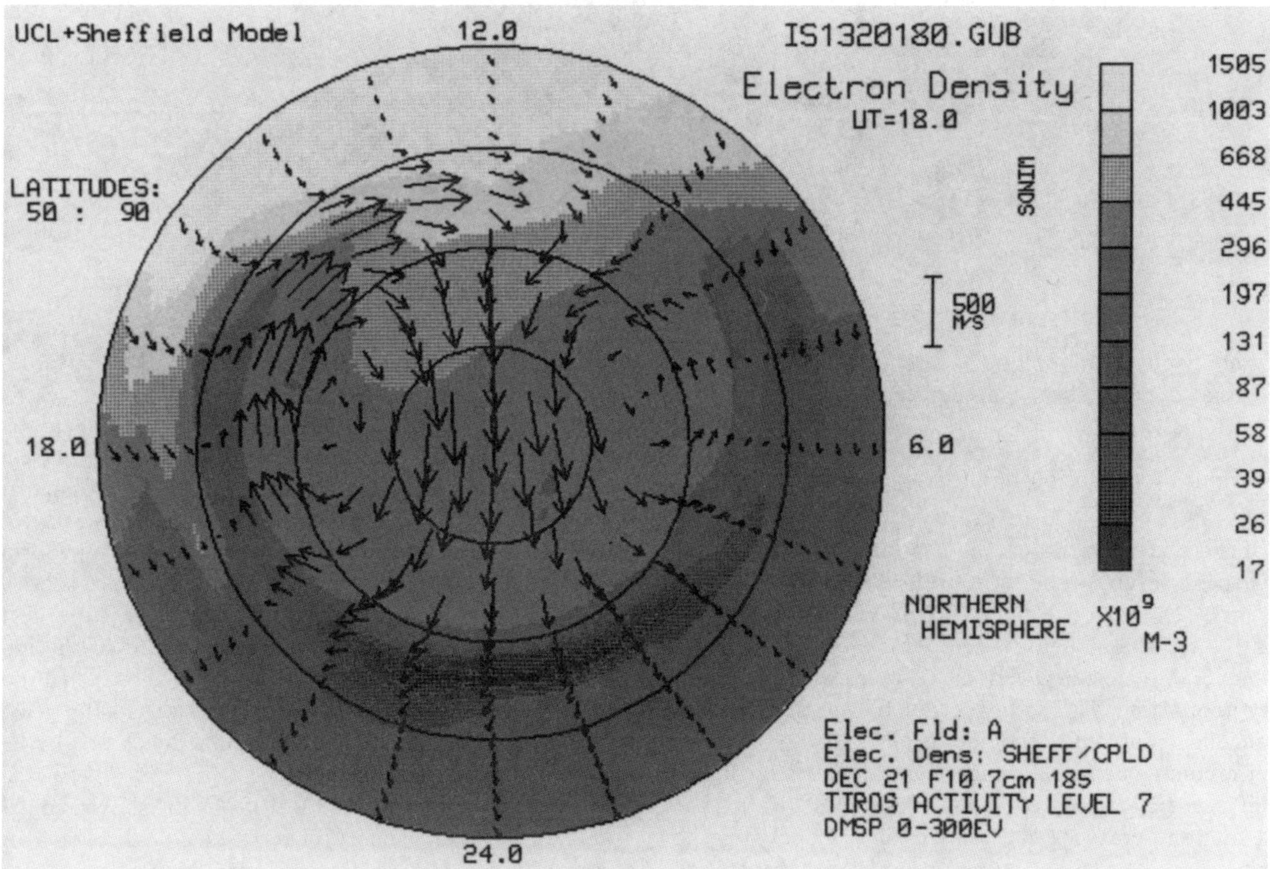

Fig. 13. Calculated electron densities near the F-peak produced by the UCL/Sheffield coupled thermosphere/ionosphere model, for steady conditions. The calculations are for the December solstice, and the plots show only a section of the northern hemisphere down to 50° geographic latitude. The circles of latitude are separated by 10°. Plot (a) is for UT = 1800 and plot (b) is for UT = 0600.

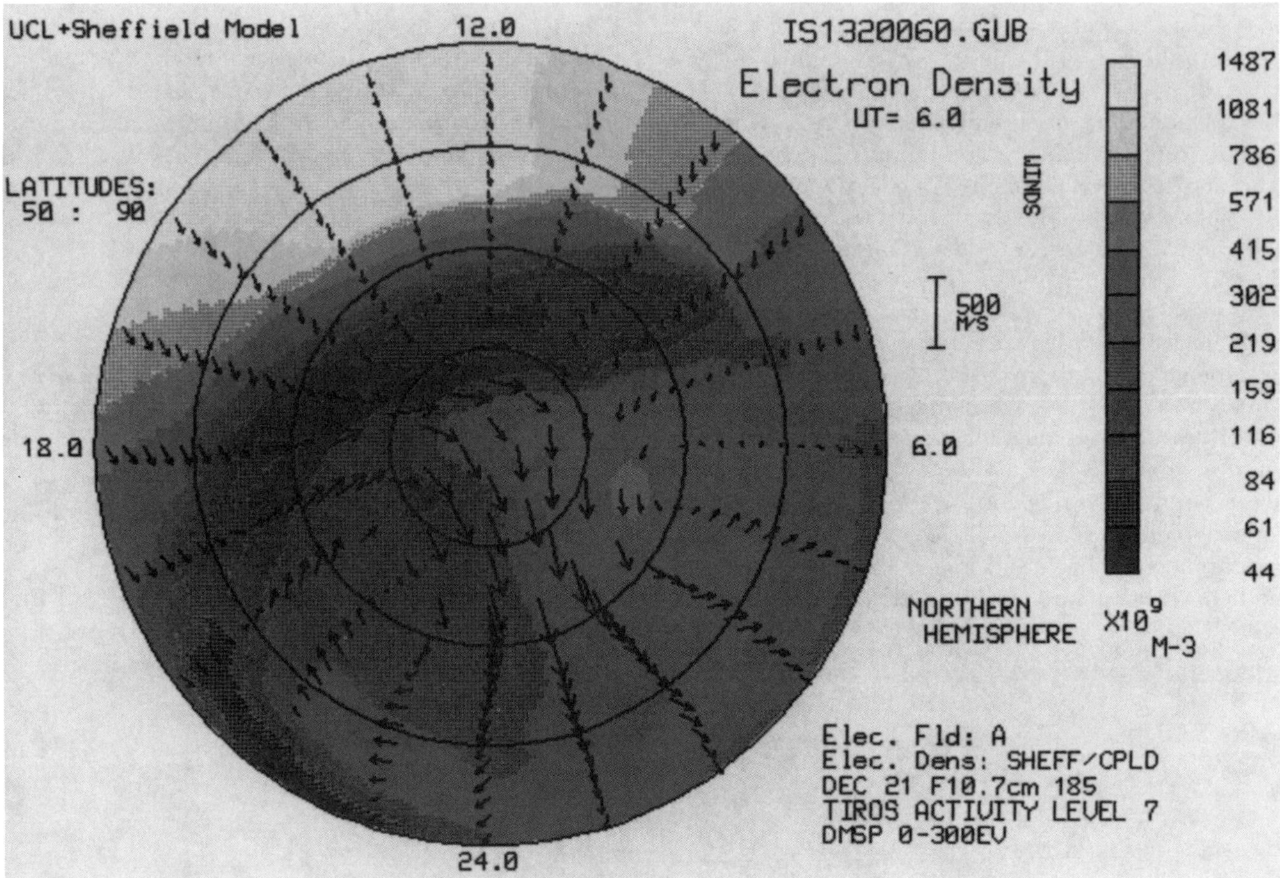

Fig. 13 (continued)

hemispheric modelling is not used, or if the field lines are not always closed (as occurs at high latitudes);

- values for collision frequencies.

All of these are external to the partial differential equations to be solved. The ion temperature is not in this list because it can be derived from the other parameters (electron and neutral temperatures, electric field and neutral atmosphere velocity), by means of the ion energy equation. This is important because of the dramatic effects of enhanced ion temperatures on ionospheric behaviour (e.g., trough formation, ion outflow), which occur particularly at high latitudes. Note that electron temperatures can be provided by solving the energy equation; this requires a model for the energy sources and sinks. Neutral atmosphere parameters can be derived self-consistently with the ionosphere at the expense of considerably increased computational effort; this has been carried out by a number of workers [Fuller-Rowell et al., 1987, 1988; Burns et al., 1989; Roble et al., 1988].

Despite these uncertainties in the input parameters, models based on this approach have been very successful in elucidating the behaviour of the ionosphere and of the coupled ionosphere/thermosphere system [Rees et al., 1987]. They are driven by inputs appropriate to a general set of conditions, such as season, solar cycle epoch, magnetic conditions, universal time (UT), etc., and the ensuing large-scale structure of the ionosphere is then derived. An example is shown in Figure 13, which is produced from calculations using a coupled thermosphere/ionosphere model developed by University College London and the University of Sheffield [Fuller-Rowell et al., 1987, 1988]. It shows contour plots of electron density near the F2-peak over a universal day in the winter northern hemisphere, calculated for steady electric field (taken from Heppner and Maynard, 1983) and particle precipitation (taken from Fuller-Rowell and Evans, 1987) conditions. The arrows on the plots show the response of the neutral wind to pressure gradients and plasma convection. Geographic latitude is indicated by circles separated by 10°.

Plot (a) is for UT = 1800, when the magnetic axis is near its maximum tilt towards the sun. The most notable features are (1) the high electron densities on the dayside due to solar photoionization; (2) a 'tongue of ionization' occupying the central polar cap, due to electric field induced plasma transport from the dayside, with plasma decay as it is transported in darkness; (3) at lower latitudes than the polar cap, a mid-latitude trough, deepest near midnight, but extending into the afternoon and morning sectors. Plot (b) is for 0600 UT. At this time the magnetic axis is tilted away from the sun, and the whole of the convection pattern is in darkness. As a result, transport of enhanced plasma density from the dayside into the polar cap does not occur. Instead, there is a marked trough feature on the dayside, and the polar cap is a region of low densities. Density enhancements due to particle precipitation occur, particularly round the morning side of the auroral oval. The mid-latitude trough is not as clear in this plot, because on the nightside it has moved to latitudes outside the range of the plot. Numerical models used in this way can produce predictions of expected ionospheric structure under any defined conditions. Comparisons with data have shown reasonable agreement [Sojka et al., 1982].

Numerical models can also be used for case studies of individual events, when there is sufficient evidence to permit realistic estimates of the input parameters. As an example (described in detail in Quegan et al., 1988), Figure 14 shows a comparison between Eiscat incoherent scatter data and model calculations. In this case, solar wind observations suggested that ionospheric conditions would be steady, and justified the use of average models of electric fields and particle precipitation. At the start of the comparison it was unclear what had been occurring in the previous period, which may explain the differences between the model calculations and the observations. As the calculation progresses, the agreement with data becomes very good. Several studies of this type using coordinated data from a variety of sources are currently in progress.

The possible impact of such models on position measurements is perhaps best illustrated by the information they provide on the dynamics of the mid-latitude trough. This depression in electron density at the F2-peak also manifests itself as a marked decrease in TEC (see Figure 2). Modelling studies have played a major role in elucidating the dependence of trough occurrence and depth on UT and hemisphere [Sojka and Schunk, 1989]. They indicate that the variation in NmF2 at a

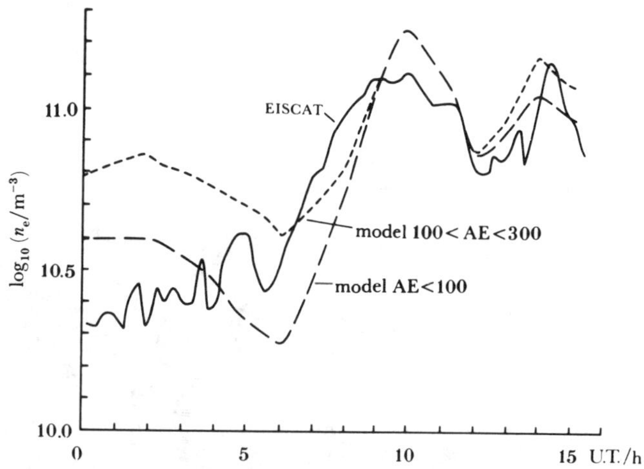

Fig. 14. Comparison between Eiscat observations of electron density and the results of a numerical model (reprinted with permission from Quegan et al., Copyright (1988), Pergamon Press Ltd.).

given position due to the mid-latitude trough is highly dependent on UT. Trough modelling studies have not up to now dealt with TEC, but if we assume that TEC varies as NmF2, this would imply significant longitudinal differences in the ionospheric delay at trough latitudes. This is, of course, a rather naive statement, since it ignores the fact that, for most applications, the propagation path is likely to cut across the magnetic field lines threading the trough. For such ray-paths, calculations of TEC need the global distribution of electron density. Examples of this type of model calculation can be found, for example, in Bailey et al. (1991).

Low Frequency Models

Though the modelling studies described above are successful in reproducing the global scale features observed in the ionosphere, it is important to realise their limitations. What they can do is to produce meaningful calculations of ionospheric densities, as long as the input parameters are realistic. They provide no direct information about these parameters, which must be supplied externally. The problem is then not one of solving the fluid equations, but of providing them with realistic inputs. This means that the large-scale numerical models have little to say directly about the day-to-day variations which are normal in observations, since these correspond to variations in the driving parameters.

This does not mean that the large-scale models cannot be applied to time-dependent behaviour. Figure 15 shows calculations from the same numerical model which produced the results in Figure 13, but for a mag-

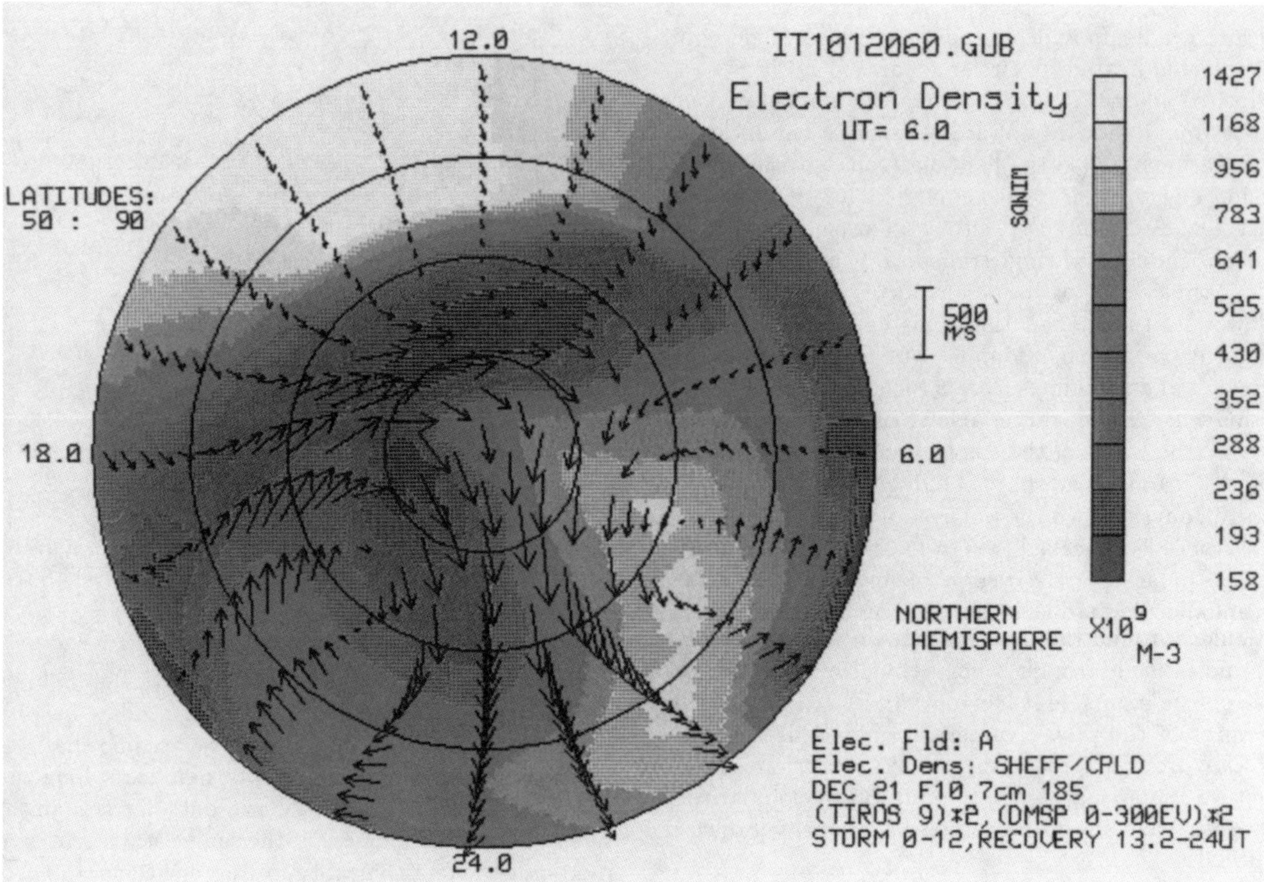

Fig. 15. As for Fig. 13, but for disturbed conditions, corresponding to enhancements of electric field strengths and precipitation intensities at UT = 2400. The Figure shows electron densities at UT = 0600, six hours after the event starts.

netically active period (enhanced electric fields and particle precipitation) which begins at 2400 UT. The Figure shows calculated electron densities at 0600 UT, and should be compared with Figure 13(b). The polar cap shows comparable densities (note that the scale is not the same on the two plots). The dayside trough feature is still present, but with enhanced densities. Marked density enhancements due to particle precipitation occur, particularly round the morning side of the auroral oval. Comparing values of electron density in Figures 13(b) and 15 indicates regions of very significant difference. For example, in the morning sector auroral zone, the density changes from 4×10^{11} to 1×10^{12} electrons m^{-3} in some places. If these values translate into equivalent variations in TEC (which is not necessarily the case), then, depending on the propagation geometry, these differences are indicative of range delay variations caused by ionospheric disturbances. (For the calculated data shown in these Figures, evaluation of field-aligned TEC was not carried out, so that stronger statements are not possible here. However, large-scale models can readily supply such information).

Case studies of individual events have also been successfully carried out, when there is enough observational evidence to construct a credible set of input parameters (see, for example, Quegan, 1989). However, it is comparatively rare to have data which is comprehensive enough to provide inputs, such as the global electric field pattern, which can be used to drive the models reliably. Essentially, our predictive capability is limited by our inability to predict electric field and particle precipitation patterns with any confidence. If we have this information, there are reasonable grounds to believe that we could model the variations of ionospheric behaviour. Recent work indicates, however, that a solar wind and interplanetary magnetic field monitor may go a long way towards filling this need [Lockwood et al., 1990].

High Frequency Models and Scintillation

Large-scale numerical models operate on grid sizes of the order of degrees (i.e., tens to hundreds of km), while irregularities have a range of scale sizes from tens of km down to metres or smaller. There are a number of excellent reviews of their morphology, underlying physics, and their effects on propagation [Fejer and Kelley, 1980; Yeh and Liu, 1982; Aarons, 1982; Tsunoda, 1988]. All that is attempted here is a description of some of the physics involved, and how it relates to the processes and calculations relevant in the large-scale numerical models.

Irregularities strong enough to cause scintillations at microwave frequencies are always present throughout the high latitude auroral zone and polar cap, and in the post-sunset equatorial zone. A great deal of experimental evidence, from scintillation measurements [Fremouw and Secan, 1984] or in-situ electron density probes [Singh et al., 1985], shows that the power spectrum of irregularities obeys a power law, i.e., over a range of spatial scales, the three-dimensional power spectrum of density fluctuations is of the form

$$\Phi_{\Delta N}(\mathbf{k}) = A|\mathbf{k}|^{-p} \qquad (20)$$

Here $\mathbf{k}$ is the three-dimensional spatial wave number, p is the spectral index and A is a constant. There is also evidence that in some regions the spectral index is different in the high and low frequency parts of the spectrum, giving a two-component power law. The ratio of the spatial correlation lengths of irregularities in the zonal and meridional directions relative to the local L-shell (which we can think of as defining irregularity shape) varies with location. Around the flanks of the auroral zones, 'sheet-like' irregularities occur, which are elongated in the zonal direction; elsewhere they tend to be 'rod-like', with no asymmetry [Rino and Vickrey, 1982].

A major effort by Fremouw, Secan and co-workers has gone into establishing an extensive data-base of irregularity occurrence and characteristics [Fremouw and Secan, 1984; Secan et al., 1987; Secan, 1991]. Using scintillation data from the Wideband, HiLat and Polar BEAR satellite beacons, this has been used to generate an irregularity model known as WBMOD. This provides the shape factors, spectral indices and integrated strength of the spatial spectrum of irregularities as functions of latitude, time (local, magnetic local and day of the year), K_p and sunspot number in three regimes, high latitude, mid-latitude and equatorial. Validation studies indicate that WBMOD predicts mean scintillation well, but there is large scatter around the mean [Secan, 1991].

WBMOD was designed not simply to provide an empirical model of irregularities, but also to support the major effort being made by a number of workers to understand the physics which gives rise to irregularities and their observed properties. Possible sources for irregularities have been identified as plasma instabilities, structured electric fields or particle precipitation, and/or neutral wind dynamic coupling with the ion gas [Fejer and Kelley, 1980; Keskinen and Ossakow, 1983]. Electric field induced motions then provide a mechanism for transporting the larger irregularities away from their source region before they decay. The convection electric field will distribute irregularities throughout the high-latitude region (see Figure 12 and Kelley et al., 1982). In the equatorial region, electric fields are equally important in determining the motions of irregularities in the post-sunset ionosphere [Aarons, 1982; Basu et al., 1986; Fejer and Kelley, 1982].

Considerable progress has been made in understanding and modelling the physical processes generating instabilities. Many different mechanisms have been proposed; an excellent and comprehensive review of these mechanisms and of the zoo of instabilities is given in Fejer and Kelley (1980). As an example, Figure 16 illustrates a form of instability known as the gradient drift instability, which is generated when a density fluctuation occurs in the presence of a density gradient and an electric field (or a neutral wind). The instability

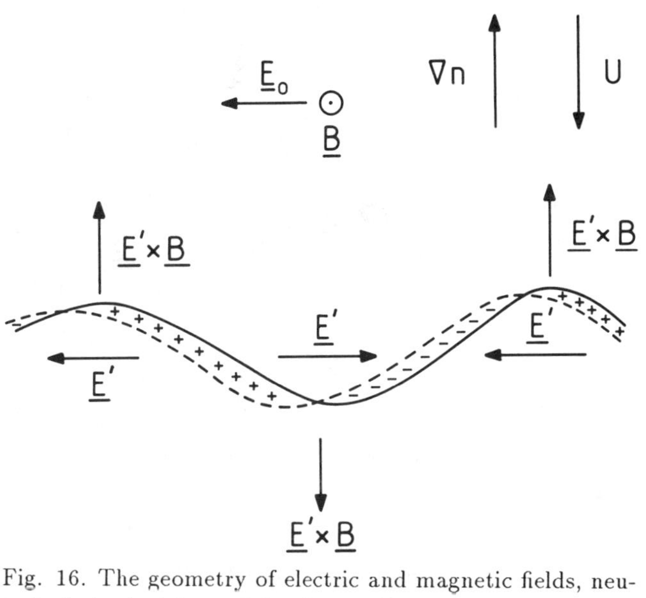

Fig. 16. The geometry of electric and magnetic fields, neutral wind velocities and density gradients which gives rise to the gradient drift instability.

arises for the geometry shown in the Figure, in which the electric field **E** is perpendicular to the density gradient ∇n, and the magnetic field **B** is perpendicular to both. A small amplitude fluctuation with wave-vector parallel to **E** leads to a charge separation; this in turn gives rise to an electromagnetic force which causes the fluctuation to grow. A similar effect would occur if, instead of an electric field, there was a neutral wind **U** directed oppositely to the density gradient.

Figure 17 shows a calculation of how such a perturba-

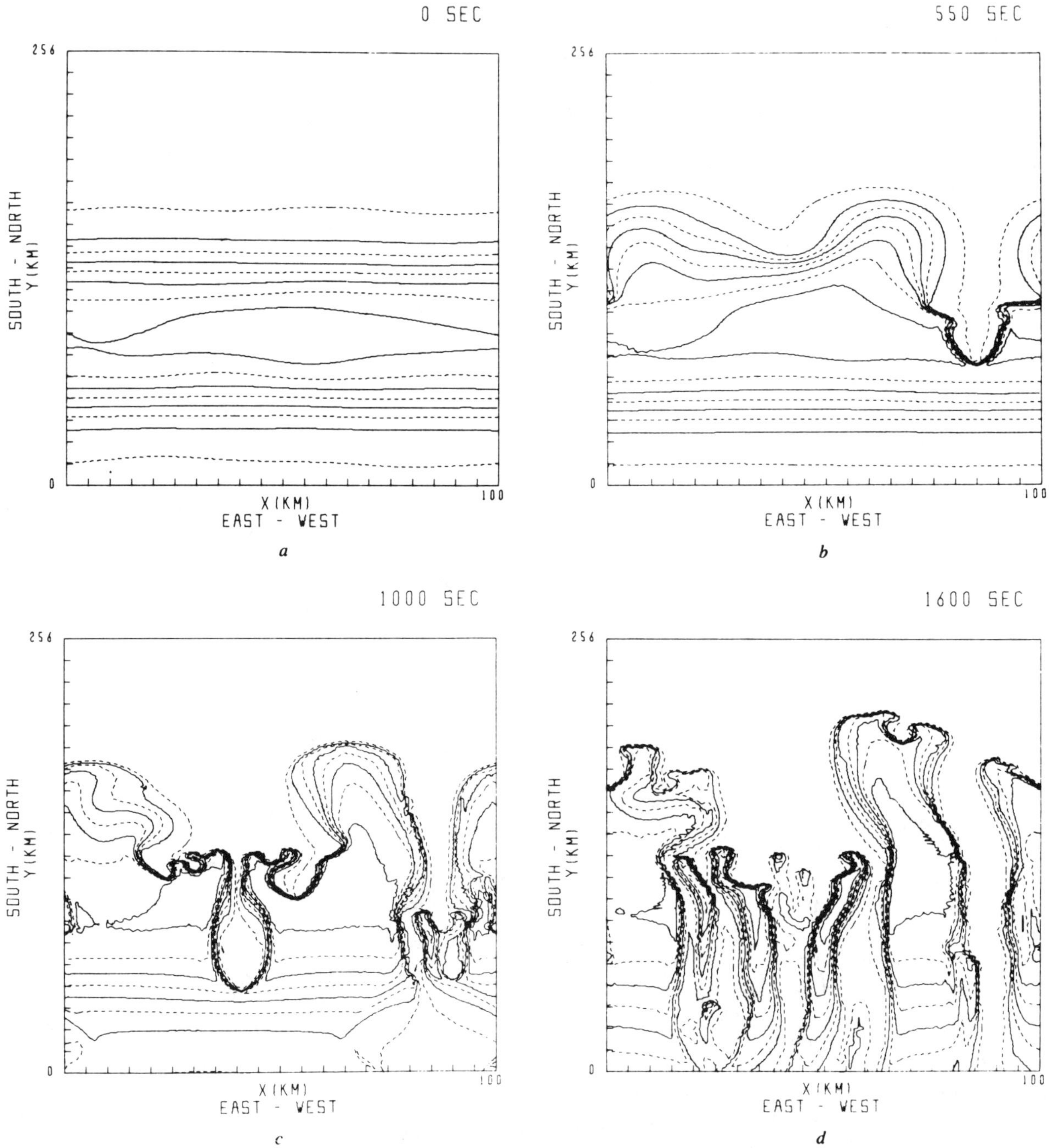

Fig. 17. Simulations of the growth of the gradient drift instability over a period of 1600 seconds for an imposed electric field. At time zero a 1% density fluctuation is imposed. The solid contours are separated by a factor of 1.25 in density (from Keskinen and Ossakow, 1982).

tion would grow in a more realistic ionospheric configuration. The plots show the development of an initial 1% perturbation of a density gradient oriented north-south, when a westward electric field is present. The solid lines are spaced by a factor of 1.25 in density. Inside half an hour, the regular initial configuration has been transformed into a highly distorted pattern with 'fingers' of enhanced and depleted density oriented north-south. Such features are often observed. Other forms of instability have been similarly modelled (see Fejer and Kelley, 1980).

Model studies have also indicated how irregularities can be generated by structured electric fields and particle precipitation, particularly in the high-latitude cusp region. Kelley et al. (1982) have shown how large-scale irregularities produced in this way can be transported across the polar cap, while instability processes occurring at their boundaries cause a cascade of energy into smaller scale sizes. (They in fact used the convection pattern of Figure 12 to quantify these effects). However, there are significant gaps in our understanding of irregularities. In particular, the relative importance of the different mechanisms which can produce irregularities is not known. As for the macro-scale models, this in part is because of our ignorance of the ambient conditions (the detailed electric field and precipitation patterns, for example). Physical models of irregularities do not provide information on their strength or the strength of the associated scintillations. These are key deficiencies when we wish to use models to predict the performance of space-based systems.

We have already indicated how scintillations can affect positional measurements (and other aspects of image quality) for space-borne SAR. The calculations leading to Figure 11 were based on severe disturbance conditions, using parameters of the type supplied by WBMOD.

Summary

Large-scale numerical models based on the equations for a multi-component fluid can reproduce ionospheric behaviour in a mean sense, or in case studies, as long as the input parameters are realistic. Difficulty in providing regular or comprehensive information on the input parameters provides the major stumbling-block to a true predictive capability on a day-to-day basis for these models. Global electric fields, particle precipitation patterns and neutral atmosphere composition and dynamics are particularly difficult to describe reliably. Coupled thermosphere/ionosphere models are becoming increasingly able to provide the third of these inputs, but the first two remain problematical. Progress is likely as we understand better the coupling of the solar wind to the earth's magnetosphere and the subsequent response of the magnetosphere/ionosphere system.

The same physical processes which generate the large-scale ionospheric structure are also responsible for the production of irregularities, which are themselves intimately related to the large-scale structures. An examination of Figure 12 makes this quite clear. Paths of convecting plasma show markedly different behaviour because of the electric field structure. This leads naturally to the density gradients needed to drive, for example, the gradient drift instability, which itself feeds off the same electric field structure. The flow paths marked illustrate how irregularities produced in the cusp will be distributed across the polar cap, if they survive long enough. Observational evidence also confirms this coupling of scales. Irregularities around the flanks of the polar cap are organised with respect to the convection pattern [Rino et al., 1983]. The cusp and auroral oval are of central importance both to ionospheric densities and to irregularity formation [Kelley et al., 1982].

There has up to present been little effort to provide this coupling of scales in model studies (exceptions include a study by Lockwood and Fuller-Rowell (1987) which made use of the University College London/Sheffield model to predict the occurrence of instabilities arising from neutral wind/ionosphere interactions, and the work by Kelley et al. (1982) on the generation and distribution of irregularities at high latitudes). There is a variety of reasons for this. Modelling of both large-scale ionosphere structure and irregularities requires specialised, sophisticated computer codes expensive on processing power and computational time, carried out at only a few centres in the world. The spatial grid required for irregularity modelling is much finer than that required for large-scale structure, and hence direct embedding of one in the other is not attractive. The communities interested in the physics and modelling of irregularities and large scale processes do not overlap much, and neither overlaps too much with the propagation community.

Nonetheless, this splitting of interests is artificial. Intrinsic to the calculations of ionospheric structure such as those shown in Figure 13 and 15 are conditions where irregularities will occur. Irregularities provide a means of observing the parameters needed by the macro-scale models, since they can act as tracers of convection and precipitation conditions. Tracking these tracers relies on probing tools, such as radar, whose performance is intimately linked to ambient ionospheric densities and to irregularity occurrence. Such obvious interactions of interests suggest that the time is becoming ripe for the advances in understanding both large and small-scale ionospheric structure (made largely in the name of basic research) to be related and to begin to have an impact on applications. We should therefore expect considerable development in coupling the physics of the two scales into a unified picture. Such a unification will provide consistent descriptions of large and small scale ionospheric structure. These can be used to support a variety of applications to propagation problems, of which positional measurements form a part. It is premature to claim that this will give a predictive capability, because of our uncertainties about the solar wind/magnetosphere/ionosphere coupling mechanisms. This is currently an area of intense research, which holds the key to any true prediction of ionospheric conditions, and the behaviour of systems using the ionosphere as a propagation path.

References

Aarons, J., Global morphology of ionospheric scintillations, *Proc. IEEE, 70*, 360-378, 1982.

Ashkenazi, V., and R. M. Sykes, Doppler translocation and orbit relaxation techniques, *Phil. Trans. R. Soc. Lond. A, 294*, 357-364, 1980.

Ashkenazi, V., and D. N. McLintock, Very long baseline interferometry: an introduction and geodetic applications, *Survey Review, 26*, 279-288, 1982.

Bailey, G. J., and R. Sellek, A mathematical model of the Earth's plasmasphere and its application in a study of He^+ at L = 3, *Ann. Geophysicae, 8*, 171-190, 1990.

Bailey, G. J., R. Sellek and N. Balan, The effect of interhemispheric coupling on nighttime enhancements in ionospheric total electron content during winter at solar minimum, *Ann. Geophysicae, 9*, 738-747, 1991.

Basu, S., S. Basu, C. E. Valladares, A. DasGupta and H. E. Whitney, Scintillations associated with bottomside sinusoidal irregularities in the equatorial F region, *Jnl. Geophys. Res., 91*, 270-276, 1986.

Burns, A. G., T. L. Killeen , G. Crowley, B. A. Emery and R. G. Roble, On the mechanisms responsible for high latitude thermospheric composition variations during the recovery phase of a geomagnetic storm, *Jnl. Geophys. Res., 94*, 16961-16968, 1989.

Egeland, A., O. Holter and A. Omholt (eds), *Cosmical Geophysics*, Universitetsforlaget, Oslo, 1973.

Elachi, C., *Spaceborne radar remote sensing: applications and techniques*, IEEE Press, 1988.

Fejer, B. G., and M. C. Kelley, Ionospheric irregularities, *Rev. Geophys. and Space Phys., 18*, 401-454, 1980.

Fremouw, E. J., and J. A. Secan, Modelling and scientific application of scintillation results, *Radio Sci., 19*, 687-694, 1984.

Fuller-Rowell, T. J., and D. S. Evans, Height-integrated Pedersen and Hall conductivity inferred from the TIROS/NOAA satellite data, *Jnl. Geophys. Res., 92*, 7606-7618, 1987.

Fuller-Rowell, T. J., D. Rees, S. Quegan, R. J. Moffett and G. J. Bailey, Interactions between neutral thermospheric composition and the polar ionosphere using a coupled ionosphere-thermosphere model, *Jnl. Geophys. Res., 92*, 7744-7748, 1987.

Fuller-Rowell, T. J., D. Rees, S. Quegan, R. J. Moffett, and G. J. Bailey, Simulations of the seasonal and universal time variations of the high-latitude thermosphere and ionosphere using a coupled, three-dimensional model, *PAGEOPH, 127*, 189-217, 1988.

Heppner, J. P., and N. C. Maynard, High latitude electric field models, AGU Chapman Conf. on Magnetospheric Currents, Irvington, Va., March, 1983.

Hothem, L. D., and C. J. Fronczek, Report on test and demonstration of macrometer Model V-1000 Interferometric Surveyor, *FGCC Rep. FGCC-LS-83-2*, Federal Geodetic Control Centre, Rockville, USA, 1983.

Kelley, M. C., J. F. Vickrey, C. W. Carlson and R. Torbert, On the origin and extent of high-latitude F region irregularities, *Jnl. Geophys. Res., 87*, 4469-4475, 1982.

Keskinen, M. J., and S. L. Ossakow, Nonlinear evolution of plasma enhancements in the auroral ionosphere 1. Long wavelength irregularities, *Jnl. Geophys. Res., 87*, 144-150, 1982.

Keskinen, M. J., and S. L. Ossakow, Theories of high-latitude ionospheric irregularities: a review, *Radio Sci., 18*, 1077-1091, 1983.

Lockwood, M., and T. J. Fuller-Rowell, The modelled occurrence of non-thermal plasma in the ionospheric F region and the possible consequences for ion outflows into the magnetosphere, *Geophys. Res. Lett., 14*, 371-374, 1987.

Lockwood, M., S. W. H. Cowley and M. P. Freeman, The excitation of plasma convection in the high latitude ionosphere, *Jnl. Geophys. Res., 95*, 7961-7972, 1990.

Lorell, J., E. Colquitt and R. J. Anderle, Ionospheric correction for Seasat altimeter height measurements, *Jnl. Geophys. Res., 87*, 3207-3212, 1982.

Marsh, J. G. and T. V. Martin, The Seasat altimeter mean sea surface model, *Jnl. Geophys. Res., 87*, 3269-3280, 1982.

Mendillo, M., and J. A. Klobuchar, Investigations of the ionospheric F region using multistation total electron content observations, *Jnl. Geophys. Res., 80*, 643-650, 1975.

Myers, W. J., R. J. Gjeldum, C. H. Liu, and K. C. Yeh, A study of ionospheric scintillations of phase and quadrature components, *Jnl. Geophys. Res., 84*, 2039-2048, 1979.

Quegan, S., G. J. Bailey, R. J. Moffett, R. A. Heelis, T. J. Fuller-Rowell, D. Rees and R. W. Spiro, A theoretical study of the distribution of ionisation in the high latitude ionosphere and the plasmasphere: first results on the mid-latitude trough and the light-ion trough, *Jnl. Atmos. Terr. Phys., 44*, 619-640, 1982.

Quegan, S., G. J. Bailey and R. J. Moffett, Regions of He^+ dominance in the high-latitude topside ionosphere, *Planet. Space Sci., 32*, 791-802, 1984.

Quegan, S. and J. Lamont, Ionospheric and tropospheric effects on synthetic aperture radar performance, *Int. Jnl. Remote Sensing, 7*, 525-539, 1986.

Quegan, S., R. S. Gill and M. Lockwood, Comparisons between EISCAT observations and model calculations of the high latitude ionosphere, *Jnl. Atmos. Terr. Phys., 50*, 1057-1076, 1988.

Quegan, S., The influence of convection on the structure of the high latitude ionosphere, *Phil. Trans. R. Soc. Lond. A, 328*, 119-137, 1989.

Ratcliffe, J. A., *An introduction to the ionosphere and magnetosphere*, Cambridge University Press, 1972.

Ratcliffe, J. A., Some aspects of diffraction theory and their application to the ionosphere, *Rep. Progr. Phys., 19*, 188-267, 1956.

Rees, D., T. J. Fuller-Rowell, S. Quegan, R. J. Moffett and G. J. Bailey, Thermospheric dynamics: understanding the unusual disturbances by means of simulations with a fully-coupled global thermosphere/high-latitude ionosphere model, *Ann. Geophys., 5*, 303-328, 1987.

Rino, C. L., and J. F. Vickrey, Recent results in auroral-zone scintillation studies, *Jnl. Atmos. Terr. Phys., 44*, 875-887, 1982.

Rino, C. L., R. C. Livingston, R. T. Tsunoda, R. M. Robinson, J. F. Vickrey, C. Senior, M. D. Cousins and J. Owen, Recent studies of the structure and morphology of auroral zone F region irregularities, *Radio Sci., 18*, 1167-1180, 1983.

Roble, R.G., E.C. Ridley, A. D. Richmond and R. E. Dickinson, A coupled thermosphere/ionosphere general circulation model, *Geophys. Res. Lett., 15*, 1325-28, 1988.

Schunk, R. W., Transport equations for aeronomy, *Planet. Space Sci., 23*, 437-485, 1975.

Secan, J. A., E. J. Fremouw and R. E. Robins, A review of recent improvements to the WBMOD ionospheric scintillation model, in *Effect of the ionosphere on communication, navigation and surveillance systems*, Proc. 1987 Ionospheric Effects Symp., edited by J. Goodman, pp. 607-616, 1987.

Secan, J. A., WBMOD model validation study, Results, Vol. 1, Equatorial region, Vol. 2, High latitude region, *Rep. NWRA-CR-91-R070*, Northwest Research Associates, Inc., Bellevue, WA., 1991.

Singh, M., P. Rodriguez, and E. P. Szuszczewicz, Spectral classification of medium-scale high-latitude F region plasma density irregularities, *Jnl. Geophys. Res., 90*, 6525-6532, 1985.

Sojka, J. J., W. J. Raitt, R. W. Schunk, F. J. Rich and R. C. Sagalyn, Observations of the diurnal dependence of the high latitude F region ion density by DMSP satellites, *Jnl. Geophys. Res., 87*, 1711-1718, 1982.

Sojka, J. J., and R. W. Schunk, Theoretical study of the seasonal behaviour of the global ionosphere at solar maximum, *Jnl. Geophys. Res., 94*, 6739-6749, 1989.

Sojka, J. J., and R. W. Schunk, A theoretical study of the global F region for June solstice, solar maximum, and low magnetic activity, *Jnl. Geophys. Res., 90*, 5285-5298, 1985.

Szuszczewicz, E. P., P. Rodriguez, M. Singh and S. Mango, Ionospheric irregularities and their potential impact on synthetic aperture radars, *Radio Sci., 18*, 765-774, 1983.

Tsunoda, R.T., High-latitude F region irregularities: a review and synthesis, *Rev. Geophys., 26*, 719-760, 1988.

Yeh, K. C. and C.-H. Liu, Radio wave scintillations in the ionosphere, *Proc. IEEE, 70*, 324-360, 1982.

Department of Applied and Computational Mathematics, University of Sheffield, Sheffield S10 2TN, UK.

The Effect of Horizontal Gradients of Ionization on Position Determination and the Availability of Relevant Data

R. LEITINGER

Institut für Meteorologie und Geophysik
Universität Graz, A-8010 Graz, Austria

In applications of the trans-ionospheric propagation of radio waves first order range errors are proportional to (slant) electron content of the ionosphere, second order errors to the square of this quantity. In many applications range differences are more important than range itself and this is especially true for all interferometric methods. Range difference measurements are influenced by differences of electron content or 'horizontal gradients'.
The role of gradients is demonstrated. Unfortunately the presently available ionospheric models give no realistic indication of horizontal gradients and therefore we have to rely on experience from measurements and on estimates from atmospheric / ionospheric physics. Three types of gradients can be distinguished: 1) gradients from stable spatial structures which appear in averaged as well as in actual data, 2) gradients from the signature of atmospheric gravity waves, 3) gradients from ionospheric disturbances and from phenomena which occur only occasionally.
The availability of observed data is discussed together with the relevance of conclusions from measurements and estimates.

INTRODUCTION

The propagation of radio waves through the ionosphere of the earth finds a wide range of applications for science and engineering. Among the scientific application areas we name ionospheric physics, radio astronomy, space geodesy, oceanography. The engineering applications span such diverse areas as communication, navigation and time dissemination via satellites. The atmosphere influences all applications because it changes the properties of the radiowaves. All measurable signal parameters are affected. The most important ones are the amplitude, the phase and the polarization of the wave components.

Considering the plasma influence only (ionosphere and plasmasphere) and omitting the so-called scintillation effects which are produced by small scale irregularities in the ionization distribution and all other effects which are outside the domain of geometric optics we have to deal with refraction problems. In this case the propagation medium is described by refractive indices which have to be 'nearly constant' over distances of one wave length. (A 10% variation is near the limit for simple geometric optics.)

Environmental Effects on Spacecraft Positioning and Trajectories
Geophysical Monograph 73, IUGG Volume 13

A refractive index n can be considered to define the velocity of phase propagation $v = c/n$ (c: free space velocity of light) for a monochromatic wave (a principal component of a monochromatic wave in the birefringence case).

For an inhomogeneous medium the refractive indices are not constant but they are spatial functions. This implies that in general there exist no analytic solutions for the propagation of monochromatic waves from source to receiver.

For frequencies above 100 MHz the following approximation for the refractive index is valid under 'nearly all' conditions:

$$n_{1,2}^2 \simeq 1 - \frac{f_p^2}{f^2 \pm f\, f_g \mid \cos\theta \mid - (1/2)\, f_g^2 \sin^2\theta}$$

f_p: plasma frequency; the square of plasma frequency is proportional to the electron density N_e: $f_p^2 = A\ N_e$ (using S.I.-units: $A = 80.6\ \mathrm{m}^3\ \mathrm{s}^{-2}$);

f_g: gyrofrequency of free electrons which is proportional to the geomagnetic induction, B;

θ : angle between the phase propagation vector and the vector of geomagnetic induction.

Because of the magnetic field of the earth the ionosphere and the plasmasphere are anisotropic media (wave propagation in a magneto-plasma). This means that a radiowave has to be split into two components. Their polarizations are the two 'principal polarizations' (birefringence). Generally the principal polarizations are (left–hand and right–hand) elliptic but in this approximation the ellipses degenerate into circles. In the geomagnetic north hemisphere the + sign corresponds to the left–hand circular component, the − sign to the right–hand circular component. The approximation is not valid for propagation nearly perpendicular to the geomagnetic field vector. (It is valid outside the interval $\theta = \pi/2 \pm \epsilon$, $\epsilon \simeq (3/2)(f_g/f)$ if $f \gg f_p$. Since in the atmosphere of the earth everywhere $f_g < 1.7$ MHz it follows $\epsilon \leq 1.5°$ for $f = 100$ MHz, and $\epsilon \leq 0.15°$ for $f = 1$ GHz.)

Since $f_g < 1.7$ MHz the birefringence can be neglected for many purposes if the signal frequency $f \geq 100$ MHz. This leads to the approximation $n^2 \simeq 1 - (f_p^2/f^2)$ or (with expansion of the square root) $n \simeq 1 - (1/2)(f_p^2/f^2) = 1 - (1/2)(1/f^2)\ A\ N_e$.

In the case of radiowave applications the primary discussion of refraction effects is concerned with phase path length ($\int n\ ds$) and group path length ($\int n_g\ ds$) (ds: ray path element; n_g: group refractive index). Both are compared with the free space distance, ρ, from the transmitter to the receiver.

Since it is not difficult to expand phase path considerations to group path estimates we restrict the following discussion of refraction effects to the phase path length and add the results for group path.

The phase path length has a direct relation to one important group of applications: the so–called Doppler navigation and Doppler geodesy. The basic quantity to be measured in these applications is range difference, $\rho_2 - \rho_1$ (the indices indicate two different positions of a satellite transmitter). Provided that the phase of a received radio signal can be observed continuously between position 1 and 2, the differences $S_2 - S_1$ in phase path length can be determined simply by multiplying the measured phase difference by a factor $(2\ \pi\ f\ /\ c)$.

The difference between the measured and the wanted quantities is called 'propagation error'. For simplicity and generality we discuss 'propagation errors' for ranges and not for range differences bearing in mind that the phase path length itself cannot be 'measured' (because we have no possibility to measure absolute phases).

Now we make use of the series expansion method for the ionospheric refractive index [Leitinger, 1974], [Leitinger and Hartmann, 1977]. Then the propagation errors are separated in terms of increasing power of $(1/f^2)$.

The following scenario is assumed: radiowaves emitted from a satellite transmitter, T, are received at an observing station, R. With the length of the phase path (optical path), S_n, and with true (geometrical) range, $S_o = \rho$, we define the propagation 'error' to be $S_n - S_o$. If we subdivide the error into two partial errors s_1 and s_2 we have $S_n = S_o + s_1 + s_2$. Using the phase path element ds and the line element ds_o (straight line RT from the receiving station to the satellite) then $S_n = \int_R^T n\ ds$ and $S_o = \int_R^T ds_o$. Error distribution is done in such a way that $s_1 = \int_R^T n\ ds_o - \int_R^T ds_o$ and $s_2 = \int_R^T n\ ds - \int_R^T n\ ds_o$. ($s_1$ comprises the first order error and those errors arising from the series expansion of the refractive index. s_2 is the refraction error in the sense that it comprises the effect of the curvature of the ray path.)

Numerical model calculations and derivations for a homogeneous ionosphere (planar as well as spherical layering) have shown that s_2 contains no terms of order zero or one [Leitinger, 1974]. Model calculations demonstrate that the birefringence (anisotropy) influence produces no errors of orders zero, one, or two. Therefore restriction to error terms up to order two (order 4 in $1/f$) allows to use the 'quasi–isotropic' approximation for the ray path, i.e., assuming the ray path to be identical with the wave normal.

s_1 can be derived immediately from the series expansion of n leading to a series expansion in $(1/f^2)$: $s_1 = (s_1)_1 + (s_1)_{3/2} + (s_1)_2 + \cdots$. s_2 is the refraction error (from the curvature of the ray path), the first term of its expansion is of order 2 in $1/f^2$ (of order 4 in $1/f$).

For frequencies > 100 MHz the 'first order error' $(s_1)_1$ is by far the largest partial error. It can be compensated in coherent dual frequency systems but at the expense of enhancing the higher order errors.

There are many single frequency applications which need good estimates of the first order error. Therefore it is necessary to relate it to observable quantities. Using $n = 1 - (1/2)\ A\ N_e$ gives $(s_1)_1 = \int_R^T N_e\ ds_o$. With the zenith angle of the straight line RT, ζ, and the height element dh we get $ds_o = dh/\cos\zeta$. Taking a mean value of $\overline{1/\cos\zeta} = 1/\cos\chi$ leads to $(s_1)_1 = \int_R^T N\ ds_o = \overline{1/\cos\zeta} \int_0^{h_s} N\ dh = (1/\cos\chi)\ N_\perp$. (h_s: height of the satellite transmitter) $\int_0^{h_s} N\ dh = N_\perp$ is the vertical electron content from the ground to the height of the transmitter (abbreviated TEC for 'total electron content').

Identifying h_i with the height of the first moment of the height distribution of electron density ('center of gravity' of the ionosphere) leads to a good approximation. A rule of thumb gives $h_i = h_m + \Delta$, with $\Delta = 50$ km, h_m being the height of the electron density maximum. For many purposes the assumption of a fixed 'mean ionospheric height' is good enough. For mid–latitudes $h_i = 400$ km has found general acceptance.

There are applications which need electron content for error assessment (e.g., single frequency altimetry), other applications need electron content differences (e.g., the so–called Doppler systems which rely on range differences). An important class of applications is based on single frequency interferometry and here again error assessment needs electron content differences. The interferometric arrangement uses a minimum of two rays from a distant source to two antennas separated by the 'baselength' of the equipment. For radioastromomical sources as well as for high orbit satellite sources the baselength Δ gives approximately the separation of the rays in the 'ionospheric height' h_i. This means that in error assessment for short baseline interferometry (small Δ, meaning $\Delta < 100$ km) we have to deal with horizontal gradients of electron content, whereas Very Long Baseline Interferometry (VLBI) has to use the differences in electron content for largely different ionospheric conditions (e.g., nighttime ionization for one ray, daytime ionization for the other, etc.).

Even some dual frequency systems with compensation of first order errors need information about TEC gradients for the assessment of residual plasma influences: the second order propagation errors are proportional to the square of electron content [Leitinger and Putz, 1988]. Therefore the residual error in dual frequency interferometry is proportional to the difference in TEC for the two rays times the average TEC.

Applications of radio wave pulses (radars, altimeters, timing systems) need information about the plasma influence on pulse travel time. In this case the group refractive index is responsible for the 'error' (ionospheric/plasmaspheric time delay). The first order contribution is again proportional to the electron content as shown in Table 1.

Table 1: Ionospheric (plus plasmaspheric) electron content and first order propagation errors.

Frequency	TEC	range error $\Delta\rho$	pulse time delay $\Delta\tau$
	$10^{15}\mathrm{m}^{-2}$	m	μs
100 MHz	100	400	1.3
1 GHz	100	4	0.013

It will be demonstrated in the following sections that realistic assessment of TEC gradients is difficult because of lack of relevant data. Presently there exist no usable theoretical or empirical models. The empirical models are based on smoothing processes which remove most of the 'gradients'.

We apply the term 'gradient' to electron content differences per unit of distance (km) or per degree of latitude, restricting the observation distance somewhat arbitrarily to 550 km or 5 degrees of latitude.

Gradient data samples

Without aiming at completeness we suggest the following list of TEC gradients:

1. Gradients from stable spatial structures which appear in averaged as well as in actual data.
2. Gradients from the signature of atmospheric gravity waves.
3. Gradients from ionospheric disturbances and from phenomena which occur only occasionally.

1. The most stable spatial structures are the general equatorward increase of TEC in mid latitudes during daytime, the west to east increase of TEC in the morning in all seasons, the east to west increase in the afternoon in winter.

All TEC models show these structures but the actual gradients can differ considerably from average values because of a substantial 'day to day variability' of the ionization even in mid latitudes. Except in winter actual diurnal curves of ionization show enhanced variation around noon, in the afternoon, and in the evening. Models can be very misleading when used to predict, say, minimal longitudinal gradients in the early afternoon in summer. There are days with two peaks and a 'bite out' in the early afternoon and others with a single peak of ionization. It is equally impossible to derive reliably from models the rate of TEC decrease for the summer afternoon or early evening.

Afternoons with a well developed Equatorial Anomaly of the F–region show a strong increase of TEC from mid latitudes toward the peak of the anomaly. Only a part of this increase appears in average models. The Equatorial Anomaly varies strongly from day to day both in its amplitude and in the time of development.

A normal feature of winter and equinox nights is the so-called 'F–region main trough' with a distinct minimum of ionization around (magnetic) midnight and in a magnetic latitude around 60°. It is comparatively narrow in latitudinal extent and the location of the minimum depends on geomagnetic activity. Furthermore the width of the main trough is highly variable. As a narrow feature it is smoothed out in models which use only a small number of coefficients for the latitude dependence of ionization. In other models the trough is nearly smoothed out because its location is not stable enough. Finally it should be mentioned that the widely used International Reference Ionosphere (IRI) [Lincoln and Conkright, 1981], [Bilitza, 1990] does not contain the main trough per definitionem: the equatorward border of the main trough is just poleward of the region of validity. A typical example for the main trough in TEC is seen in Figure 1.

2. Atmospheric gravity waves produce signatures in the F–region ionization which are usually seen as 'Travelling Ionospheric Disturbances' (TIDs) ([Hines, 1960], see [Yeh and Liu, 1974], [Francis, 1975], [Gossard and Hooke, 1975], for more recent reviews). It is useful to distinguish between 'large' (LSTIDs), 'medium' (MSTIDs) and 'small' scales. The latter can be considered as background noise. In mid latitudes the MSTIDs produce TEC gradients which are a limiting factor for short baseline interferometry [Spoelstra and Kelder, 1984], [Mercier, 1984], [Mercier et al., 1989]. On the other hand, radio interferometric observations can be used to derive a climatology for MSTIDs [van Velthoven, 1990]. An other important source of data for MSTIDs are HF Doppler measurements (see, e.g., [Waldock and Jones, 1986]). It is assumed that MSTIDs have sources in the troposphere or in the stratosphere (but compare [van Velthoven, 1990] for a discussion of the difficulties with straightforward gravity wave propagation from lower into F–region altitudes).

LSTIDs can be detected in many electron content records, both in the time dependence as observed by means of signals from geostationary satellites, and in the latitude dependence as observed by means of polar orbiting satellites. Figure 3

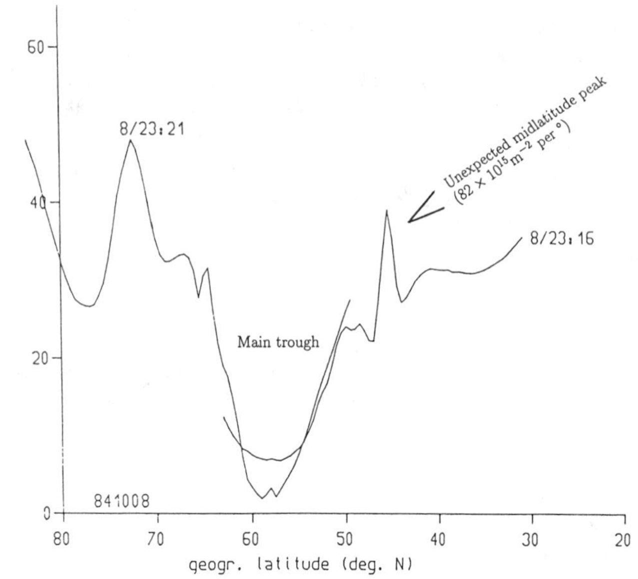

Fig. 1: Electron content in units of 10^{16} m^{-2} vs. geogr. latitude in °N. Data gained at the stations Uppsala (59.8°N, 17.6 E)(left hand curve) and Graz (47.1°N, 15.5°E)(right hand curve) from Diff. Doppler on the signals of an NNSS satellite on 8 October, 1980, around 23 LT. Trough minimum near 58°N, sharp mid-latitude gradients near 43°N.

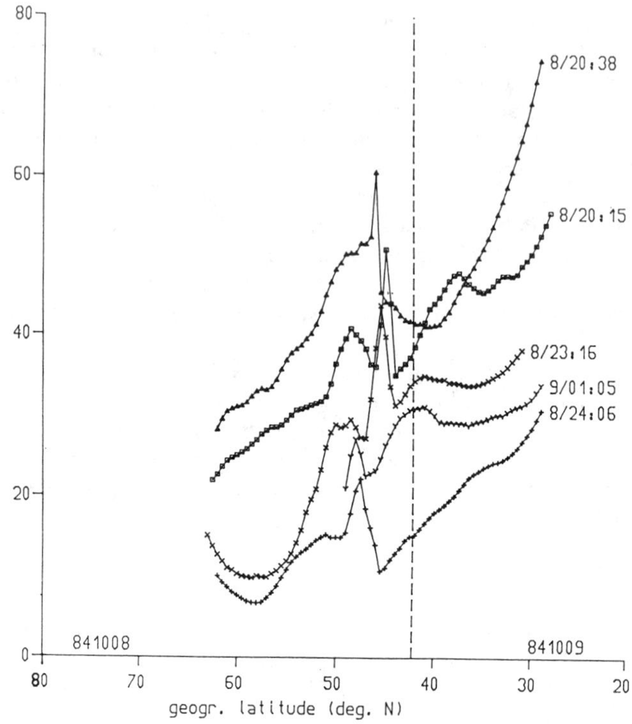

Fig. 2. Sharp mid latitude gradients as in Fig. 1. Electron content in units of 10^{16} m^{-2} vs. geogr. latitude in °N. Data gained at the station Graz (47.1°N, 15.4°E), calibrated with data from Lindau/Harz (51.6°N, 10.1°E) from Diff. Doppler on the signals of an NNSS satellite during the night 8/9 October, 1980. Time marks in CET (UT + 1 hour). (TEC gradient about $82 \times 10^{15} m^{-2}$ per °.)

gives an example for the latter. There is no doubt that most LSTIDs as observed in mid latitudes have sources in the high latitude ionosphere. It is important to note that in many cases a simple point source is not a good model: TEC observations indicate the interaction of waves from different sources and complicated radiation patterns (compare Figure 3). This means that the amplitude of LSTIDs can vary strongly with longitude and one has to be very careful when TID observations at one station are used to predict TEC gradients for an other location [Leitinger and Putz, 1991].

3. Ionospheric disturbances which produce stronger than usual gradients are in general coupled to geomagnetic activity. Therefore it is recommended to check the level of geomagnetic activity by using published indices and reports of geomagnetic storms (e.g., from Solar–Geophysical Data prompt reports).

An other possibility is to use the station to station variability of ionosonde data as an indicator for disturbed conditions. In low and mid latitudes this works often better: the F–region of the ionosphere can be quite disturbed even in the case that the geomagnetic activity is rather low and even a strong geomagnetic storm is not a sure indicator for globally strong ionospheric disturbances. (In some regions of the world the ionosonde network is not dense enough to be used for a check on regional ionospheric disturbances.)

Small scale and short duration disturbances and some phenomena which occur only occasionally are missed by ionosonde observations. Figures 1 and 2 give an example of a phenomenon for which no link could be found to known geophysical processes [Leitinger et al., 1988]. The enhancements with strong gradients occurred clearly in mid latitudes (both geographically and geomagnetically, compare the main trough in Figure 1), a region of the ionosphere which is considered to be 'smooth' compared with lower and higher latitudes. The observations shown in Figures 1 and 2 are a lucky coincidence: the Diff. Doppler evaluations give a realistic image of small scale structures only in the vicinity of the observing station [Leitinger et al., 1975]. Nothing conspicuous can be seen in simultaneous data from the observing station Lindau/Harz which is only about 4° North of Graz. From the existing data material it is impossible to make any statement about the probability of occurrence of similar phenomena in mid latitudes. Our observations should be used as a warning that occasionally TEC gradients can be much higher than assumed from models or from observations made nearby.

In higher latitudes strong gradients have to be expected as normal (see, e.g., [Klobuchar et al., 1986], [Buchau et al., 1988]). Even in the region of the main F–layer trough which is sometimes (wrongly) thought to be a mid latitude feature one has to be aware of phenomena which produce strong TEC gradients, namely 'trough region enhancements'. Examples are shown in Figures 4 and 5. (Compare [Leitinger et al., 1982], [Leitinger et al., 1985].)

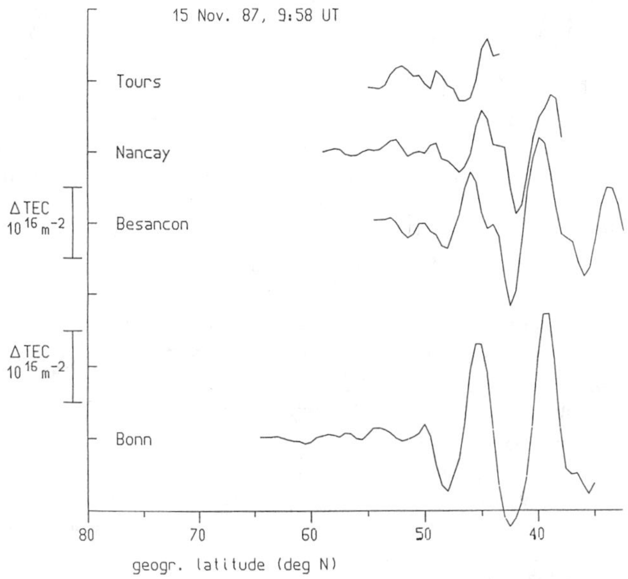

Fig. 3. Example for gradients from TIDs (signatures of atmospheric gravity waves) in electron content. High pass filtered TEC data in units of 10^{16} m^{-2} vs. geogr. latitude in °N. Data gained at the stations Tours (47.4°N, 0.7°E), Nançay (47.3°N, 2.2°E), Besançon (47.3°N 6.0°E), Bonn (50.5°N, 6.9°E) from Diff. Doppler on the signals of an NNSS satellite on 15 November, 1987 around 10 UT. Maximal gradient for Bonn: 16×10^{15} m^{-2} per °of latitude.

Discussion: Availability of Gradient Information

The following typical example is used to demonstrate the differences between model values and observed values.

Situation: northern slope of the equatorial anomaly (increase in TEC from mid latitudes toward the crest of the anomaly). Geographic coordinates: 31°N, 13.5°E, date: 22 February 1991, time: 17:39 UT. Actual sunspot number: 214, monthly mean sunspot number: 168, actual 10.7cm solar flux: 309 units, monthly mean: 243 units. The observed difference of vertical electron content over 1° of latitude was 189×10^{15}m^{-2} whereas the International Reference Ionosphere (IRI 1986) gives 21×10^{15}m^{-2} (February average, highest available mean sunspot number 145, TEC difference for 35°N minus 30°N divided by 5). Linear scaling of the IRI value to the actual sunspot number would give a TEC difference of 31×10^{15}m^{-2}, scaling by using solar flux would give the same value.

The main F–region trough is smoothed out in most empirical models using spherical harmonics or other large scale functions for the dependence of the modeled data on the geographic coordinates. Investigations have shown that trough gradients are highly variable especially in times of high solar activity under geomagnetically disturbed conditions and for troughs appearing in the local time interval 06–14 hours (see [Putz and Leitinger, 1988]). A gradient around 50×10^{15}m^{-2} per degree of latitude can be considered as a typical value for high solar activity conditions but an interquartile range from about 30 to about 130×10^{15}m^{-2} per degree of latitude

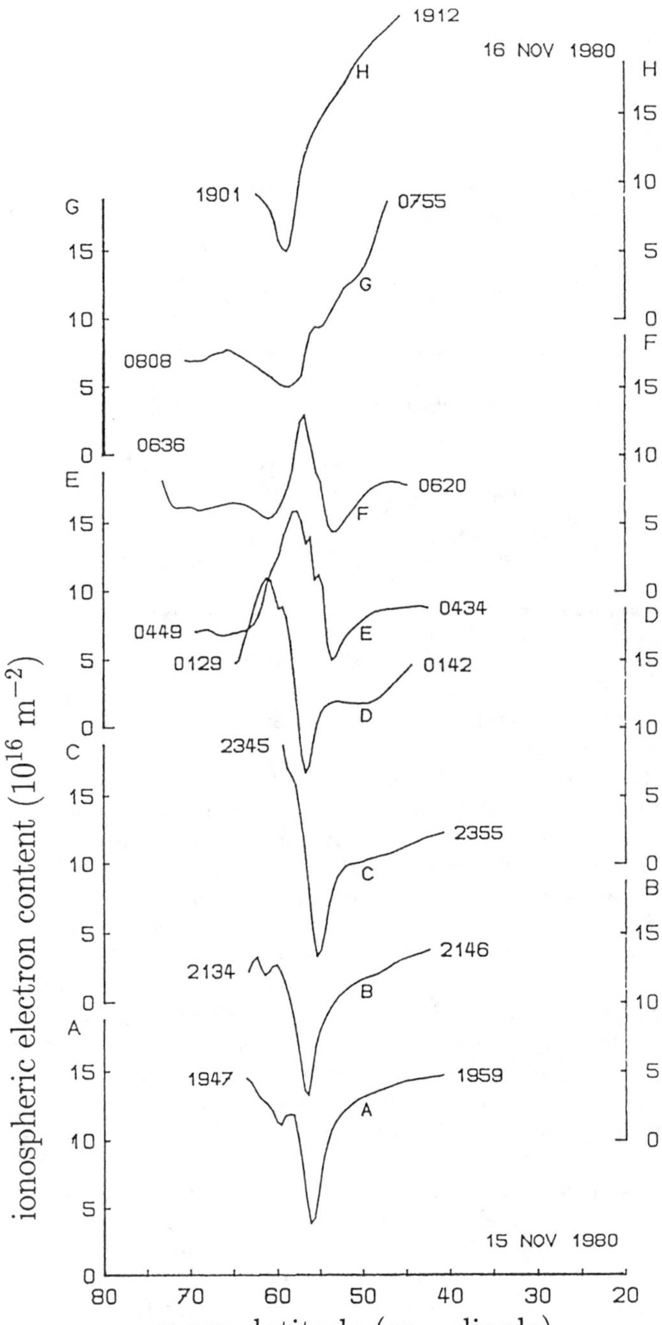

Fig. 4: Example for trough region enhancements. Electron content from Diff. Doppler in units of 10^{16} m^{-2} vs. geomagnetic latitude (eccentric dipole) in °N. Data gained on 16 November, 1980 at the station Uppsala (59.8°N, 17.6 E). Time marks: in CET (= UT + 1 hour). (From [Leitinger et al., 1985]. Reproduced with permission of Pergamon Press.)

was observed for high solar activity under geomagnetically disturbed conditions (Ap > 17) in the local time interval 06–14 LT.

These examples show that one has to be careful even with normal features of the electron content morphology and that

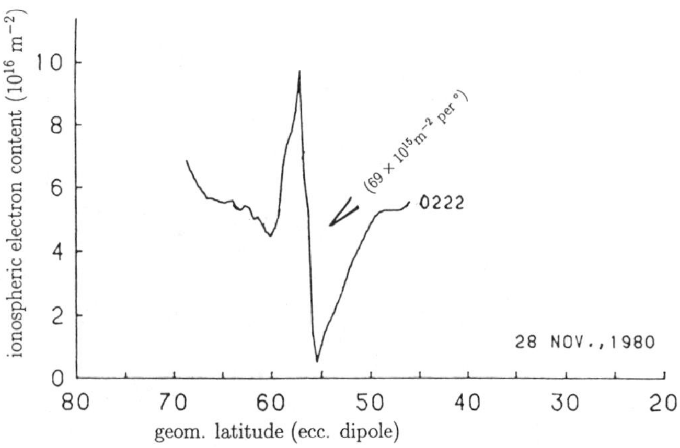

Fig. 5. Example for a very sharp trough region enhancement (strong gradients). Electron content from Diff. Doppler in units of 10^{16} m^{-2} vs. geomagnetic latitude (eccentric dipole) in °N. Data gained on 28 November, 1980 around 01:20 UT at the station Uppsala (59.8°N, 17.6 E). Simultaneous observations made from Lindau/Harz (51.6°N, 10.1°E) prove that the Uppsala observations are not an artifact: the enhancement was seen from Lindau too but with a reduced amplitude and with smoothed peaks (spatial filtering effect, see [Leitinger et al., 1975]).

for precise assessment of TEC gradients it is necessary to use measured data.

As demonstrated ionospheric models are not suited to derive reliable gradient information. Because of the smoothing processes involved in the construction of both empirical and theoretical models they give average gradients only. Even the average can be totally misleading because the smoothing processes decrease even stable gradients and stable but 'movable' structures are smoothed out.

Gradients should be derived directly from observations. For latitudinal gradients one reliable source is the observed latitude dependence of ionospheric electron content from carefully calibrated Differential Doppler observations on the signals of the NNSS satellites (Navy Navigation System, formerly TRANSIT, with 5 to 7 active satellites in nearly polar and nearly circular orbits, height around 1000 km). An other possibility is the use of navigation instruments on polar orbiting satellites, a good example is DORIS (Doppler Orbitography and Radiopositioning Integrated by Satellite) for which a receiver network of about 50 ground stations exists and which was used on the French satellite SPOT2 [Nouël et al., 1991], [Escudier et al., 1991].

Because of the projection from oblique to vertical the 'true' gradients are seen near the latitude of the receiving station only and small scale spatial structures with strong gradients are likely to be missed.

For longitudinal gradients the present prospects are not good: the appropriate source of data are observations of the Faraday effect on the VHF signals of geostationary satellites with a suitable polarization. Unfortunately the VHF beacons are disappearing because there are no operational reasons left to keep them in operation. The geostationary beacons can be replaced partly by the use of GPS beacons (Global Positioning System – satellites in 12 hour orbits) provided that some evaluation problems are solved and provided that the system is not degraded for the unprivileged user. Partly we will have to rely on past experience when assessing longitudinal gradients. There is an important role for theoretical models: The critical local time for longitudinal gradients is sunrise in the production height (around 150 km): at about this time we expect the strongest quiet time longitudinal gradients. For some experimental scenarios even satellite beacon measurements cannot provide good gradient information around sunrise and one should add information from computed electron density behavior. (In general the ray path to a geostationary or to a GPS satellite is not meridional and therefore the sunrise gradients are decreased by smoothing.)

Very good information on latitudinal gradients could be gained by means of a closely spaced meridional chain of receiving stations for the NNSS signals: close spacing would increase the chance to detect small scale structures with steep gradients. Suitable evaluation procedures exist, one possibility is to used weighted latitudinal means which ensure that the small scale structure is taken from the vicinity of the nearest receiving station. Of course a necessary condition would be that the NNSS system is kept in operation.

Coordinated efforts should be made to collect information about TEC gradients with the aim to construct gradient models. With the cautions indicated above the Global Positioning System (GPS) could provide a good basis for long term studies. This has do be done in close cooperation between the beacon satellite community which is interested in ionospheric research and the GPS user groups involved in geodesy related research. Presently good quality TEC data suitable to derive gradients have to be gained by means of the highest quality 'geodetic' dual frequency / dual code receivers. The so-called 'ionospheric' receivers lack the necessary resolution.

References

Bilitza, D. (Editor), *International Reference Ionosphere 1990* (NSSDC/WDC-A-R&S 90–22), pp. 11–22, 1990.

Buchau, J., B. W. Reinisch, D. N. Anderson, E. J. Weber, and C. Dozois, Polar cap plasma convection measurements and their relevance to the modeling of the high latitude ionosphere. *Radio Sci. 23*, 521–536, 1988.

Escudier, P. N., N. Picot, and O. Z. Zaniffe, Altimetric ionospheric correction using DORIS Doppler data, *IUGG Vienna UNION Program and Abstracts*, p. 159, 1991.

Francis, S. H., Global propagation of atmospheric gravity waves: A review, *J. Atmos. Terr. Phys. 37*, 1011–1054, 1975.

Gossard, E. E., and W. H. Hooke, *Waves in the atmosphere. Atmospheric infrasound and gravity waves. Their generation and propagation. (Developments in Atmospheric Science 2)*, Elsevier, Amsterdam–New York, 1975.

Hines, C. O., Internal atmospheric gravity waves at ionospheric heights, *Can. J. Phys. 38*, 1441–1481, 1960.

Klobuchar, J. A., D. N. Anderson, G. J. Bishop, and P. H. Doherty, Measurements of trans-ionospheric propagation parameters in the polar cap ionosphere, in *Proceedings of the Inter-*

national Beacon Satellite Symposium, URSI/COSPAR, part II (edited by A. Tauriainen), pp. 59–67, University of Oulu, Oulu, Finland, 1986.

Leitinger, R., Der Einflußionosphärischer Ausbreitungsfehler bei der geodätischen Anwendung von Navigationssatelliten, *Kleinheubacher Berichte 17,* 321–332, 1974.

Leitinger, R., G. Schmidt, and A. Tauriainen, An evaluation method combining the differential Doppler measurements from two stations that enables the calculation of the electron content of the ionosphere, *J. Geophysics (Zs. Geophysik) 41,* 201–213, 1975.

Leitinger, R., and G. K. Hartmann, Zeit– und Breitenabhängigkeit von transionosphärischen Ausbreitungsfehlern, *Kleinheubacher Berichte 20,* 267–276, 1977.

Leitinger, R., G. K. Hartmann, W. Degenhardt, A. Hedberg, P. Tanskanen, The electron content of the ionosphere and the southern boundary of diffuse aurora, *J. Atmos. Terr. Phys. 44,* 369–374, 1982.

Leitinger, R., P. H. G. Dickinson, A. Dumbs, G. K. Hartmann, A. Hedberg, and A. Ranta, The state of the F–region during the Energy Budget Campaign, *J. Atmos. Terr. Phys. 47,* 49–60, 1985.

Leitinger, R., P. Wilkinson, and R. Hanbaba, The ionosphere in mid–latitudes during the SUNDIAL campaign. *Ann. Geophysicae 6,* 59–68, 1988.

Leitinger, R., and E. Putz, Ionospheric refraction errors and observables, in *Atmospheric effects on geodetic space measurements,* (edited by F. K. Brunner), *Monograph 12,* School of Surveying, University of New South Wales, Australia, pp. 81–102, 1988.

Leitinger, R., and E. Putz, Untersuchung der Begrenzung von F–Schicht–Prognosen durch den Einflußatmosphärischer Schwerewellen, *Kleinheubacher Berichte 34,* 535–542, 1991.

Lincoln, J. V, and R. O. Conkright (Editors), *International Reference Ionosphere — IRI 79,* Report UAG–82, WDC A for Solar–Terrestrial Physics, NOAA, Boulder, CO, USA (Report of URSI Working Group G.4, chairman K. Rawer), 1981.

Mercier, C., Observations of gravity waves by radio interferometry, *J. atmos. terr. Phys. 48,* 605–624, 1984.

Mercier, C., F. Genova, and M. G. Aubier, Radio observations of atmospheric gravity waves. *Ann. Géophys. 7,* 195–202, 1989.

Nouël, F. J. P. Berthias, A. Guitart, A. Piuzzi, and C. Valorge, Drag effect on SPOT satellite and its estimation with DORIS tracking system, *IUGG Vienna UNION Program and Abstracts,* p. 160, 1991.

Putz, E., R. Leitinger, A proposed model for the TEC main trough in winter and equinox nights, in *Investigation of the ionosphere by means of beacon satellite measurement* (edited by Cao Chong), International Academic Publishers (Pergamon), pp. 209–216, 1988

Spoelstra, T. A. TH., and H. Kelder, Effects produced by the ionosphere on radio interferometry, *Radio Sci. 19,* 779–788, 1984.

van Velthoven, P. F. J., *Medium scale irregularities in the ionospheric electron content,* Phil. Diss. (Ph. D. thesis) University of Eindhoven, the Netherlands, Drukkerij Elinkwijk bv, Utrecht, 1990.

Waldock, J. A., and T. B. Jones, HF Doppler observations of medium–scale travelling ionospheric disturbances at mid–latitudes, *J. atmos. terr. Phys. 48,* 245–260, 1986.

Yeh, K. C., and C. H. Liu, Acoustic–gravity waves in the upper atmosphere, *Rev. Geophys. Space Phys. 12,* 193–216, 1974.

R. Leitinger, Institut fur Meteorologie und Geophysik, Universitat Graz, A-8010 Graz, Austria.

Prediction of Average Transionospheric Radio Propagation Delay for the Latitudinal Zone from 15° to 35° N

YINN-NIEN HUANG[1]

Telecommunication Training Institute, Ministry of Transportation and Communications Panchiao, Taipei, Taiwan, Republic of China

The differential Doppler data obtained from the polar orbiting NNSS satellites and the Faraday rotation data obtained from the geostationary satellite, ETS-2, at Lunping (25.00° N, 121.17° E) have been used to construct the contour charts of trans-ionospheric radio propagation delay in a latitude versus local time coordinate system for each day from October 1, 1985, to November 30, 1990. Monthly mean contour charts are obtained from these daily contour charts. The monthly mean delay, T, at a given local time and monthly mean sunspot number, R, are related by a formula T=a+bR. Contour charts for each month for R=10 and 100 are constructed for prediction use.

INTRODUCTION

The U.S. Navy navigation satellite system (NNSS) has been widely used not only for satellite navigation systems but also geodetic and geophysical purposes. In the NNSS positionings, most of the error sources are due to instrumentation such as antenna; receiver; clock; and oscillator and atmospheric refraction including tropospheric and ionospheric refractions. Tropospheric refraction delays are generally smaller than ionospheric ones. Radio waves which pass through the ionosphere propagate more slowly than in free space due to group retardation along the ray path which is proportional to the total number of free electrons integrated along the ray path. Ranging systems such as satellite detection radars or satellite navigation systems will have their ranging accuracy affected by this ionospheric group path delay (Johanson et al., 1978). Under conditions of high electron density such as are encountered during solar maximum periods, especially near the equatorial anomaly crest region (Aravindan and Iyer, 1990; Balan and Iyer, 1983; Dabas et al., 1984; Huang et al., 1989; Rastogi and Alex, 1987), the degradation of ranging system accuracy can be significant. At VHF, ionospheric range errors of greater than one mile are possible. Satellite detection radars operating at UHF can have range errors of several hundred meters; and at L band, such as Global Positioning System (GPS), range error can exceed 200 feet. A two frequency ranging system can be used to correct for the ionospheric time delay. However, for those systems or users to which the two frequency system is impractical, indirect means are required to correct for the time delay due to the ionosphere.

The purpose of present study is to construct the monthly mean contour charts of trans-ionospheric radio propagation delay in a latitude (from 15° N to 35° N) versus local time coordinate system for low (mean sunspot number, R=10) and high (R=100) solar activity periods by use of differential Doppler frequency shift data from the NNSS data and Faraday rotation data from the geostationary satellite, ETS-2, at Lunping (25.00° N, 121.17° E). The charts can be used to predict the average transionospheric radio propagation delay for a given time and month at a place located within 15° to 35° N if a monthly mean sunspot number of the month is known. It is hoped that these charts can be used to correct for the average time delay due to the ionosphere for those systems or users to which two frequency system is impractical.

THEORY

The ionospheric travel time for a radio wave signal transmitted from a satellite to the ground or vice versa is represented by

$$t = \int_0^s \frac{1}{Vg}\, ds \tag{1}$$

where ds is the differential distance along the trans-

---

[1]Now at Department of Posts and Communications, MOTC, 2 Chang Sa Street, Sec. 1, Taipei, Taiwan 10001, Republic of China

Environmental Effects on Spacecraft Positioning and Trajectories
Geophysical Monograph 73, IUGG Volume 13

mission path, and Vg is the group velocity of the radio signal and is related to the refractive index, n, of the medium by

$$Vg = cn \qquad (2)$$

where c is the free space velocity (3×10^8 m/sec). In terms of n, Equation (1) becomes

$$t = \frac{1}{c}\int_0^s \frac{ds}{n} \qquad (3)$$

The refractive index in the ionosphere can be expressed by the relationship

$$n = \sqrt{1 - \frac{Ne^2}{\varepsilon_o m \omega^2}} \qquad (4)$$

where N is the electron density (electron/m^3), e is the electron charge (1.6×10^{-19} coulomb), m is the electron mass (9.1×10^{-31}Kg), ε_o is the electric permittivity of free space (10^{-9} /36π F/m) and ω ($=2\pi f$) is the angular frequency of the radio wave (rad/sec). It should be noted that the ionospheric refractive index is also a function of both the electron collision frequency and the earth's magnetic field. However, for frequencies in VHF band and above, and at altitudes greater than 80 km, the effect of collision frequency term on the refractive index is negligible (Davies, 1990).

For a frequency, f, larger than the plasma frequency of the ionosphere, Equation (3) becomes

$$t = \frac{s}{c} + \frac{e^2}{8\pi^2 c\, \varepsilon_o m f^2}\int_0^s N\, ds \qquad (5)$$

The first term on the right side of Equation (5) is the transmit time of the signal in free space while the second term is the additional time delay introduced by the ionosphere. It follows from Equation (5) that, for frequencies in the VHF band and above , the increase in propagation delay time, $\triangle t$, due to the ionosphere, can be expressed by

$$\triangle t = \frac{e^2}{8\pi^2 c\, \varepsilon_o m f^2}\int_0^s N\, ds$$

$$= \frac{e^2}{8\pi^2 c\, \varepsilon_o m f^2}\int_0^s N \sec\theta\, dh \qquad (6)$$

where θ is the angle between the ray path and the zenith and dh is the differential distance in the vertical direction. The parameter, $\sec\theta$, is the obliquity factor accounting for the slant path geometry. For vertical propagation, Equation (6) becomes

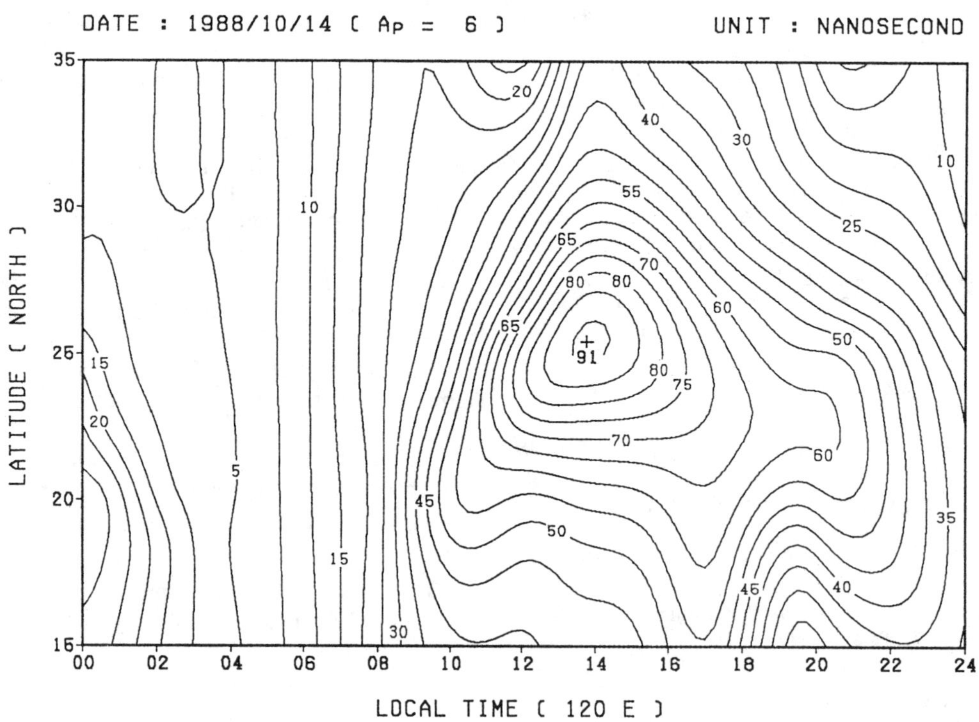

Fig 1. A contour chart of transionospheric radio propagation time delay for October 14, 1988.

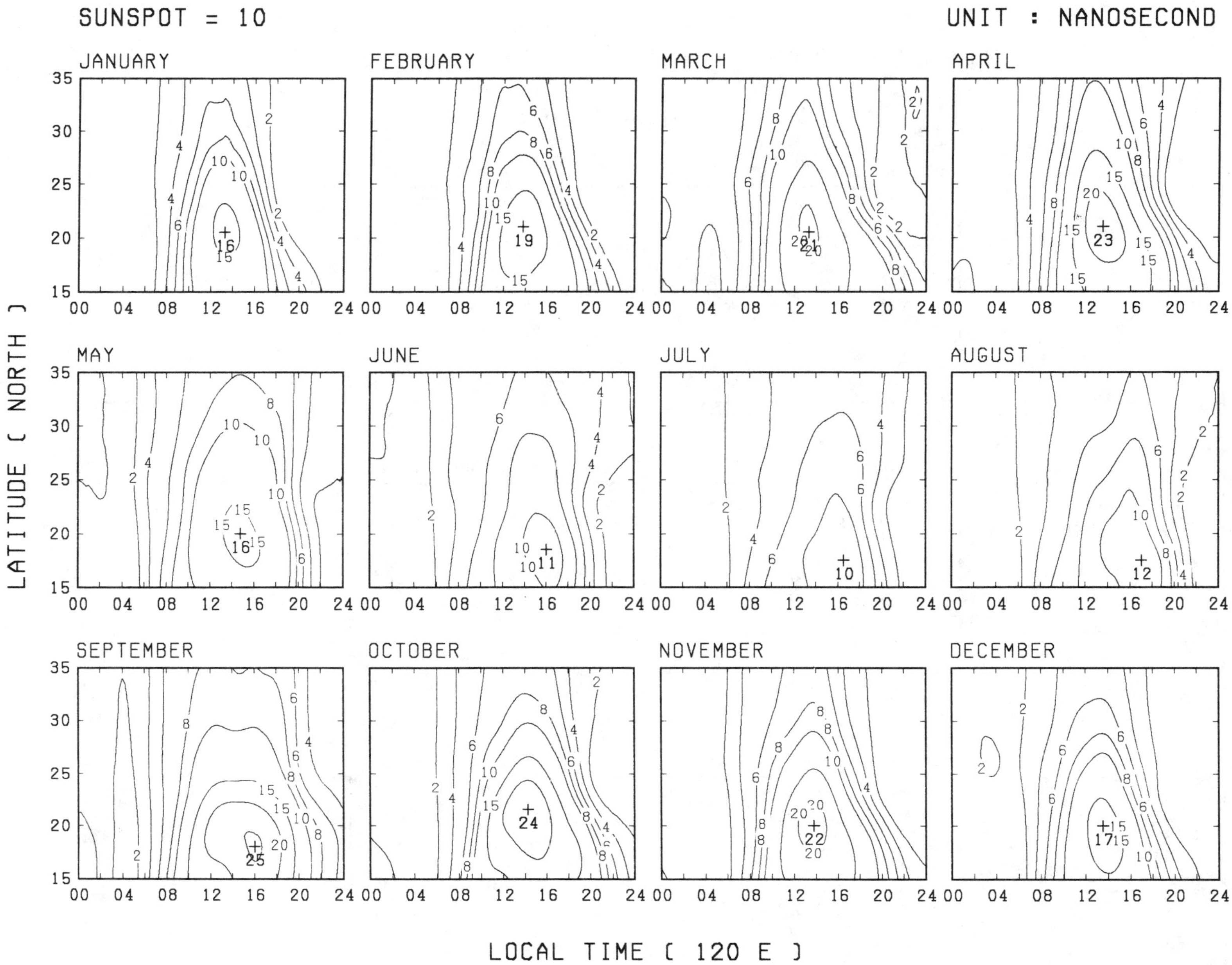

Fig 2. Monthly mean contour charts of transionospheric radio propagation time delay for each month for low solar activity, R=10.

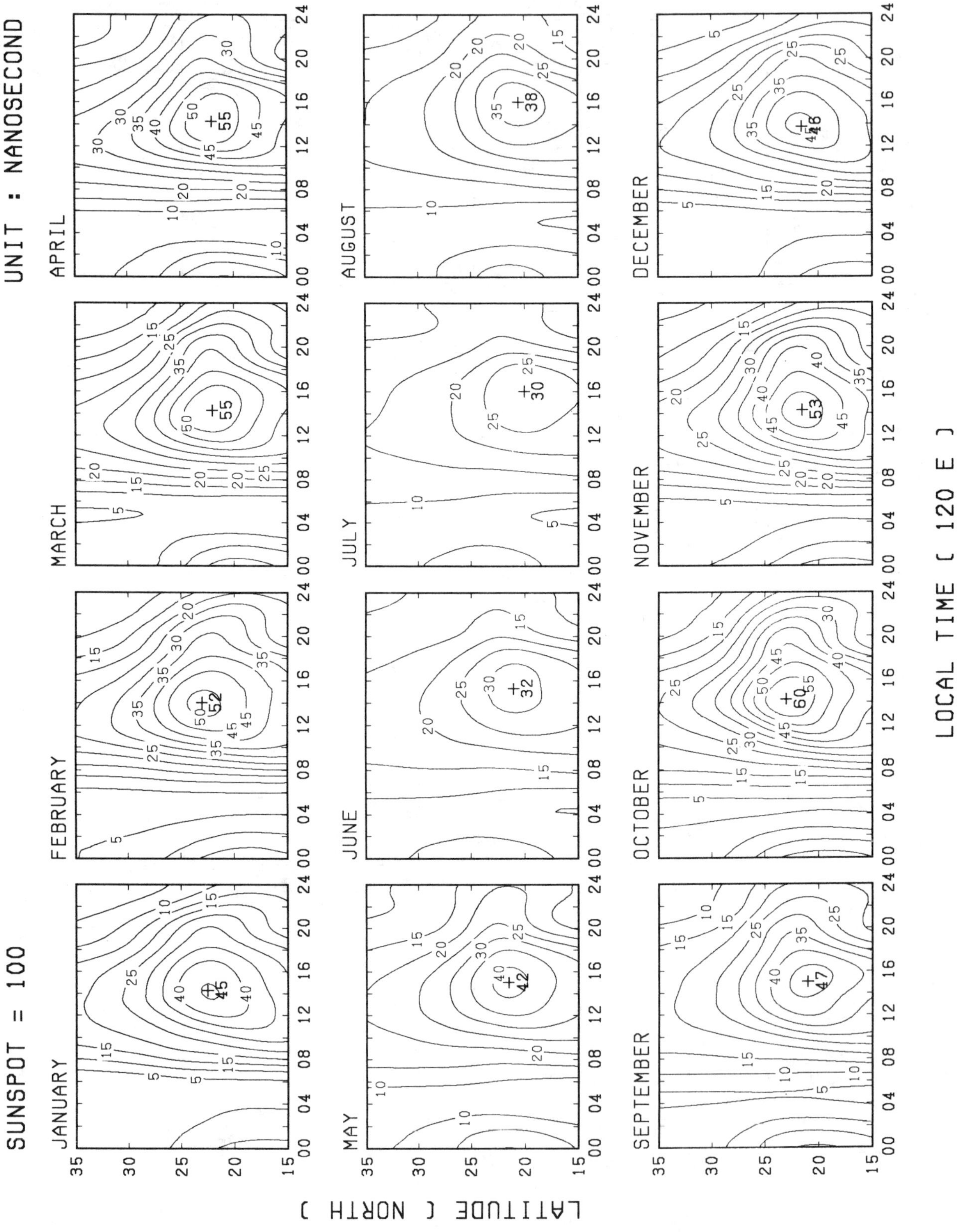

Fig 3. Monthly mean contour charts of transionospheric radio propagation time delay for each month for high solar activity, R = 100.

$$\triangle t = \frac{e^2}{8\pi^2 c \varepsilon_o m f^2} \int_0^s N\, dh$$

$$= \frac{e^2}{8\pi^2 c \varepsilon_o m f^2} I$$

$$= 1.3436 \times 10^{-7} \frac{I}{f^2} \text{ (sec)} \quad (7)$$

where

$$I = \int_0^s N\, dh \quad (8)$$

is the vertical total electron content (TEC) which can be obtained by measuring the Faraday rotation angle of the VHF beacon signal transmitted from a geostationary satellite or the differential Doppler frequency shift of two coherent radio signals from NNSS satallites.

Construction of Daily Time Delay Contour Chart

In order to observe the latitudinal variations of the ionospheric total electron content , TEC, the coherent radio signals at frequencies of 150- and 400-MHz transmitted from the polar orbitted satellites of U.S. Navy Nevigation Satellite System have been received to obtain the differential Doppler frequency shift of the coherent radio signals for each passage of NNSS statellite at Lunping observatory since March 1984 (Huang et al., 1987). By combining the differential Doppler frequency shift data with Faraday rotation data from geostationary satellite, ETS-2, the latitudinal distribution of ionospheric total electron content can be obtained for each passage of the satellite and used to construct TEC contour chart in latitude versus local time coordinates (Huang et al., 1987). This TEC contour chart can be converted to vertical trans-ionospheric radio propagation delay time ($\triangle t$) contour chart by use of Equation (7) and (8) at a given frequency. Figure 1 shows one example for f = 1.57542GHz obtained on Oct. 14, 1988. It can be used to estimate the delay time, $\triangle t$, at a given local time and a given latitude within 15° to 35° N for f = 1.57542GHz.

In this paper, a daily delay time contour chart is constructed for each day from October 1, 1985, to November 30, 1990. Monthly mean contour charts are obtained from these daily contour charts. The monthly mean delay time, T(t,m), at a given local time, t; month, m; and monthly mean sunspot number, R(m) are linearly correlated and can be expressed by

$$T(t,m) = a(t,m) + b(t,m)R \quad (9)$$

where the parameters, a(t,m) and b(t,m), are determined by use of least mean square method. Once a(t,m) and b(t,m) are determined, they are used to construct monthly mean delay time contour charts for each month at R = 10 and R = 100. Figures 2 and 3 show the results obtained for each month for R = 10 and R = 100 respectively. These chart can be used to predict the monthly mean delay time for a desired local time; month and latitude if the mean sunspot number, R, of the month can be predicted by some means.

Conclusions

The latitudinal distributions of the total electron content were obtained by observing differential Doppler frequency shifts data from NNSS satellite combined with Faraday rotation data from geostationary satellite, ETS-2. These latitudinal distributions of TEC obtained in one day were used to construct daily contour chart of trans-ionospheric propagation delay time. Monthly mean delay time contour charts are obtained from these daily contour charts and used to obtain contour chart at low solar activity (R = 10) and high solar activity (R = 100). Using these low and high solar activity contour charts, time delay at any local time, month and location within 15° to 35° N can be predicted if the monthly mean sunspot number can be predicted. It is hoped that these contour charts at R = 10 and 100 can be used to correct for the time delay due to ionosphere for those systems or users to which two frequency system is impractical.

References

Aravindan, P., and K. N. Iyer, Day-to-day variability in ionospheric electron content at low latitudes, Planet. Space Sci., 38, 743, 1990.

Balan, N., and K. N. Iyer, Equatorial anomaly in ionospheric electron content and its relation to dynamo currents, J. Geophys. Res., 88, 10259, 1983.

Dabas R. S., P. K. Bhuyan, T. R. Tyagi. R. K. Bharadwaj, and J. B. Lal, Day-to-day changes in ionospheric electron content at low latitudes, Radio Sci., 19, 749, 1984.

Davies, K., Ionospheric Radio, IEE Electromagnetic Waves, series 31, 1990.

Huang Y. N., K. Cheng, and S. W. Chen, Daily observations of the development of the ionospheric equatorial anomaly by means of differential doppler shift method, Radio Sci., 22, 433 1987.

Huang Y. N., K. Cheng, and S. W. Chen, On the equatorial anomaly of the ionosopheric total electron content near the northern anomaly crest region, J. Geophys. Res., 94, 13515, 1989.

Johanson, J. M., M. J. Buonsanto, and J. A. Klobuchar, The variability of ionospheric time delay, in Beacon satellite measurements of plasmaspheric and ionospheric properties, Proc. Symp. of COSPAR Satellite Beacon Group, Firenze, Italy, May, 1978.

Rastogi, R. G., and S. Alex, Day-to-day variability of ionospheric electron content, J. Atmos. Terr. Phys., 49, 1133, 1987.

Y. -N. Huang, Telecommunication Training Institute, MOTC. 168 Minchu Road, Panchiao, Taipei Hsien, Taiwan 22077, Republic of China.

Table 1. Specifications of TECMETER

Antenna	:	directional
System noise	:	$2 \times 10^{16} el/m^2$
Measurement interval	:	1-20 min (4 min typically)

Mapping GPS measurements to any other directions as well as an interpolation in the time domain is, therefore, necessary not only for a comparative study between TEC measured by two techniques, GPS and VLBI, but also for an application of GPS measurements to the ionospheric correction of a single band VLBI. If GPS measurements are distributed fairly densely both in a celestial hemisphere and in time, we can easily get the TEC for a certain direction by a simple interpolation using the adjacent data; of course, the receiver and transmitter offsets must be considered in advance. However, the actual sky coverage of GPS measurements for a few hours is sometimes limited in a restricted area, because the number of GPS satellites visible in the sky is at most 4 or 5. At one time, there were periods of several hours with no satellites visible but now the number of GPS satellites has been increased so that we are always able to see at least one satellite in the sky at any time although strong interference signals sometimes result in a lack of GPS measurements. To make possible the mapping of N_{GPS} to other directions and over a period of at least 24 hours (the typical session duration of geodetic VLBI) under the unfavourable condition of limited sky and time coverage, we use the concept of the "ionospheric shell" proposed by Lanyi and Roth [1988]; trajectories of GPS satellites are represented by intersections of the lines of sight with the ionospheric shell, and the vertical TEC is expressed as a function of longitude and latitude of the shell. In this paper, for convenience sake on the application to the VLBI ionospheric corrections, universal time (UT) is used as the

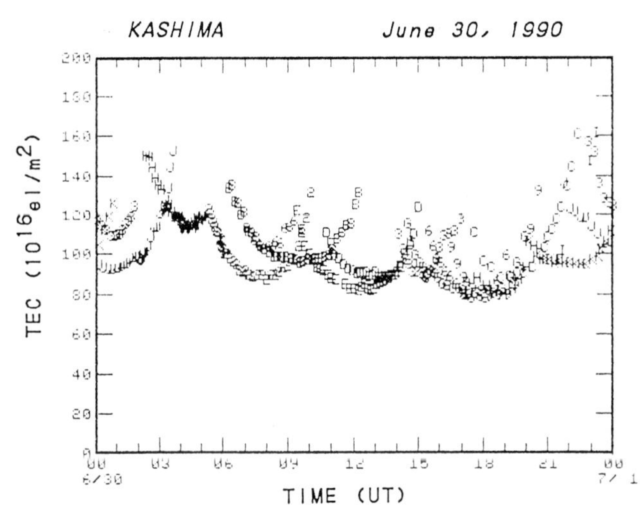

Fig.2. An example of measurement of total electron content obtained by the GPS dual band receiver "TECMETER" at Kashima on June 30, 1990.

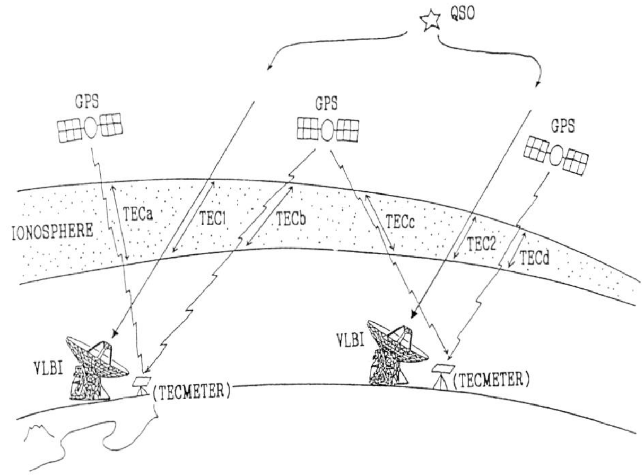

Fig.3. A concept of calibration of ionospheric excess delay using TEC data measured by GPS receivers. The direction of radio star is usually different from the GPS direction. Therefore TEC observed for GPS directions must be projected to the radio star direction.

one of shell coordinates instead of longitude . Lanyi and Roth adopted a two- or three-dimensional polynomial to model the vertical TEC distribution on the shell. Their model is adopted in our study. Besides the polynomial model, we also model the vertical TEC variation in time by the Fourier series to investigate the possibility of interpolation when no GPS measurements are made for more than several hours.

Assumptions regarding the ionosphere used in the study are basically similar to those in Lanyi and Roth's work and can be summarized as follows,

a). the ionosphere is represented by a spherical shell of infinitesimal thickness at an altitude of 300 km, and

b). the global structure is stable in a frame referred to the earth-sun axis, but allows a gradual variation on a time scale longer than one day.

4.1 UT-Latitude Coordinates

Trajectories of GPS measurements can be represented in terms of intersections of the lines of sight with the ionospheric shell (see Figure 4), i.e., the trajectories are given by UT and latitude coordinates of the intersection points with the shell. To take account of longitude difference between receiver position and intersection, we use the "corrected" UT defined as,

$$t = UT_O + \Delta\phi/360^\circ \qquad \text{(day)} \tag{5}$$

where UT_O is universal time (in day) when a GPS measurement made and $\Delta\phi$ is the difference between the shell longitude of intersection ϕ_P and a receiver longitude ϕ_O in degree $(\phi_P - \phi_O)$, i.e., the longitude difference is converted into the local- time-like difference. From now on, we will use the term, "UT" or "time" to mean the corrected UT.

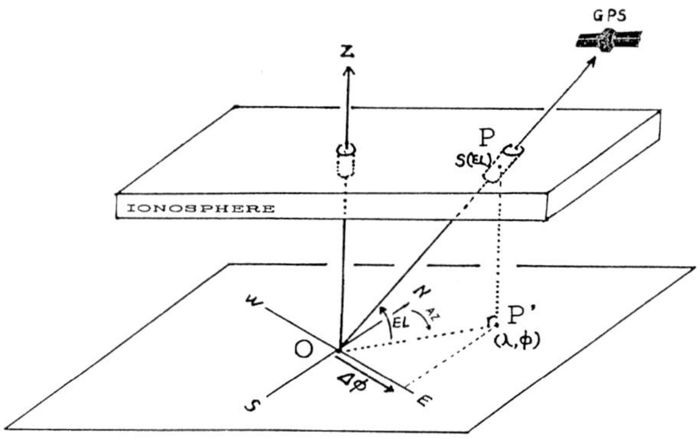

Fig.4. An intersection of GPS satellite with the ionospheric shell. The position of GPS satellite can be represented by a position P where the signals from GPS satellite intersects the ionosphere.

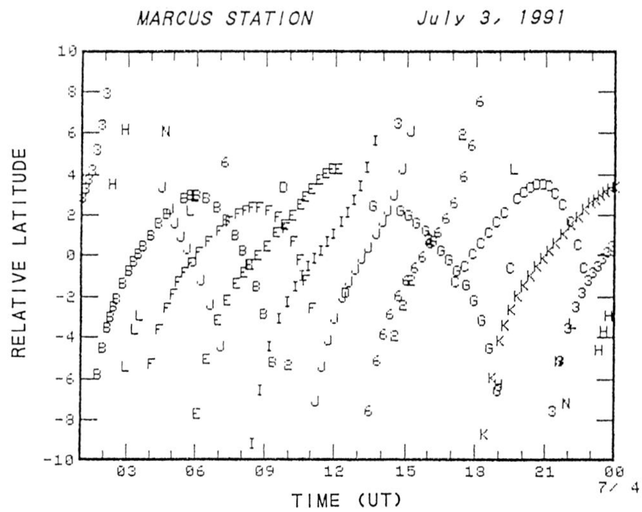

Fig.5. Trajectories of GPS satellites on the ionospheric shell for the data measured at Kashima on July 3,1991. Latitude means relative latitude from the receiver position on ground. Time is "corrected UT" (see text). Modeling of TEC distribution on the UT-latitude plane is the first step of projection of GPS measurements to any other desired direction at an arbitrary time.

Notice that here we define the ionospheric shells referred to each receiver (station), therefore, the shells (i.e., UT-latitude plane) may be different between distant receivers. Figure 5 demonstrates the trajectories of GPS satellites on the ionospheric shell, hence on UT(t)-latitude(λ) plane, referred to Kashima (35.9°N, 140.7°E), Japan, on July 3, 1991. In the figure each data point represents one TEC measurement by the TECMETER. TEC distributions on the $t-\lambda$ plane will be expressed by two kinds of mathematical models in this study (details will be described in Section 4.3).

4.2 OFFSET ESTIMATION

As already mentioned in Section 3, the TEC measured by a GPS receiver is inherently accompanied by both receiver and satellite transmitter offsets. Transmitter offsets vary with the satellites and can be up to 10×10^{16}el/m^2 [Coco et al., 1991]. Thus satellite dependent offsets must be considered in a mapping model utilizing multi satellite data. However no distinction is made here among the satellites for the following two reasons. One is to reduce the number of parameters to be fitted, because too many parameters compared to the number of data points will lead to a meaningless result in the least squares estimation. The other and principal reason from a practical point of view, is that transmitter offsets depend on the particular satellite and are thought to be canceled out as follows. In our comparative study between GPS measurements and VLBI ionospheric corrections, the distance between VLBI stations is at most about 2000 km as seen in the later section. In this case, the trajectories of GPS satellites on $t-\lambda$ planes referred to each station resemble each other. This means the satellite offsets affect almost the same regions on the $t-\lambda$ planes and tend to be canceled out by calculating differential TEC to produce the VLBI corrections. Thus we do not consider each transmitter offset explicitly. However an offset averaged over satellites is taken into the model as a part of the receiver offset. This offset is determined for each receiver from GPS measurements in advance of the estimation of TEC distribution.

The TEC measured by the TECMETER can be expressed for an intersection by

$$N_{GPS} = g(\lambda, t)S(EL) + O_t \qquad (6)$$

where $g(\lambda, t)$ is a vertical TEC at the intersection, $S(EL)$ is a ratio of slant path length to zenith path, a slant factor, as a function of elevation angle EL, and O_t is an offset including both a receiver and averaged transmitter offsets. For the ionospheric shell, from assumption a), $S(EL)$ is simply given by

$$S(EL) = 1/\cos\{\sin^{-1}[R_E \cos EL/(R_E + h)]\} \qquad (7)$$

where R_E and h denote the earth's radius (6371.2 km) and the altitude of the ionosphere. In this study h is fixed to be 300 km. To determine O_t effectively, hence to reduce the number of parameters, we assume the uniform ionosphere around the station but allow a gradual change in time, i.e., $g(\lambda, t)$ is thought to be a constant ($= N_0$) in an area with a narrow UT range (a sub-UT area) on the ionospheric shell. N_0 and O_t are determined for every sub-UT area by means of the least squares estimation using GPS measurement data. Figure 6 shows examples of O_t derived this way, where the ionospheric shell is divided into a number of sub-UT areas each two hours in width and with adjacent areas overlapped by one hour. As shown in the figure, the estimated O_t displays large daily and sub-daily variations. These variations are thought to be partly caused by an imperfect de-coupling between parameters O_t and N_0 due to non-uniform distribution of vertical TEC and partly caused by potential transmitter offsets. Though some problems remain in the determination of offset, we consider O_t averaged over one day or more as an actual observation offset applicable to TEC distribution estimation.

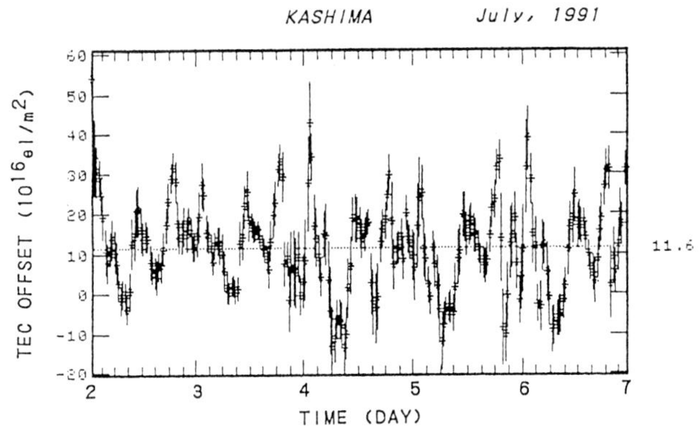

Fig.6. An example of estimated TEC offset from the data taken during July 2-6, 1991 at Kashima. Offsets are estimated using the data spanning two hours with start time of data shifting by one hour step. Averaged offset is drawn by a dotted line.

4.3 TEC Distribution Models

Two mathematical models are proposed to represent the vertical TEC distribution on the $t-\lambda$ plane. One is a Fourier approximation model that describes TEC variation in time by a summation of different periodical components. The other one is a two dimensional polynomial model like that adopted by Lanyi and Roth [1988]. The parameters of each model are determined by means of a least squares estimation using TEC data measured by the TECMETER (an offset due to the receiver and transmitters is removed through the process described in Section 4.2).

Fourier Model

The variation of vertical TEC with respect to time is modeled in a narrow latitudinal zone ($\lambda - \Delta\lambda \sim \lambda + \Delta\lambda$) as

$$g(t)_{\text{at }\lambda} = a_0 + \sum_{m=1}^{4}(a_m \cos 2\pi mt + b_m \sin 2\pi mt) + c_0 t \quad (8)$$

where t is the corrected UT (in days) as defined by eq.(5), a_0, a_m, and b_m are coefficients of the Fourier series and the last term is introduced to model any longer term TEC trend. Fourier coefficients are independently determined for every latitudinal zone. Thus this model gives the TEC distribution continuously in the time domain but discontinuously in latitude. The final distribution of TEC on the $t-\lambda$ plane is derived by using a smoothing filter (we use a Hanning filter) in the latitudinal direction. The width of each latitudinal zone is conventionally taken to be 2°, and adjacent zones are overlapped over each other by 1°.

Polynomial Model

This model describes the distribution of TEC on the $t-\lambda$ plane by a two dimensional third-order polynomial. A polynomial is, in general, inadequate to describe a periodic variation like a TEC daily change. Therefore we divide the $t-\lambda$ plane into sub-UT areas with 4 hours width in time domain. Since the latitudinal coverage of the $t-\lambda$ plane is at most about 20°, it is not necessary to consider periodic variations in the latitudinal direction. In each sub-UT area, TEC distribution is modeled as

$$g(\lambda, t) = \sum_{n=1}^{3}\sum_{m=1}^{3} p_{nm}\lambda^n t^m \quad (9)$$

where p_{nm} are polynomial coefficients which are determined for every sub-UT area by the least squares fitting. As this gives a discontinuous TEC distribution at a boundary between sub-UT areas, we apply the Hanning smoothing filter again for the time direction to get a continuous TEC distribution.

These two models are thought to produce the similar results when the GPS measurements on the $t-\lambda$ plane are fairly dense and uniform. However the actual data coverage is, as described before, often biased and is sometimes entirely lacking in some area, particularly in the time domain. This was often the case with data obtained in the past. Under these unfavourable conditions the Fourier model is expected to give a good approximation to the actual TEC distribution because a priori knowledge of the TEC daily variation is implicitly taken into account by assuming a daily component and its harmonics up to the 4-th order (see eq.(8)).

Figure 7 shows examples of contour plots of TEC distributions obtained by these two models for Marcus station (24.3°N, 154.7°E) on July 3, 1991.

Once $g(\lambda, t)$ is determined for the GPS measurements, we can compute the slant TEC, N_s, for any desired direction and time from,

$$N_s = g(\lambda, t)S(EL) + O_t \quad (10)$$

5 Model Evaluation

We compare the "ionospheric corrections" actually observed by the VLBI ($\Delta\tau_{ion-x}$) with those calculated from the models based on GPS measurements ($\Delta\tau_{GPS}$), in order to evaluate the validity of models proposed in the previous section. This comparative study is thought to give us the most objective evaluation. We have been carrying out simultaneous observations of the VLBI and TECMETER since 1989 on domestic baselines in Japan (Figure 8). At each VLBI station TEC observations were made by the TECMETER for a period longer than one day in which a 24-hour VLBI session was entirely included.

The comparison is carried out as follows. First we estimate the offset of each TECMETER including average transmitter offsets according to the procedure described in Section 4.2. Next a vertical TEC distribution on the ionospheric shell referred to each station is estimated using the two models proposed in Section 4.3, then the vertical TEC at the intersection of line of sight to radio star of VLBI observation is derived from the estimated TEC distribution on the ionospheric shell.

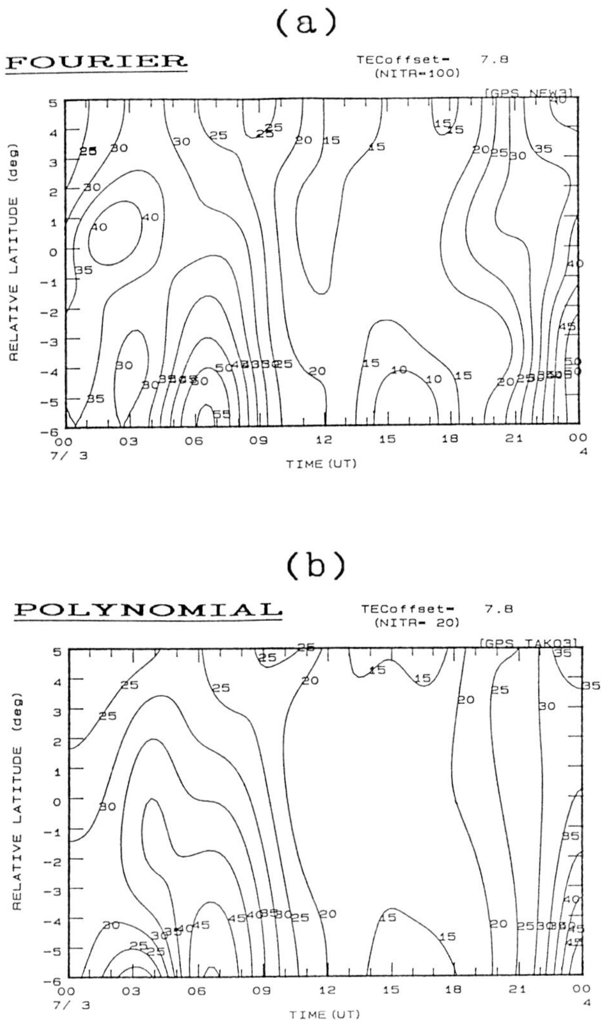

Fig.7. Contour plots of TEC distributions obtained by (a): Fourier model and (b): Polynomial model, for Marcus station on July 3, 1991. Latitude is relative latitude and unit of TEC contour is 10^{16}el/m^2.

This vertical TEC is multiplied by a slant factor there to project to the radio star direction. Lastly slant TECs at two stations are differentiated and converted into X band time delays to produce the VLBI ionospheric corrections. The ionospheric corrections calculated this way from GPS measurements can be directly compared with those observed by S and X band VLBI.

Figure 9 represents examples of comparisons between $\Delta\tau_{GPS}$ and $\Delta\tau_{ion-x}$ together with standard deviations (SD) of the residuals, $\Delta\tau_{GPS} - \Delta\tau_{ion-x}$, for an experiment made on the baseline between Kashima and Marcus station (about 2000 km length) on July 3, 1991. Each data point corresponds to one scan (observation) of the VLBI (one geodetic VLBI session usually consists of more than 150 scans). Error bars in the figure denote one sigma formal error in the estimation of $\Delta\tau_{GPS}$.

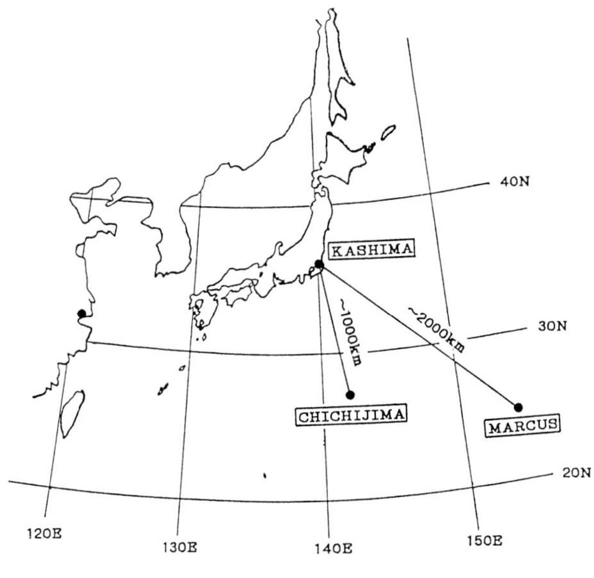

Fig.8. Locations of VLBI stations, Kashima, Chichijima and Marcus.

Errors of $\Delta\tau_{ion-x}$ calculated from the signal to noise ratio of VLBI data are usually very small compared with those of $\Delta\tau_{GPS}$, and consequently are not displayed on the figure. Offsets seen in the difference between $\Delta\tau_{GPS}$ and $\Delta\tau_{ion-x}$ are mainly due to a differential instrumental delay between S and X bands in a VLBI system and are not essential in this study. The comparison results are summarized in Table 2. Residuals are scattered by about 0.2-0.3 ns ($10 - 15 \times 10^{16}$el/m^2), and are about two times larger than the requirement of 0.1 ns (5×10^{16}el/m^2) for the precise geodetic VLBI. As judged by the scatter of the residuals, the Fourier model gives the ionospheric corrections better than those obtained with the polynomial model at the present time, but the differences are not significant.

6 DISCUSSION

The comparison study between the GPS measurements and the dual band VLBI observations shows that both the Fourier and the polynomial models give the ionospheric corrections with an accuracy of about 0.2-0.3 ns in terms of X-band delay ($10 - 15 \times 10^{16}$el/m^2). At the present time the Fourier model seems to give slightly better corrections than those of the polynomial model, even though the difference is not very significant. This is thought to be due to an uneven distribution of the GPS measurements on the $t - \lambda$ plane. As mentioned in Section 4.3, in the case that the data distribution of TEC observed by a GPS receiver on the $t - \lambda$ plane is not uniform, i.e., data are lacking in some area especially in the UT domain, the Fourier model is expected to give a good approximation to the actual TEC distribution because the a priori information about the TEC daily variation is implicitly taken into account in the model. According to an improvement in

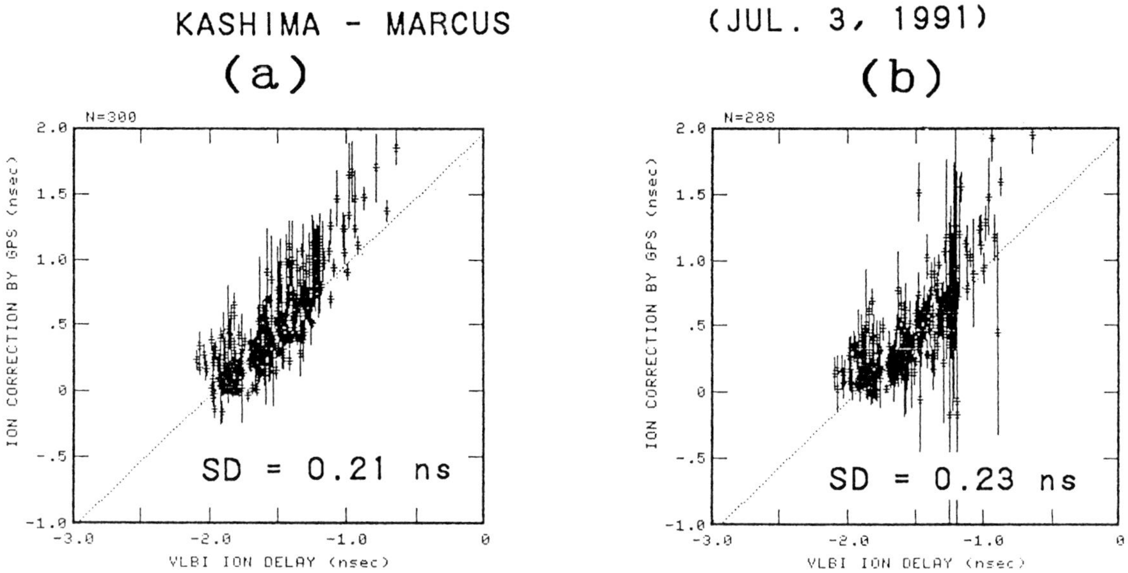

Fig.9. Comparison of calculated ionospheric excess delays and observed ones for an experiment made on the Kashima-Marcus baseline on July 3, 1991. An ordinate is excess delay at X band calculated from the GPS measurements and abscissa is that observed by a dual band VLBI. (a):Fourier model. (b):Polynomial model.

the data distribution on the $t - \lambda$ plane in accordance with the increase of the number of GPS satellites, the polynomial model is expected to produce ionospheric corrections as good as those from the Fourier model.

In VLBI observations dedicated to a precise geodetic measurement, excess delays due to the ionosphere must be corrected with an accuracy better than 0.1 ns in terms of X band excess delay (5×10^{16}el/m^2 in TEC). However the best estimatates from the GPS measurements are scattered by 0.2 ns with respect to actual VLBI corrections (see Figure 9) which is about two times larger than the requirement. Therefore the model proposed here must be further improved. To investigate a cause for the scatter of the residuals, we performed the so-called "elevation cut-off test". In the elevation cut-off test data in a limited range of elevation angle are utilized for the comparative study. Figure 10 shows an example of this elevation cut-off test ($EL > 30°$) in the case of the Fourier model for the same experiment shown in Figure 9. As shown in the figure, SD is drastically improved from 0.21 ns to 0.10 ns. It is therefore suggested that a refinement of $S(EL)$ especially in low elevation angle may improve the fitting. In this paper, the ionosphere is assumed to be a spherical shell with infinitesimal thickness. A slant factor calculated on the basis of this assumption will have a considerably large error for a low elevation path when the plasma density has a large gradient in the horizontal direction such as at dawn and at dusk. In this case, thickness of ionosphere must be taken into account along with a TEC horizontal distribution.

Also we must consider errors due to differential GPS transmitter offsets which are neglected here as they are canceled out. According to recent works [e.g., Coco et al., 1991], transmitter offsets can vary by 10×10^{16}el/m^2 (0.2 ns in X band delay) with satellites. Our elevation cut-off test demonstrates the scatter of residuals is less than 0.2 ns, i.e., the offsets are thought to be actually canceled out to some extent. As mentioned earlier, trajectories of GPS satellites on a $t - \lambda$ plane do not differ much from those referred to a neighbouring station, i.e., the same satellite is measured from both stations in almost the same direction. Thus offsets attributed to transmitter could be common to both stations in the estimation of TEC distribution on $t - \lambda$ planes and could be canceled out through the calculation of differential TECs to form VLBI ionospheric corrections. In this way, transmitter offsets are thought not to affect seriously the calculation of VLBI ion corrections on a short baseline (< 2000 km). However it should be taken into the model to apply to ionospheric corrections on a long baseline such as an intercontinental baseline. Of course, in an application to research on the ionosphere itself,

Table 2. Summary of model comparison

Baseline	Date	Standard deviation of residuals (ns)	
		Fourier Model	Polynomial Model
Kashima-Chichijima	Dec. 6,1989	0.28	0.29
Kashima-Marcus	Jun.30,1990	0.21	0.25
Kashima-Marcus	Jul. 3,1991	0.21	0.23

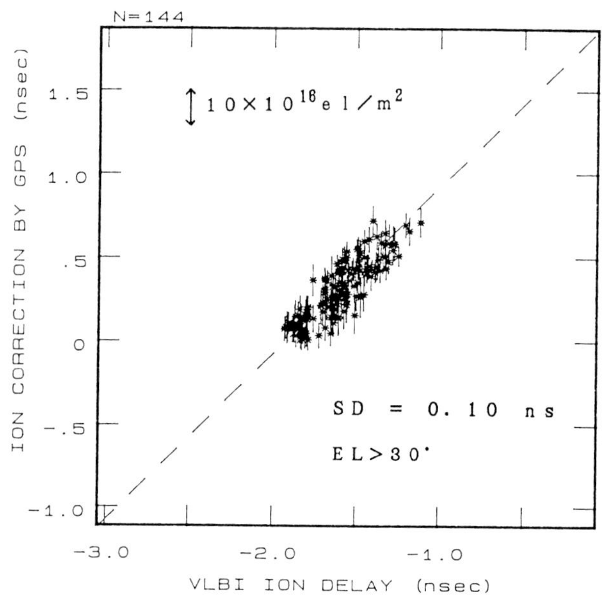

Fig.10. An example of elevation cut-off test for the same experiment data used in Fig.9. Data are restricted in the region of elevation angle larger than 30°. Note that standard deviation of residuals is improved by factor about two, i.e., 0.1 ns in terms of X band delay (5×10^{16}el/m^2 in TEC).

we must carefully consider the effect of satellite offsets.

7 CONCLUSIONS

We have proposed two projection models (Fourier and polynomial) that calculate the ionospheric excess delays, which are necessary for a geodetic VLBI made at a single frequency band, using TEC data measured by GPS dual band receivers. An evaluation of the models has been made by comparing the differential excess delays calculated from the models with those actually observed by a dual band VLBI. The comparison shows that both the Fourier and the polynomial models give almost same results but are slightly in favor of the Fourier model. The scatter of residuals is 0.2 ns in terms of X band delay and is about two times larger than the requirement (0.1 ns) from a precise geodetic measurement. We also made an elevation cut-off test to search a cause of large scatter after fitting seen in the comparison study. This test suggests that a refinement of function calculating slant factor in particular at low elevation angle will improve the whole projection model. Even though residuals are scattered more than the requirement from X band (8 GHz) VLBI, the current models can be applied to an ionospheric delay correction of VLBI planned at a higher fre- quency band such as a 22 GHz band, because the ionospheric effects will reduce in proportion to the inverse square of the frequency, i.e., 0.2 ns at 8 GHz corresponds to 0.03 ns at 22 GHz and it will satisfy the requirement.

Lastly the projection models proposed in this paper can be applied to monitoring the ionosphere. The current model yields a TEC distribution on the UT-latitude plane. This will be a helpful tool for monitoring the ionosphere. Furthermore it is possible to cover wider region by spaced GPS receivers separating a few hundreds of kilometers along a longitude line. In the application to ionospheric study, it is important and essential to remove the effect arising from the GPS transmitter offsets.

Acknowledgments. The VLBI experiment on the baseline between Kashima and Chichijima was carried out in cooperation with the Geographical Survey Institute (GSI) of Japan. We would like to express our thanks to GSI VLBI staff members for the close cooperation.

REFERENCES

Clark, T. A., B. E. Corey, J. L. Davis, G. Elgered, T. A. Herring, H. F. Hinteregger, C. A. Knight, J. I. Levine, G. Lundqvist, C. Ma, E. F. Nesman, R. B. Phillips, A. E. E. Rogers, B. O. Ronnang, J. W. Ryan, B. R. Schpler, D. B. Shaffer, I. I. Shapiro, N. R. Vandenberg, J. C. Webber, and A. R. Whitney, Precision geodesy using the Mark-III very long baseline interferometer system, *IEEE Trans. Geosci. Remote Sensing, GE-23*, 438-449, 1985.

Coco, D. S., C. Coker, S. R. Dahlke, and J. R. Clynch, Variability of GPS satellite differential group delay biases, *IEEE Trans. Aerosp. Electron. Syst., AES-27*, 931-938, 1991.

Colombo, O. L. and M. M. Watkins, Satellite positionning, *Rev. Geophys. Suppl.*, 138-147, 1991.

Imae, M., T. Kondo, J. Amagai, A. Kaneko, and C. Miki, An application of GPS dual frequency codeless receiver for ionospheric delay to single frequency band VLBI observation, *IEEE Trans. Instrum. Meas., 40*, 208-211, 1991.

Imae, M., W. Lewandowski, C. Thomas, and C. Miki, A dual frequency GPS receiver measuring ionospheric effects without code demodulation and its application to time comparison, in *Proc. 20th PTTI meeting,* 1988.

Lanyi, G. E. and T. Roth, A comparison of mapped and measured total ionospheric electron content using global positioning system and beacon satellite observations, *Radio Sci., 23,* 483-492, 1988.

Royden, H. N., R. B. Miller, and L. A. Buennagel, Comparison of NAVSTAR satellite L band ionospheric calibrations with Faraday rotation measurements, *Radio Sci., 19,* 798-804, 1984.

Soicher, H., Ionospheric and plasmaspheric effects in satellite navigation systems, *IEEE Trans. Antennas Propag., AP-25,* 705-708,1977.

T. Kondo and M. Imae, Kashima Space Research Center, Communications Research Laboratory, Kashima, Ibaraki 314, Japan.

Altimetric Ionospheric Correction Using DORIS Doppler Data

P. ESCUDIER and N. PICOT

Centre National d'Etudes Spatiales, Toulouse, FRANCE

O.Z. ZANIFE

CLS/ARGOS, Toulouse, FRANCE

The DORIS precise positioning system [Dorrer et al., 1991] provides dual-frequency Doppler measurements over a global network of about 50 transmitting beacons. These measurements are used to estimate a global map of the vertical Total Electron Content in the ionosphere at the satellite local time. We have applied this technique to measurements obtained with the DORIS receiver flying on board the SPOT2 satellite. Based on comparisons with an empirical model and with estimates from GPS observations, the ionospheric correction from DORIS appear to be sufficiently precise to apply to single-frequency radar altimetry data. During the TOPEX/POSEIDON mission [Fu et al., 1991] dedicated to ocean topography observation, this method will be used to correct the POSEIDON single-frequency radar altimeter for the effect of the ionosphere.

INTRODUCTION

The presence of free electrons in the ionosphere modifies the propagation of radio waves (see Figure 1 for a typical electron density profile as a function of altitude).

To the first order the effect on range measurement is proportional to the total number of electrons along the ray path and inversely proportional to the square of the frequency of the radio wave. For satellite radar altimeters intended to perform space measurements of the distance between the satellite and the sea surface height using 13 GHz frequency radio waves, this effect is typically of order of 20 cm. This magnitude is comparable to the amplitude of large oceanic signals. In terms of spatial signature, the ionospheric correction is mainly correlated with geomagnetic latitude (showing maxima in the tropics). The temporal evolution is correlated with local time (maxima around 2 pm), season and solar or geomagnetic activity.

Figure 2 illustrates a typical variation of the ionospheric correction in cm as a function of latitude during medium solar activity at 10.30 am local time.

Callahan (1984), Monaldo (1990), Musman (1990) and others have analyzed the ionospheric effect on altimeter measurements.

The main spatial variations are on large scales. Monaldo (1990) indicates that for scales less than 1200 km variations of sea surface height are larger than the apparent height changes induced by the ionosphere.

To achieve its goal of an rms less than 2 cm in altimetry, TOPEX/POSEIDON mission will require a precise ionospheric correction be applied, especially for large scales.

STRATEGIES FOR IONOSPHERIC CORRECTIONS

To correct for the ionospheric effect one may use semi-empirical models such as the one proposed by Llewellyn and Bent (used to correct ERS-1 altimeter) which predicts the Total Electron Content (TEC). These models have been computed using historical data, they represent a mean state of ionosphere and cannot display any day to day variations. The prediction error in these models is typically 25 percent of the TEC [Llewellyn, 1973], with the possibility of larger errors during high solar activity periods. The spectrum of the model errors is not random and unfortunately, their spatial signature can be confused with an oceanographic signals.

TOPEX/POSEIDON will carry the NASA dual frequency altimeter. By using both frequencies and the known dispersion curve, this instrument will calculate its own ionospheric correction.

On the other hand, the CNES POSEIDON Solid State Altimeter is a mono-frequency radar altimeter. This instrument will fly on the same platform, it will share the same antenna and will be used exclusively about 10% of the time. In order to compute the ionospheric correction it was suggested to use the bifrequency measurements from the DORIS system that will fly on the same platform. Preliminary studies [Fleury et al., 1989 and 1991] have indicated that the method is feasible due to the density of the DORIS network and because DORIS and POSEIDON measurements are quasi simultaneous.

Environmental Effects on Spacecraft Positioning and Trajectories
Geophysical Monograph 73, IUGG Volume 13

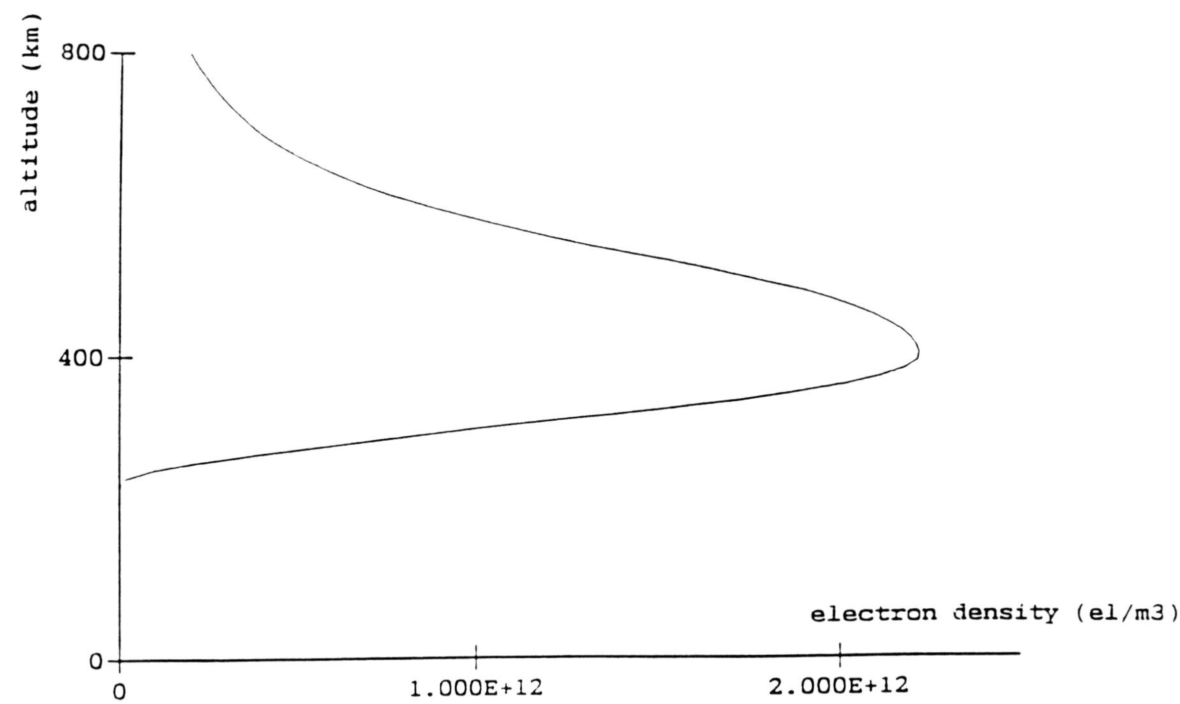

Fig. 1. Typical Electron Density Profile based on the BENT model.This simulation is for a location at 0 degree latitude and 0 degree longitude and for a local time of 10.5 a.m..

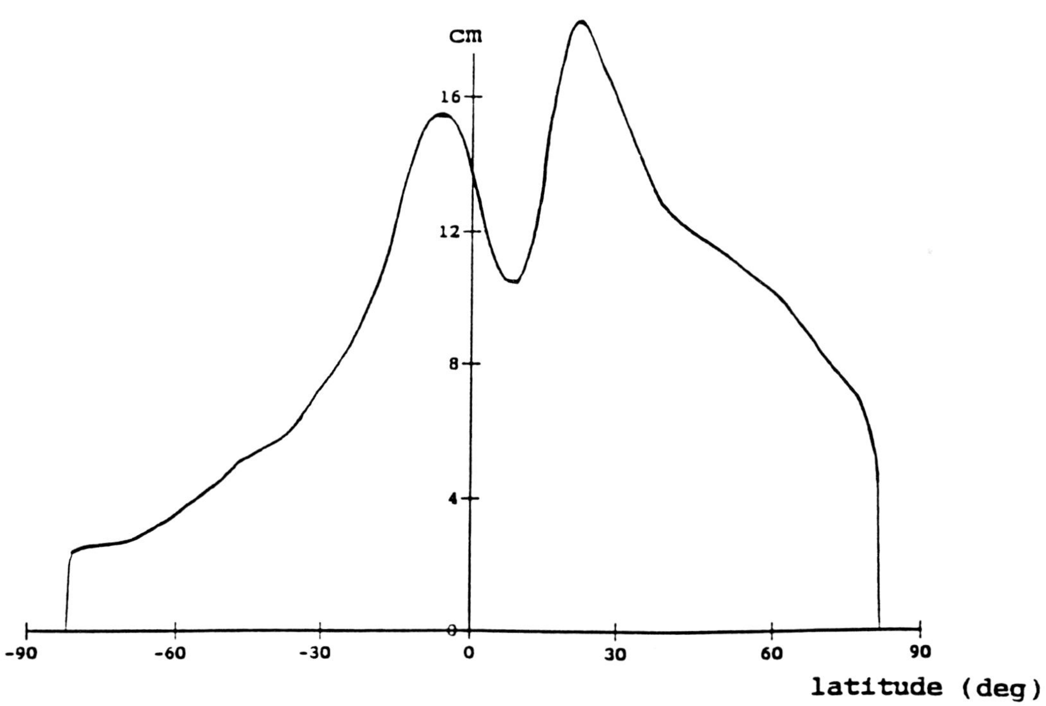

Fig. 2. Typical Altimeter Ionospheric Correction based on the BENT model. This simulation is for a location at 0 degree latitude and for the SPOT2 local time (see Fig. 11)

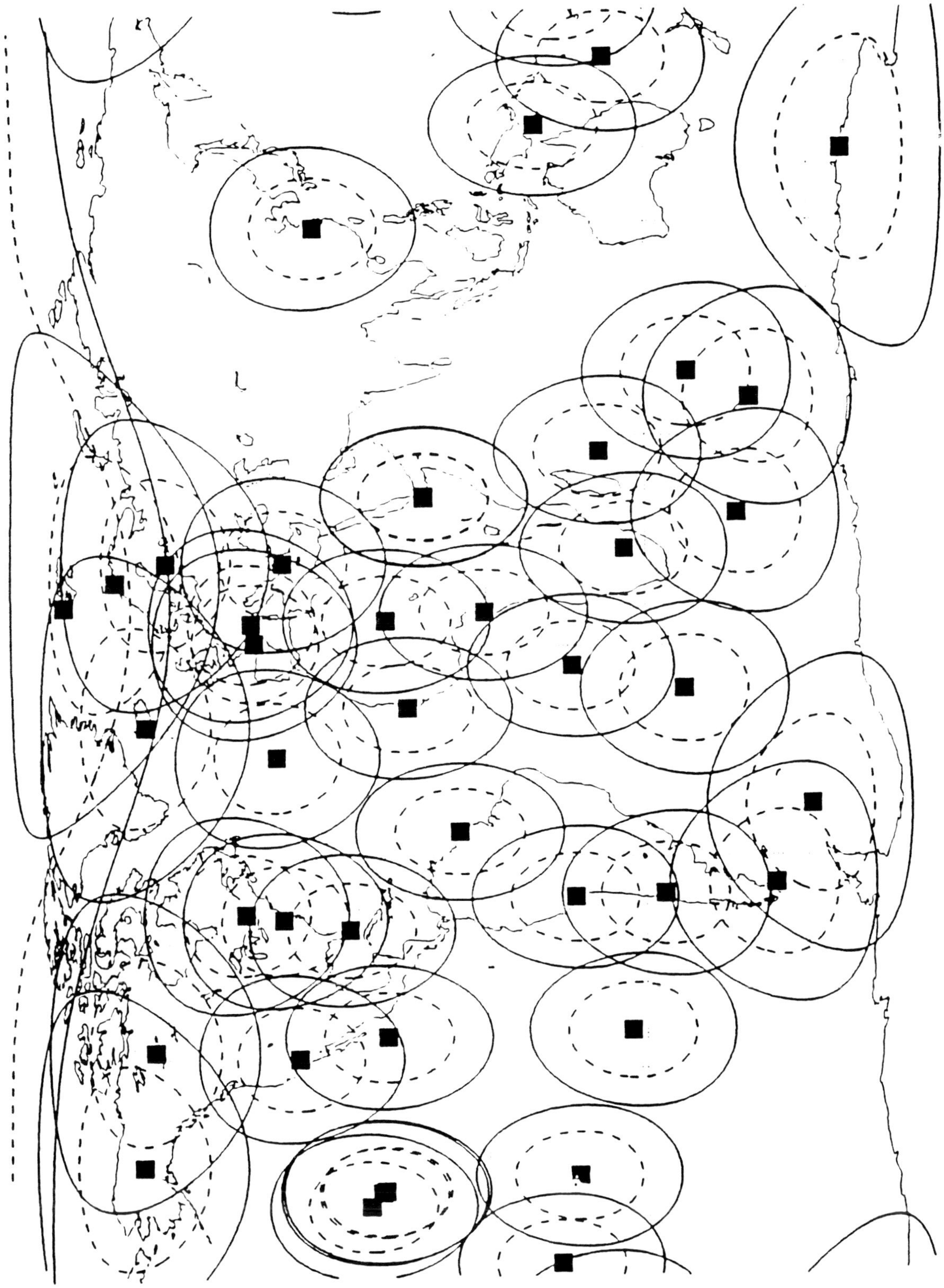

Fig. 3. DORIS Network in use October 30, 1990. The station locations are represented by a square, solid lines are the visibility circles at SPOT2 altitude (830 km), dash lines are the visibility circles at subionospheric altitude.

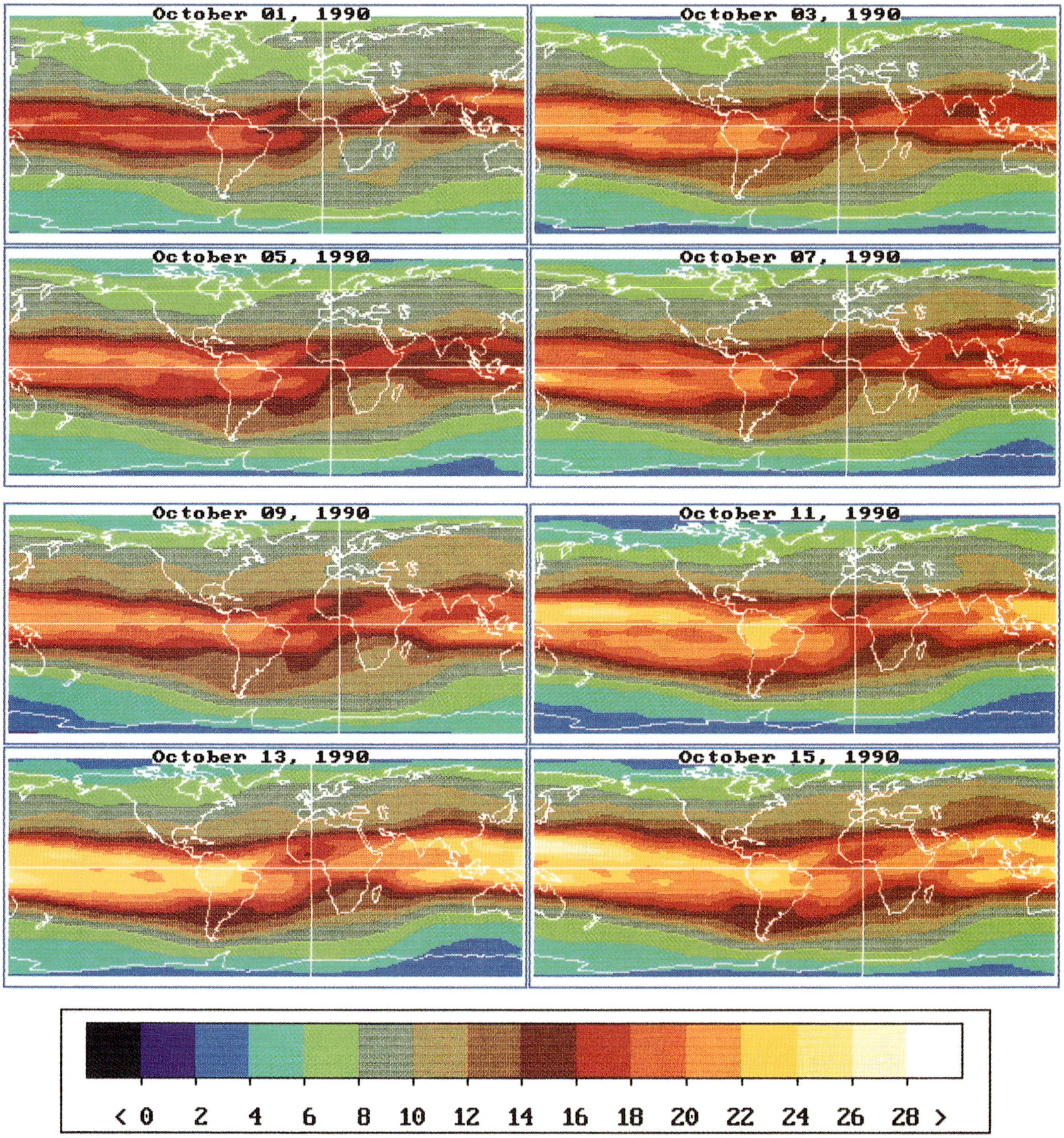

Fig. 4. DORIS computed altimeter ionospheric corrrection in centimeters between October 01, 1990 and October 15, 1990. Note the day-to-day variability and the effect of the magnetic field.

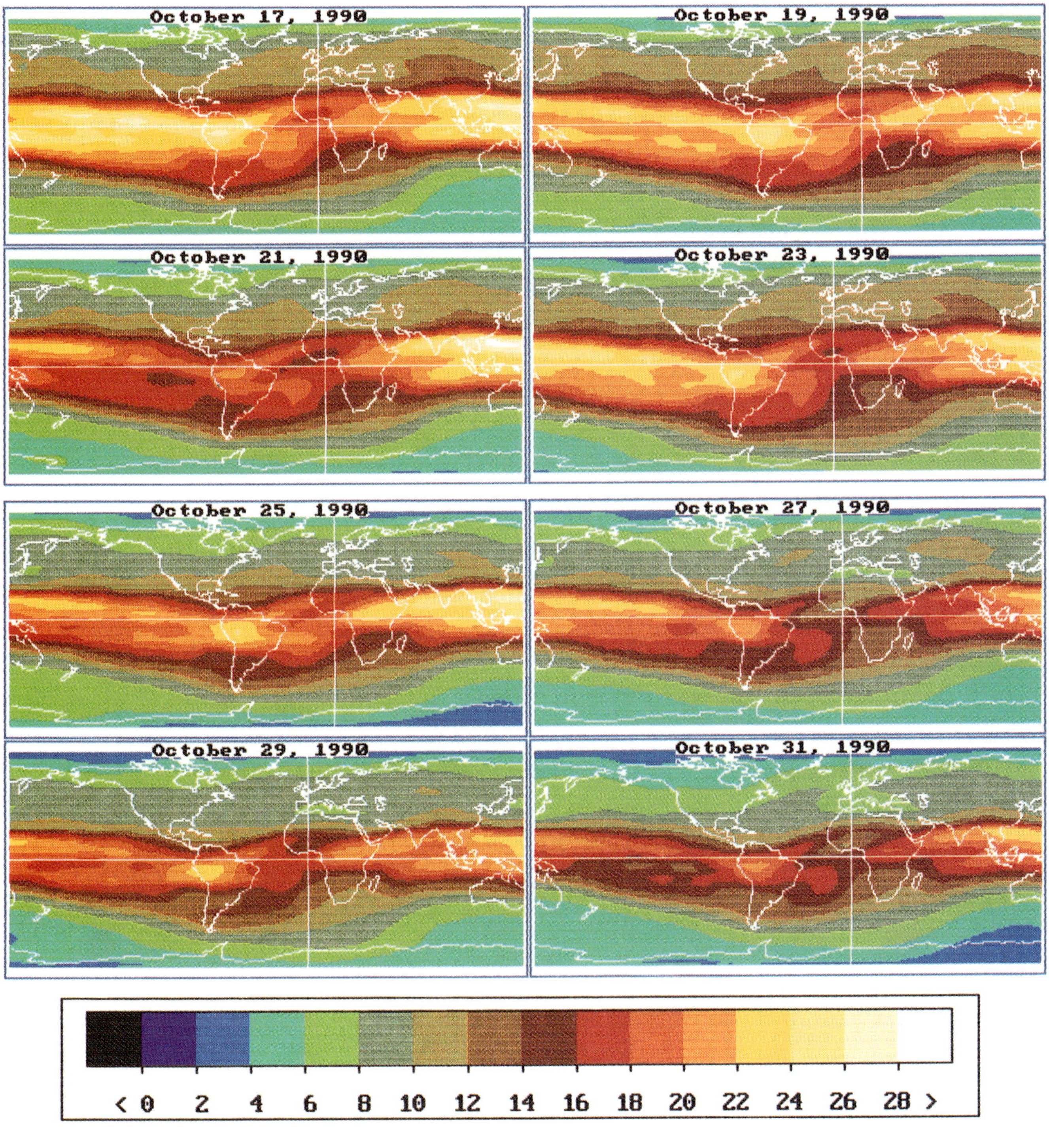

Fig. 5. DORIS computed altimeter ionospheric corrrection in centimeters between October 17, 1990 and October 31, 1990

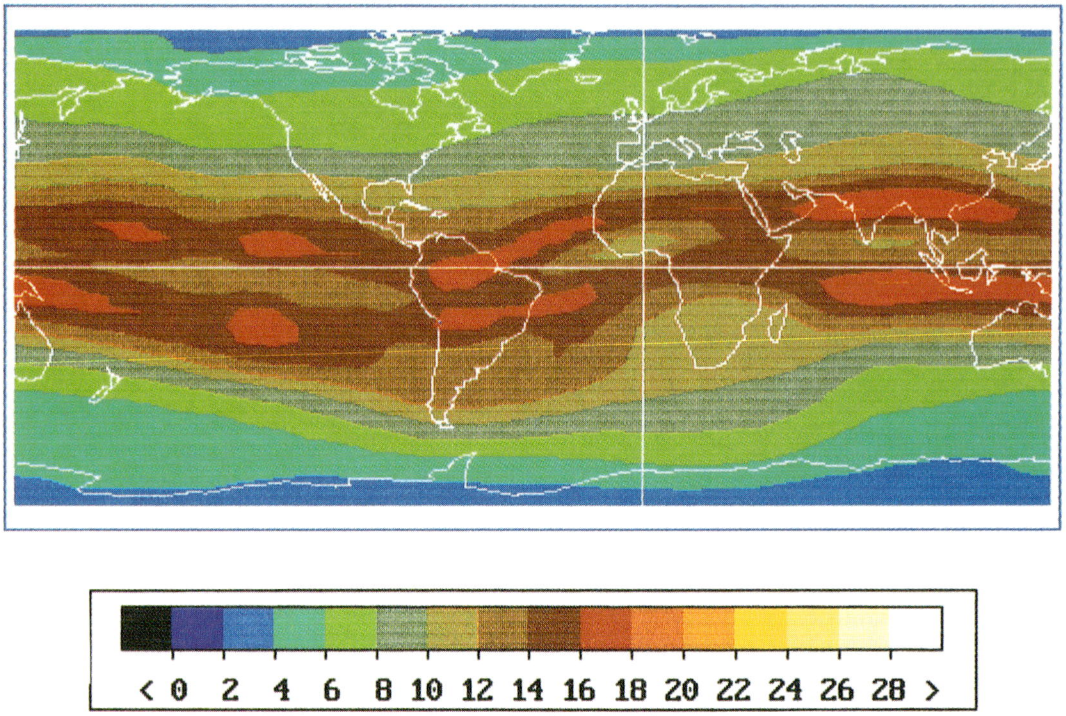

Fig. 6. Bent model altimeter ionospheric correction in centimeters computed for October 1990 period.

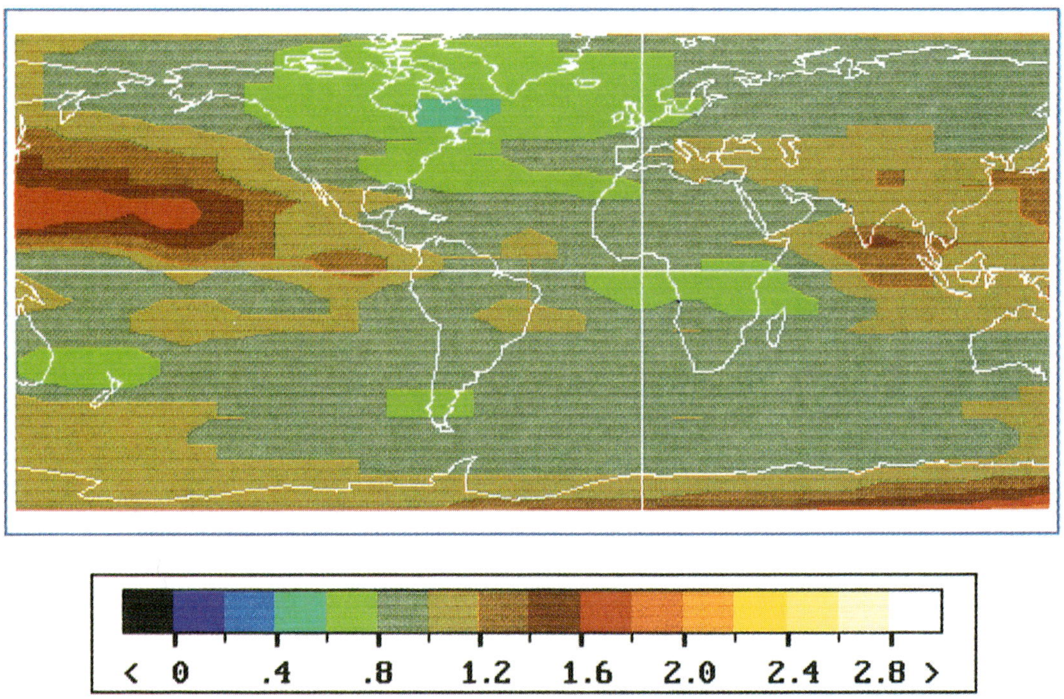

Fig. 7. Predicted uncertainty for DORIS computed ionospheric correction.

DORIS Use For Ionosphere Correction

DORIS

DORIS is a positioning system that was designed to perform precise orbit determination for missions such as TOPEX/POSEIDON and to carry on the precise positioning of beacons (DORIS/SPOT mission). The system includes a network of about 50 transmitting beacons spread all over the world, receivers flying on board satellites and a control and processing system on the ground. Each beacon transmits a signal on two frequencies (2 GHz and 400 MHz). The on-board receiver measures the Doppler count over a counting period of about 10 seconds. This system has been in use since February 1990, when the first DORIS flew on board the SPOT2 satellite.

DORIS ionospheric measurements

DORIS data.

At each frequency f_i (2 GHz and 400 MHz) DORIS measures N_i, T_{1i}, T_{2i}. The equation of measurement is :

$$N_i = \int_{T_{1i}}^{T_{2i}} (f_{ri} - f_{oi})\, dt \qquad (1)$$

with:

N_i	:	Doppler count
T_{1i}	:	beginning of Doppler count
T_{2i}	:	end of Doppler count
f_{ri}	:	$f_i + dop_i$
dop_i	:	Doppler shift
dop_i	:	$- f_i \cdot vr / c$
vr	:	radial velocity (derivative of satellite beacon distance) including propagation effects
c	:	light celerity
f_i	:	transmitted frequency
f_{oi}	:	on board reference frequency

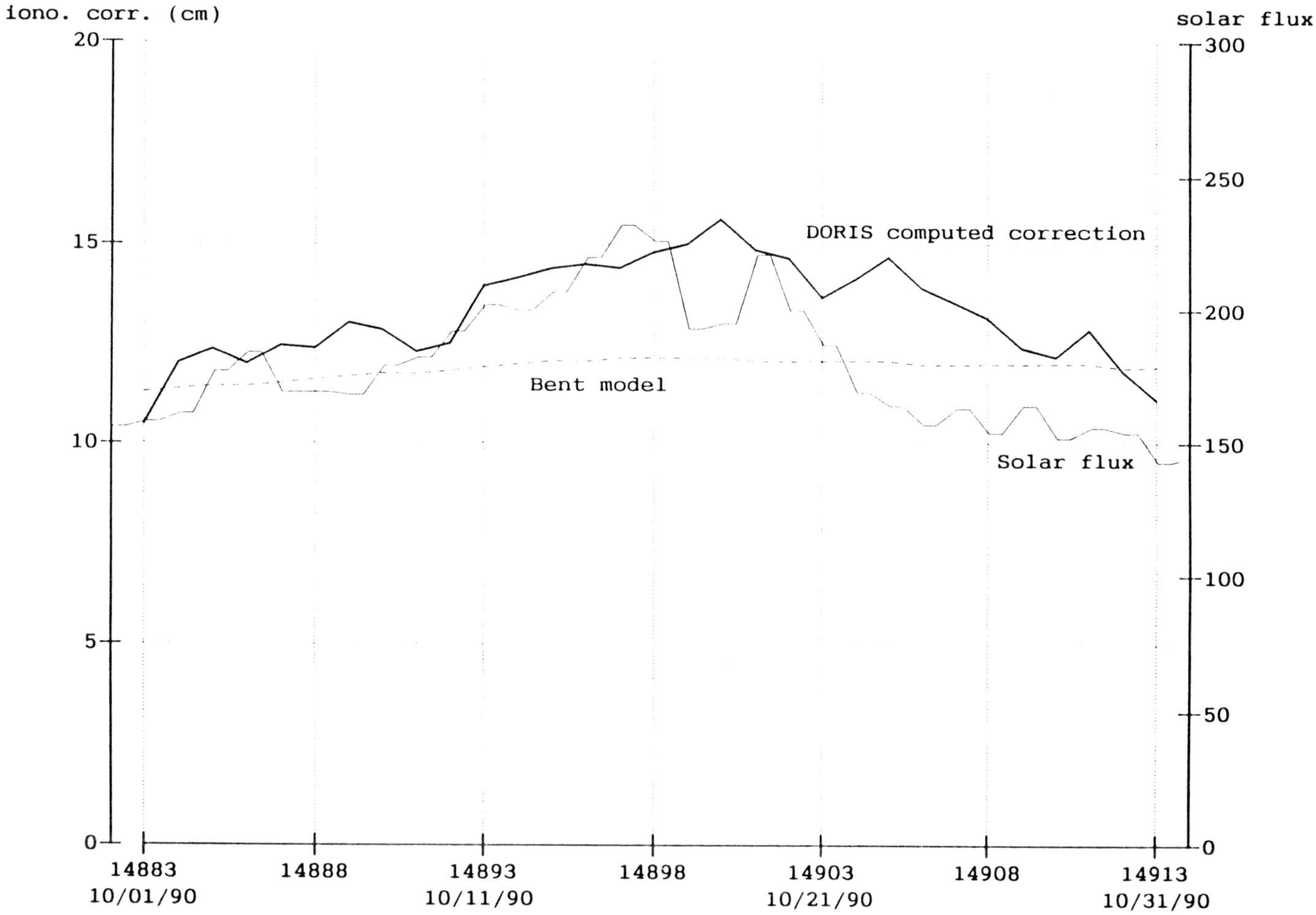

Fig. 8. Comparison between mean ionospheric correction and solar activity. The solid line is the daily solar flux, boldf line is the daily mean DORIS computed correction, dash line is the daily mean BENT modelled correction.

The Doppler measurement Do_i, in Hz, is deduced from the measurements at frequency f_i by :

$$Do_i = \frac{N_i}{(T_{2i} - T_{1i})} - (f_i - f_{oi}) \qquad (2)$$

and it is related to the physical parameters by :

$$Do_i = \frac{f_i}{c} \cdot \frac{(r_{2i} - r_{1i})}{(T_{2i} - T_{1i})} \qquad (3)$$

with

r_{ji} : satellite beacon distance at time T_{ji} including propagation effects

The shift between the counting periods at both frequencies is smaller than 12.5 µsec.

Observation equation

Assuming that Doppler measurements at both frequencies are equal except for the ionospheric effect which is supposed to be inversely proportional to the frequency squared, one can deduce from these measurements the ionospheric correction *m* to be applied to the 2 Ghz Doppler measurement Do_1.

$$m = -\frac{f_2^2}{(f_1^2 - f_2^2)} \cdot (Do_1 - \frac{f_1}{f_2} \cdot Do_2) \qquad (4)$$

m can be related to the slant electron content by :

$$m = \frac{A}{c \cdot f_1} \cdot \frac{(I_2 - I_1)}{(T_{21} - T_{11})} \qquad (5)$$

where :

A : 40.22 $m^3/(el*sec^2)$

I_j : Slant Electron Content at time T_{j1} (el/m^2)

We follow the sub-ionospheric point assumption in which the ionospheric effect is presumed to be concentrated at the sub-ionospheric altitude h_{sub} and the slant electron content I_j is related to the vertical TEC at the sub-ionospheric point TEC_j :

$$I_j = K_j \cdot TEC_j \qquad (6)$$

with

$$K_j = \frac{(R_e + h_{sub})}{\sqrt{((R_e + h_{sub})^2 - (R_e \cdot \cos S_j)^2}} \qquad (7)$$

and

Re : earth radius

h_{sub} : sub-ionospheric altitude (assumed to be 400 km in this study)

S_j : elevation angle at time T_{j1}

Then each value m, computed from the DORIS data is related to the TEC by :

$$m = \frac{A}{c \cdot f_1} \cdot \frac{(K_2.TEC_2 - K_1.TEC_1)}{(T_{21} - T_{11})} \qquad (8)$$

The ionospheric correction to be applied to the altimeter range measurement, cor, is related to the TEC by :

$$corr = -A \cdot \frac{TEC}{f_{alt}^2} \qquad (9)$$

with

f_{alt} : altimeter frequency

To recover the TEC, we perform an inversion of the values of *m* as measured by DORIS.

Inversion method

The model we use to represent the TEC is an interpolation over a latitude by longitude grid. The grid steps in latitude and longitude are parameters which can be adjusted depending on the required resolution and on the measurement density. The interpolation scheme uses cubic polynomials in both latitude and longitude.

The TEC values at every grid points are estimated using a least square fit over the whole set of measurements (two days of DORIS measurements are presently used for each fit). Measurements are weighted according to measurement noise as computed by pre-processing of the data. Two different set of TEC values are estimated, one for ascending passes and the other for descending passes to account for the difference in local time between them.

The local time of the computed TEC is the sub-satellite local time. It varies with latitude and is constant (for a sunsynchronous orbit) or slowly varying for the TOPEX orbit from one revolution to the next. The latitude parameter of the grids represents latitude plus local time variations. Figure 11 shows the dependance between latitude and local time along the SPOT orbit.

In a similar way, the longitude parameter represents longitude plus absolute time variations. Successive satellite passes are shifted in longitude due to the earth rotation, 360 degrees in longitude are covered within one day. To separate points at the same location, but at different days, we use a modified longitude which varies secularly with time. This parametrization implies that the grids cover more than 360 deg in longitude, depending on the duration of the measurements set. For example it is 720 deg for a measurement set spanning two days.

The grid is regular in latitude and longitude. In areas where few DORIS beacons are present, one could have a local divergence because the TEC values in such places would not be well constrained. To avoid divergence we constraint the variations in slope such that they are minimized at every grid point, but nonetheless compatible with the data.

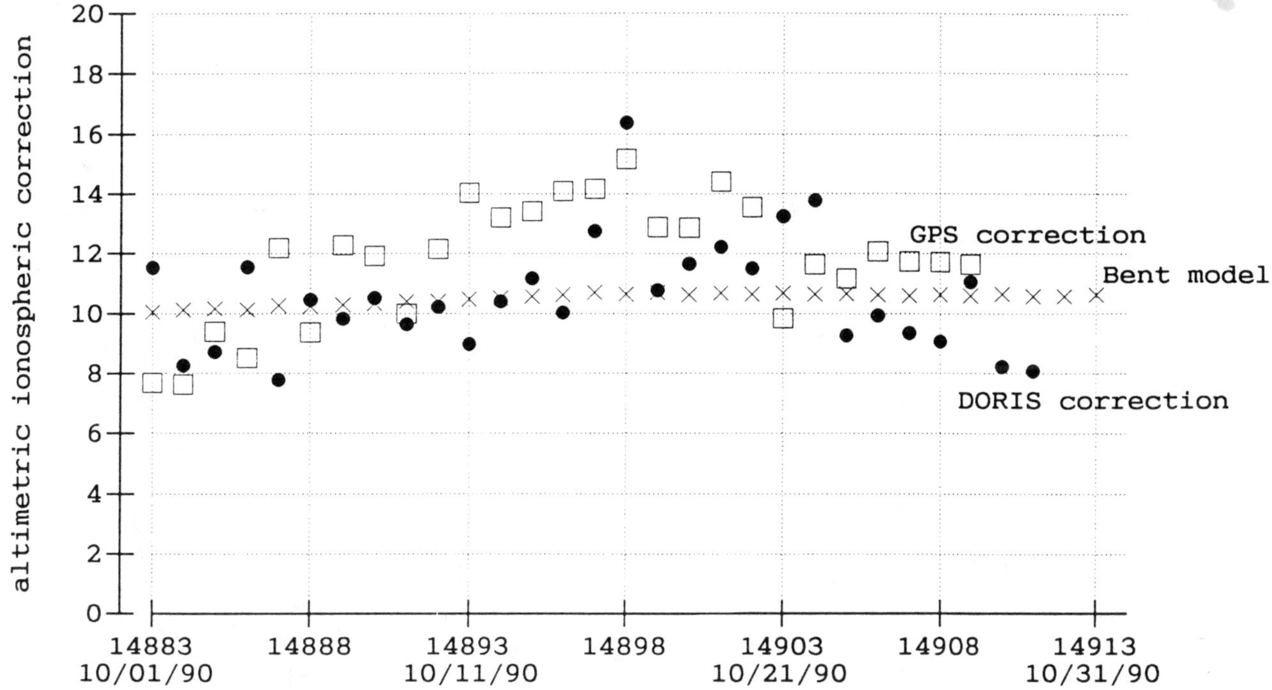

Fig. 9. Altimeter ionospheric corrections for the month of October 1990 at TSUKUBA GPS station, 10.70 local time.

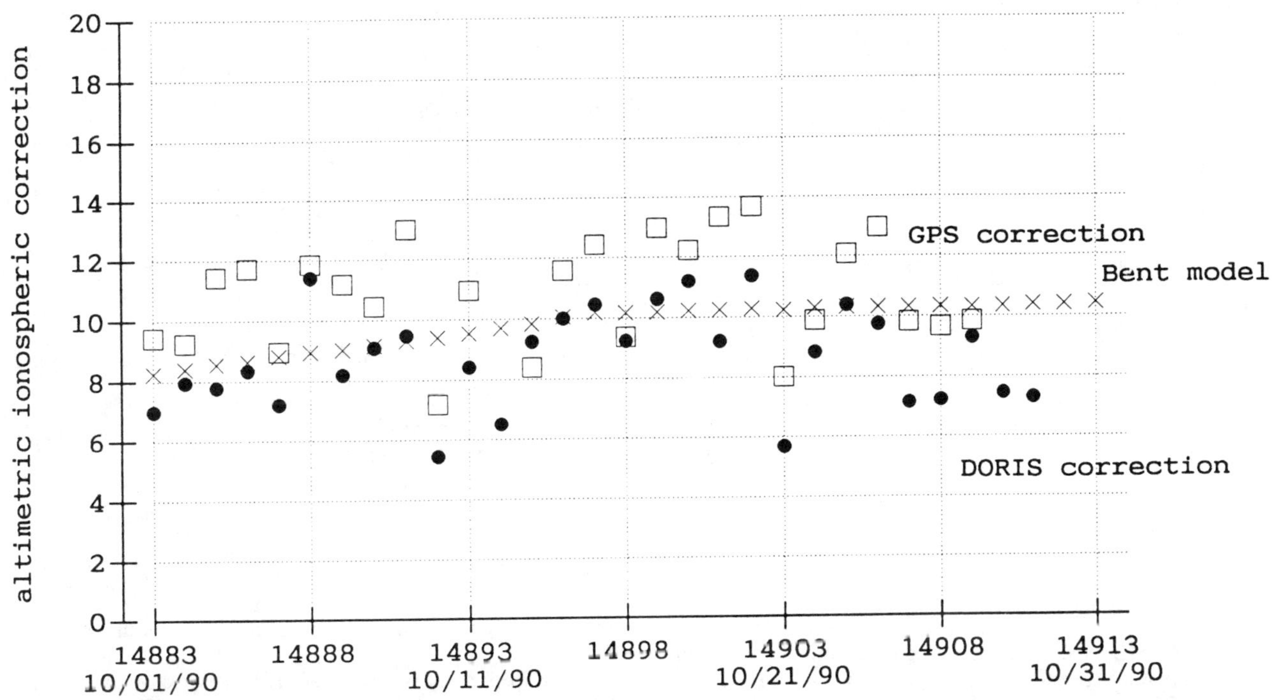

Fig. 10. Altimeter ionospheric correction for the month of October 1990 at WETZELL GPS station, 11.00 local time.

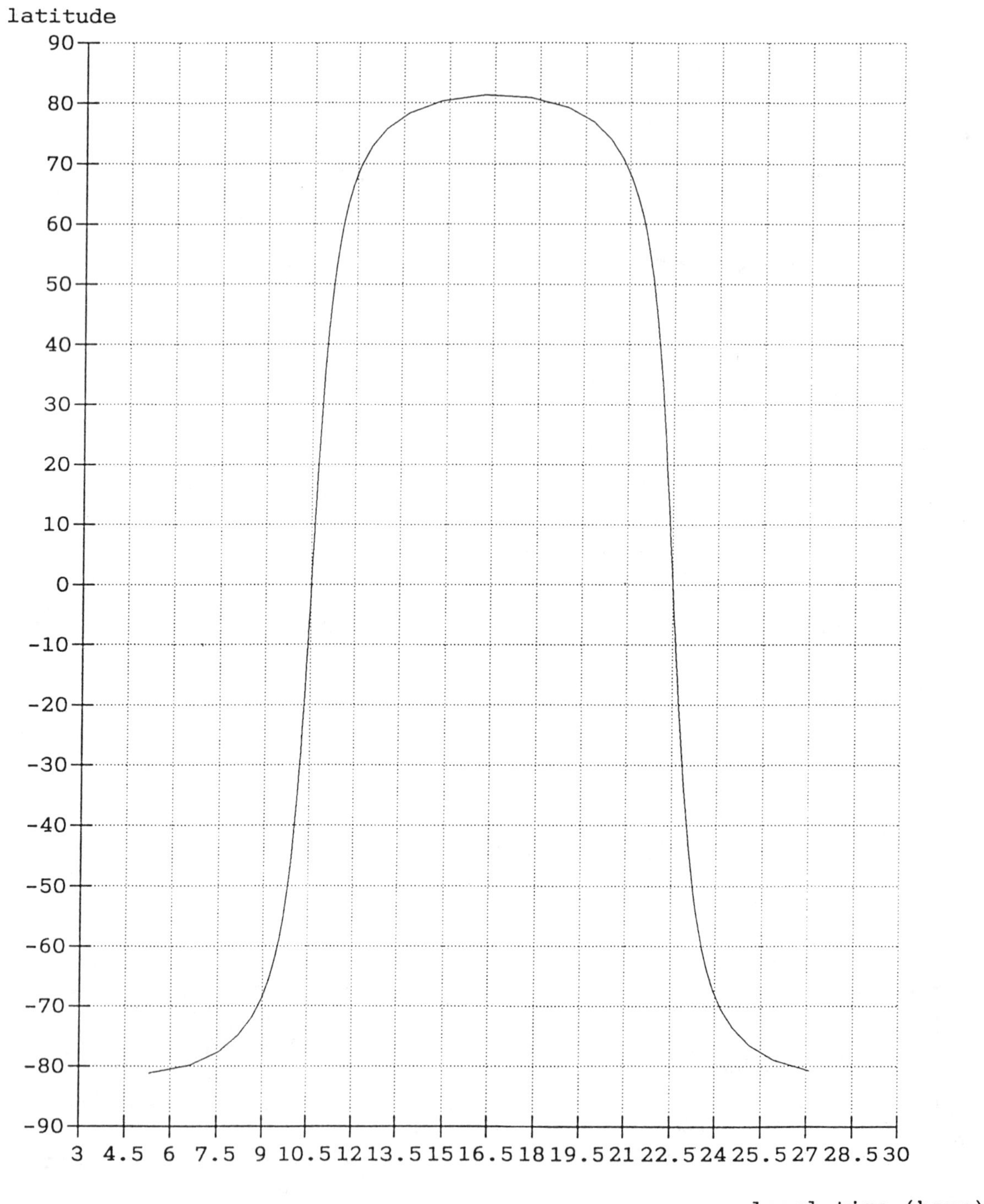

Fig. 11. SPOT2 local time variations.

Results

DORIS/SPOT2 measurements collected during the month of October 1990 have been processed. During this period the available DORIS network is shown on figure 3. The solid lines represent the visibility circles at satellite altitude (830 km) and 10 deg elevation, the dashed lines represent the visibility circles at sub-ionospheric altitude. The number of measurements is about 4500/day.

One fit per day is computed using the measurements of this day plus half a day before and half a day after. The grid step is 5 deg in latitude and 60 deg in longitude.

The results are expressed in terms of altimeter ionospheric

correction (cm). They can be converted in term of TEC evolution (TEC integrated to the satellite altitude which is about 830 km) using the correspondence :

1 cm of correction (at 13 GHz frequency) = $4.63\ 10^{16}\ el/m^2$

Figures 4 and 5 represent several daily computed ionospheric correction maps (daytime value). These maps show the contour lines for the altimeter ionospheric correction at the sub-satellite local time. SPOT flies on a sunsynchronous orbit and the local time evolution with latitude is plotted on figure 11. On each figure we can remark the spatial signature of the ionosphere which is mainly correlated with geomagnetic latitude (effect of the magnetic field). And comparing all figures we can note the day-to-day variability due to solar and geomagnetic activity evolutions.

For comparison, figure 6 displays the predicted altimeter ionospheric correction using Bent model for the October 1990 period taking into account SPOT2 local time and measured solar index for this period. The agreement between the BENT model and the DORIS estimate is quite good. The rms difference over the daily estimates for October 1990 is about 4 cm which is consistent with the known accuracy of such models as discussed previously

Figure 7 presents the estimated a priori uncertainty of the fit which, as espected, is related to the DORIS network plotted in figure 2. New DORIS beacons already installed in areas such as Manilla, Colombo (Sri Lanka), Galapagos should improve the fit by filling the holes.

A check on the internal consistency of the fit is the rms of differences between successive maps in the overlapping region. For the daytime values estimated for each day in October, the rms is 0.8 cm.

These results are globally in agreement with what is known about the physics of ionosphere (the TEC and consequently the ionospheric correction should increase with solar activity). On figure 8 are plotted the daily solar flux (10.7 cm radio flux) and the mean value of the estimated correction over the grid. It shows a close correlation.

Comparison with GPS data

GPS CIGNET network measurements were provided by NGS (National Geodetic Service). The GPS bi-frequency carrier phase measurements have been processed using the modelling proposed by Lanyi (1988).

Figure 9 and 10 represent the ionospheric correction estimated at SPOT local time and at two GPS receiving stations by using DORIS data, GPS data and the Bent empirical model.

A global statistical analysis of DORIS and GPS differences over 7 GPS station indicates a mean difference of 1 cm and a rms difference of 1.5 cm during this October month.

The correction using DORIS is smaller than the correction computed with GPS measurements due to the difference in altitude between DORIS and GPS orbit. GPS correction integrates the electron content up to 20 000 km. The 1 cm offset on the correction is consistent with the satellite altitude difference

The main DORIS technique error source is probably the local time difference between the station and the sub-ionospheric point. For example at a latitude of 40 degrees, this difference may be of twenty minutes and this represents a variation of about 1 cm in the altimeter correction.

Concerning GPS one should note that the number of available passes when SPOT2 flied overhead the GPS beacons is low during this October period. The improved GPS satellite constellation now available should improve this situation and allow more precise comparisons.

Conclusion

DORIS measurements allow us to estimate a global ionospheric model at sub-satellite local time. Processing the data from several DORIS receivers flying on different platforms should permit estimates of the daily variations of the TEC.

The association of DORIS with a single-frequency altimeter proves to be a powerful set of instruments to correct the altimeter measurements from ionosphere. In addition to the precise orbit determination DORIS provides a precise ionospheric correction.

A detailed evaluation of this correction will be performed during TOPEX/POSEIDON mission by comparing the dual frequency TOPEX correction with the DORIS one. For oceanographic studies, the goal is to decorrelate the ocean signal from any modelling error effect. It is for this reason that the spectral characteristics of the errors must be well characterized in order to be able to filter them. In particular, the error spectra of the two ionospheric correction schemes are different:

- while the dual frequency TOPEX correction increases the instrument noise, the DORIS correction filters the high frequency effect of the ionosphere.
- the magnitude of the DORIS estimate error will be correlated with DORIS network local density while the dual frequency correction may suffer some error due to unmodeled Electro Magnetic bias dependency on frequency, this error would be correlated to sea-state and then will have a different geographical signature.
- DORIS errors will decrease as long as solar activity will go down.

REFERENCES

Callahan, P.S.,Ionosphere variations affecting altimeter measurements, *Marine Geodesy*, 8, pp 249-263, 1984.

Dorrer, M., Deschamps, P., Laborde, B., DORIS (Doppler Orbitography and Radiopositionning Integrated from Space): System assessment results with DORIS on SPOT 2, *Acta Astronautica* Vol. 25, No 8/9, pp. 497-504, 1991.

Fleury, R., Foucher R., Lassudrie-Duchesne P., Correction of ionospheric effects for the precise orbit determination of satellites, AGARD-CP-453, 1989

Fleury R., Foucher F., Lassudrie-Duchesne P., Global TEC measurements capabilities of the DORIS system, *Adv. Space Res.* Vol. 11, No 10, pp 1051-1054, 1991.

Fu, L., Christensen, E., Lefebvre, M., TOPEX/POSEIDON: The Ocean Topography Experiment, *EOS*, Vol. 72, No 35, Aug. 27, 1991.

Lanyi, G.E., Roth, T., A comparison of mapped and measured total electron content using global positioning system and beacon satellite observations, *Radio Sci*, 23, pp 486-492, 1988.

Leitinger, R., Schmidt, G., Tauriainen, A., An evaluation method combining the differential Doppler measurements from two station that enables the calculation of the electron content of the ionosphere, *J. Geophys. Res.*, 41 , pp 201-212, 1975.

Llewellyn, S.K., Bent, R.B., Documentation and description of the Bent ionospheric model, AFCRL-TR-73-0657, 1973.

Monaldo, F.,Ionospheric variability and the measurement of ocean circulation with a spaceborne radar altimeter, JHU/APL SIR90U-010 1990.

Musman, S., Drew, A., Douglas, B., Ionospheric effects on GEOSAT altimeter observations,, *J. Geophys.* Res., 95, 2965-2967, 1990.

P. Escudier and N. Picot, RE/SU/TE/TP CNES, 18 Av. E. Belin 31055 Toulouse Cedex, FRANCE.

O.Z. Zanife, CLS/ARGOS, 18 AV. E. Belin 31055 Toulouse CEDEX, FRANCE

The Determination of Tropospheric Conditions (especially H_2O) Affecting the Accuracy of Position Measurements

G. K. HARTMANN

Max-Planck-Institut für Aeronomie, Postfach 20, W-3411 Katlenburg-Lindau, Germany

The lower part of the earth's atmosphere (troposphere and stratosphere) in which the constituents of the neutral gas are responsible for a refractive index > 1 is well separated in altitude and frequency from the upper atmosphere (ionosphere, magnetosphere) where ionization produces refractive indices < 1. Incomplete knowledge of the refractive index as a function of space, time and frequency gives rise to errors (inaccuracies) in the electrical path length of radio waves propagating through the atmosphere. The concept of refractive index is reviewed and the accuracy of the determination of refractive index by model calculations is discussed. Currently, the limit of path delay correction due to water vapor for the atmosphere is about 1 cm. If correction limits are to be improved and refined then more measurements and improved ray tracing calculations are required.

1. INTRODUCTION

In geodesy, as well as in various radio navigation systems, electromagnetic waves – propagating through the atmosphere – are the tools for measuring range or range rate. In some cases range rate measurements guarantee sufficient accuracy for geodetic applications. However, there are a number of cases where range measurements are required if the highest possible accuracy is wanted. The choice between the two depends, for example, on the accuracy with which the satellite orbit can be measured, on the geometry of the orbit, and on the selection of the radio frequencies (beacons) aboard the satellite. The higher the desired accuracy of the measurements for geodetic applications, the more the need for a detailed study of these two possibilities. Due to the very high standard of current technology, accurate range rate (range difference) measurements can be performed using the Doppler effect. It is much more difficult to obtain range data with the same degree of accuracy. Since all errors inherent in the "system" contribute to the total error budget, only a minor fraction of the total error is caused by measurement inaccuracies, which are called technical errors. The larger portion of the errors is not related to measurement techniques and is due to our incomplete knowledge of the physical characteristics of the propagation path. For realistic error budgets this needs to be considered. This is scarcely done in literature.

Environmental Effects on Spacecraft Positioning and Trajectories
Geophysical Monograph 73, IUGG Volume 13

2. THE REFRACTIVE INDEX AS A DESCRIPTIVE PARAMETER FOR THE PROPAGATION OF ELECTROMAGNETIC WAVES

In the approximation of geometrical optics the refractive index describes the propagation of electromagnetic waves through a medium. For an isotropic medium (e.g., the troposphere for the usable frequency domain from radio waves to light waves apart from Zeeman and Stark effects of some atmospheric constituents) the dielectric constant is a scalar and the refractive index is independent of the direction of wave propagation and wave polarization. In an anisotropic medium (e.g., the ionosphere for radio waves up to frequencies of a few GHz) the dielectric constant is a tensor and two different refractive indices are defined for two "characteristic polarizations". The values of the refractive indices and the characteristic polarizations depend on the direction of wave propagation.

A refractive index, n, can be considered to define the velocity of phase propagation $v = c/n$ (where c is the free space velocity of light) for a monochromatic wave (the component of a monochromatic wave with a characteristic polarization in the case of anisotropy or "birefringence"). For an inhomogeneous medium the refractive indices (the refractive index in the isotropic case) are spatial functions which implies that in general there exist no analytic solutions for the propagation of monochromatic waves from source to receiver. If the propagation medium ("the system") approaches the area of deterministic chaos the refractive index n is no longer a reasonable characterization of the medium. These conditions may occur at locations of phase transitions in the atmosphere and can be described by the fact that the rms (root mean square) variations of the refractive index – in time and/or space – are comparable or even larger than the temporal or spatial mean values of the refractive index.

In the hierarchy of strategies an approximate solution by the application of "geometrical optics" is by far the simplest case. One possible formulation of geometrical optics is the eikonal equation which is especially suitable for problems of phase propagation [*Born and Wolf*, 1970].

Let E_k be the "optical path". The function E_k is often called the eikonal, and the relation

$$(grad\ E_k)^2 = n^2 \tag{1}$$

is known as the eikonal equation. It is the basic equation of geometric optics. The surfaces E_k = const may be called the geometrical wave surfaces or the geometrical wave fronts.

It is clear that a necessary condition for the application of the eikonal equation is the existence of well-defined surfaces of constant phase (wave surfaces). In the presence of random (stochastic) or turbulent fluctuations in the medium which can be described by adding a stochastic part to a "slowly varying" refractive index, one may distinguish two cases: **a)** an "average" wave surface can be defined on which stochastic phase fluctuations are superimposed or **b)** the wave surface is completely disrupted.

Geometrical optics can be applied for the "average" wave surface of case a), whereas case b) cannot be treated with the geometrical optics approximation. Fortunately, it is comparatively easy to detect fluctuations in the medium and to distinguish case a) from case b) for electromagnetic waves propagating through the atmosphere of the earth. When irregular amplitude and phase fluctuations occur the strength of the fluctuations can be used to distinguish case a) from case b).

Another necessary condition for the applicability of geometrical optics is that the relative variations of the refractive index and of the quantities describing the wave field (e.g. the electric field strength of the wave, $\vec{E}$) are small over distances of one wavelength. This condition is violated, for example, at boundaries of shadows, in the vicinity of caustics, because across such regions the amplitude of the wave changes rapidly. Expressed in another way: steep gradients in the refractive index violate the condition for geometrical optics. Partial reflection, focussing, diffraction, multipath propagation can occur; and it is not possible to calculate signal phase from the simple geometrical optics approximation. Other cases which cannot be treated with the eikonal equation are scattering processes. Propagation of waves in inhomogeneous media has be considered by [*Tatarski*, 1961 and 1971].

Finally we stress that the simplicity of the geometrical optics model arises essentially from the fact that, in general, the field behaves locally as a plane wave.

3. THE REFRACTIVE INDEX OF GASES

The refractive index describes the bulk electromagnetic properties of matter [*Handbook of Chemistry and Physics*, 1969–1970; *Landolt-Börnstein*, vol. II, part 6, 1959; *Landolt-Börnstein*, 1962; *Landolt-Börnstein*, vol. I, 1965]. Because of the frequency dependence of refractive index we observe the phenomenon of dispersion. We distinguish normal dispersion where the refractive index increases with frequency and anomalous dispersion where the opposite behavior is true; in general anomalous dispersion is due to resonance effects. In most applications the two effects are treated separately, and the transition region between the two is usually neglected. In the transition region the refractive index is poorly known.

In the region of anomalous dispersion [*Sommerfeld*, 1959, p. 86 and *Debye*, 1925]: $\underline{n} = n(1+i\kappa)$, the complex refractive index, is given by:

$$\underline{n}^2 = n_m^2 + \frac{N_m e^2/m\epsilon_o}{\omega_o^2 - \omega^2 - ig\omega_o\omega} \tag{2}$$

where n_m is the contribution of all other resonances averaged in the frequency domain around ω_o. ω is the frequency at which the measurement takes place, ω_o is the line center frequency, and g is a factor due to the attenuation, e is the electron charge, m is the electron mass, N_m is the number-density of the m_{th} species of the gas. All quantities in MKS units. We obtain from eq. (2)

$$n^2(1-\kappa^2) = n_m^2 + \frac{N_m e^2}{m\epsilon_o}\,\frac{\omega_o^2 - \omega^2}{(\omega_o^2-\omega^2)^2 + g^2\omega_o^2\omega^2} \tag{3}$$

$$2n^2\kappa = \frac{N_m e^2}{m\epsilon_o}\,\frac{g\omega_o\omega}{(\omega_o^2-\omega^2)^2 + g^2\omega_o^2\omega^2} \tag{4}$$

In regions of normal dispersion the denominator in eq. (2) consists only of the first two terms and $\underline{n}^2$ is real. ($\kappa = 0$; $\underline{n}^2 = n^2$), $\kappa \neq 0$ implies that electromagnetic waves suffer absorption.

4. THE PROPAGATION OF ELECTROMAGNETIC WAVES IN THE ATMOSPHERE

One can distinguish two regions in the atmosphere of the earth where the refractive indices differ significantly from the free space value: the lower atmosphere (troposphere, lower stratosphere) in which the constituents of the neutral gas are responsible for a refractive index > 1 and the upper atmosphere (ionosphere, magnetosphere) where the ionization produces refractive indices < 1. The two regions are well separated in altitude which allows one to deal with propagation effects of the ionized part and the neutral gas separately. The errors introduced by the propagation effects – through a non vacuum, namely the atmosphere – are dealt with in detail in the section *Error Considerations and Conclusions*.

4.1. *Ionized part of the atmosphere*

The refractive indices can be found by means of the "magneto-ionic theory" [*Budden*, 1985; *Davies*, 1965, 1990; *Rat-*

cliffe, 1959; *Rawer and Suchy*, 1967]. For frequencies above 100 MHz the "characteristic polarizations" are, to a very good approximation, the right-hand and the left-hand circular polarizations except for propagation perpendicular to the geomagnetic field (and deviations of some tenths of a degree from this direction).

The latter exception has no practical implications. Furthermore, at very high frequencies, absorption can be neglected. In the geometrical optics approximation the phase of a characteristic polarization component follows from the exponential of the wave equation and is given, if the time dependent term $-\omega t$ is omitted, by

$$\phi_{\ell,r} = \int_{\vec{r}_o}^{\vec{r}} Re\{\vec{k}_{\ell,r}(\vec{r})\} d\vec{r}_{\ell,r} \doteq \frac{\omega}{c} \int_{S}^{R} n_{\ell,r} ds_{\ell,r} \tag{5}$$

where $\vec{k}$ is the wave vector, S is the source (transmitter), R is the receiver, ds is the scalar element of ray the path (the indices ℓ and r refer to the left-hand and the right-hand circular components respectively).

4.2. *Nonionized part of the atmosphere*

The troposphere and the stratosphere are not birefringent, and thus the phase of an electromagnetic wave of arbitrary polarization follows from the eikonal equation

$\phi = \frac{\omega}{c} \int_S^R n \, ds$

The neglect of absorption is justified for frequencies below about 10 GHz. (For optical frequencies absorption can be neglected for most phase propagation problems, but one should be aware that the neglect is not justified in all cases. Rayleigh scattering has an attenuation effect. Whether a possible effect on phase propagation must be taken into consideration or not should be investigated carefully from case to case.)

5. THE REFRACTIVE INDEX OF THE NON-IONIZED ATMOSPHERE

5.1. *Below* 5×10^9 *Hz (5 GHz)*

For more details about the troposphere and stratosphere see *Hopfield* [1971] and *Bean* [1962]. If n is the refractive index, we define the refractivity N as

$$N = 10^6(n-1) \tag{6}$$

If geometrical optics is valid we get the empirical frequency independent refractivity formula [*Kruspe*, 1965]:

$$N_o = (n_o - 1) \times 10^6 = K_1 \frac{p}{T} + K_2 \frac{e}{T} + K_3 \frac{e}{T^2} + K_4 \frac{p_{CO_2}}{T} \tag{7}$$

where p is the total atmospheric pressure [mbar], e is the partial pressure of water vapour [mbar], p_{CO_2} is the partial pressure of CO_2 [mbar], and T is the temperature [K].

While N is in general frequency dependent, N_o is frequency independent. See also eq. (8). Remark: Since this formula (7) was derived before the SI pressure unit kPa was accepted, the constants K_1 to K_4 are related to the pressure measured in mbar.

If we disregarded the fourth term with CO_2 we would get an error of about $\pm$ 0.02% [*Bean*, 1962]. It should be remarked that since as most authors state the whole formula is not better than $\pm$ 0.5%, this can be justified. See also Table 1 for K_1, K_2, and K_3, as well as Table 2. The references given in tables 1 and 2 can be found in *Hartmann and Leitinger* [1984] and in *Landolt-Börnstein* [1962]. The value of 0.5% implies that p, T, and e are claimed to be known better than this value at the location where N_o is measured. This error – or a larger one – must be added to any error introduced by the measuring equipment.

5.2. *Above 5 GHz and below 1000 GHz*

For frequencies above 1 GHz and below 1000 GHz a practical model (designated program code: MPM) was formulated by *Liebe* [1983, 1985, 1989]. This model simulates the complex refractive index for the propagation calculation of electromagnetic waves through the earth's atmosphere. (According to a private communication from H. Liebe this program is available on diskette for execution on PCs.) It may be obtained from Dr. H. J. Liebe, NTIA/ITS, Broadway, Boulder, Colorado 80303, USA.

The complex frequency-dependent refractivity $\underline{N}$ is given by Liebe as:

$$\underline{N}(f, p, T, A_{con}, ds) = N_o^\star + N'(f) + jN''(f) \tag{8}$$

where f is the frequency in gigahertz (10^9 Hz), p is the pressure in kilopascal [kPa], T is the temperature in Kelvin, A_{con} is the density [g/m^3] of the gaseous constituents of the atmosphere (H_2O; O_2; etc.), and ds is the ray path element between the transmitter and receiver [m].

The refractivity consists of a frequency independent term $N_o^\star$ plus frequency dependent terms for refractive dispersion $N'(f)$ and absorption $N''(f)$. In radio engineering it is customary to express the imaginary part as power attenuation rate α, and the real part as the propagation delay rate β (with reference to vacuum) as follows:

$$\alpha = 0.1820 f\ N''(f)\ [dB/km] \doteq 4.1907 \times 10^{-2} f N''(f) [1/km] \tag{9}$$

and phase dispersion of

$$\begin{aligned} \beta &= 1.2008 \times (N_o^\star + N'(f))\ [deg/km] \doteq \\ &\doteq 3.336 \times (N_o^\star + N'(f))\ [picosec/km] \end{aligned} \tag{10}$$

For easier comparisions with the references the original terminology developed by H. Liebe was used in this review.

The Kramers–Kronig relation (also called the Kramers–Heisenberg relation or simply the dispersion relation) is a general theorem that describes the interrelationship between the real part and the imaginary part of a complex quantity, e.g. the complex refractive index. Thus, under the assumption that either the absorption or the dispersion is fully

TABLE 1. Values of constants for the empirical refractivity formula (7)
Table 1 gives the values of the constants K_1, K_2, and K_3 as derived by various authors

	year	K_1 (mb^{-1} degree)	K_2 (mb^{-1} degree)	K_3 (mb^{-1} degree^2)
Barrell and Sears	1938	77.478	- 12.786	—
Schelleng, Burrows, Ferrell	1933	79.0	- 11.5	$1.35 \cdot 10^5$
Englund, Crawford, Mumford	1935	79.1	- 10.8	$3.81 \cdot 10^5$
Waynick	1940	79.0	- 10.5	$3.72 \cdot 10^5$
Smith–Rose, Stickland	1943	79.0	- 11.0	$3.8 \cdot 10^5$
Burrows, Attwood, NDRC	1946	79.0	- 11.0	$3.8 \cdot 10^5$
Met. Factors in Propagation	1946	79.0	- 11.0	$3.8 \cdot 10^5$
Crain	1948	77.1	—	—
Essen, Froome	1951	77.64	- 12.96	$3.718 \cdot 10^5$
Kerr	1951	79.0	(0)	$3.8 \cdot 10^5$
Smith, Weintraub	1952	77.6	- 5.6	$3.75 \cdot 10^5$
Essen	1953	—	—	$3.68 \cdot 10^5$
Saito	1955	77.26	- 9.76	$3.77 \cdot 10^5$
Battaglia, Boudouris, Gozzini	1957	77.6	- 5.6	$3.75 \cdot 10^5$
Crain, Whiteford	1958	77.5	- 12.5	$3.70 \cdot 10^5$

References cited in tables 1 and 2 are given in *Hartmann and Leitinger* (1984).

known over the spectral range, the other quantity can be calculated. This is not the case for eq. (8). Figure 1 shows specific H_2O attenuations α over the frequency range of 5 to 1000 GHz for five phase states of water according to the MPM model. The figure shows the very complex behaviour of the atmosphere for any ground based observation. Besides the H_2O absorption it also shows the strongest oxygen absorption. See also *Waters* [1976]. The center frequencies of the strongest lines for H_2O and O_2 above 1 GHz are displayed in Table 3. In the range up to 1000 GHz (1 THz) at least 10000 lines must be considered which implies more sophisticated model calculations that are very storage intensive. The values for α of figure 1 are not calculated from an analytical expression but from a fitted curve through measured points at some frequency ranges – for example around 60 GHz and 140 GHz – i.e. by a polynomial obtained by interpolation and extrapolation. According to H. Liebe (private communication) the accuracy of these values is at best 1% – close to the center frequencies of the resonance lines – and at best 10% in the so called "window" areas – e.g.

TABLE 2. Refractive index in the optical and microwave domain

Table 2 gives a comparison between the refractive index in the optical domain and in the microwave domain

Dependence of refractive index from water vapour at $t = 288°K$ and $p = 1013$ mb for different constants in the optical domain and in the microwave domain

		n opt	n MW
Barrell, Sears	1938	$n = 1.000273 - 0.45 \cdot 10^{-7}e$	—
Smith–Rose, Stickland	1943	$n = 1.000278 - 0.4 \cdot 10^{-7}e$	$n = 1.000278 + 45.4 \cdot 10^{-7}e$
Essen, Froome	1951	$n = 1.000273 - 0.45 \cdot 10^{-7}e$	$n = 1.000273 + 44.4 \cdot 10^{-7}e$
Smith, Weintraub	1952	$n = 1.000273 - 0.2 \cdot 10^{-7}e$	$n = 1.000273 + 45.0 \cdot 10^{-7}e$
Crain, Whiteford	1958	$n = 1.000273 - 0.45 \cdot 10^{-7}e$	$n = 1.000273 + 44.15 \cdot 10^{-7}e$

[e in mb]

Remark: Because of the missing "Debye factor" the water vapour contribution to the refractive index is much smaller in the optical domain than in the microwave domain.

the absorption minimum between two lines. (This might be – perhaps by a factor of two – too optimistic. However, there are no independent measurements available that could supply more detailed information.) This is about a factor of five better than earlier models which quoted 5 % for the line center and 50 % for the window areas [*H. Liebe*, 1991 (updated version of his MPM from 1989); *Pilewskie and Twomey*, 1987]. According to the error discussion for eq. (7) we must mention that p, T, and the density of the various A_{con} along the ray path of s interest must be known significantly better than the error margin given above. This holds true also for the technical measuring errors. Since the knowledge of the atmospheric parameters does not normally satisfy this criterian the errors can be significantly larger than quoted. A very recent model for the complex permittivity of ice below 1×10^{12} Hz (1 THz) was published by *Hufford* [1991]. In the same journal *Liebe et al.* [1991] published an extension of the MPM model up to 30 THz.

Since the main emphasis of this review is on the effects of H_2O and since in the frequency range considered – with very few exceptions, e.g. in the O_2 resonance bands – the influence of H_2O is dominant for all trans-tropospheric propagation paths and since H_2O exhibits very strong temporal and spatial variations, we will consider from now on – due to the limited space – only the H_2O influences. We will therefore continue with some remarks about H_2O in the atmosphere.

6. H_2O IN THE ATMOSPHERE

6.1. *General aspects*

Gantham et al. [1985], *Hartmann et al.* [1992 to be published], *Remsberg et al.* [1989] supply more detailed information about these general aspects.

Water is the most important substance on earth but is poorly understood even though it is the most researched fluid and possibly the strangest component in the universe.

Water, a familiar substance composed of the chemicals hydrogen and oxygen exists in vapor, liquid and several solid forms in our three geospheres: atmosphere, hydrosphere/cryosphere and lithosphere/pedosphere which are the basis of the biosphere. Water is essential to terrestrial life, participating in virtually every process that occurs in living organisms. Although the molecules of water are simple in structure, the physical and chemical properties of the compound are extraordinarily complicated.

The physical nature of water, its role in life processes, and its occurrence, distribution, and turnover present a wide spectrum of problems to physical and life scientists and engineers. Although each specialist is normally conversant with such issues as can be formulated in the particular terminology of his own discipline, few people have a full appreciation of the influence which water has had, and will have, on ecological, sociological, and industrial developments.

6.2. *Water vapor*

Hartmann et al. [1992] supply more details about this topic.

During recent years there has been an enormous growth of interest in the formation and properties of water vapor clusters formed by the attachement of molecules to ions, as well as those resulting from the clustering of neutral molecules into weakly bound complexes. This has been due in part to a recognition that work in this field is important in understanding many problems in the atmospheric sciences. Included among these are questions concerning the fate of the products of ion-cluster recombination, the role of small clus-

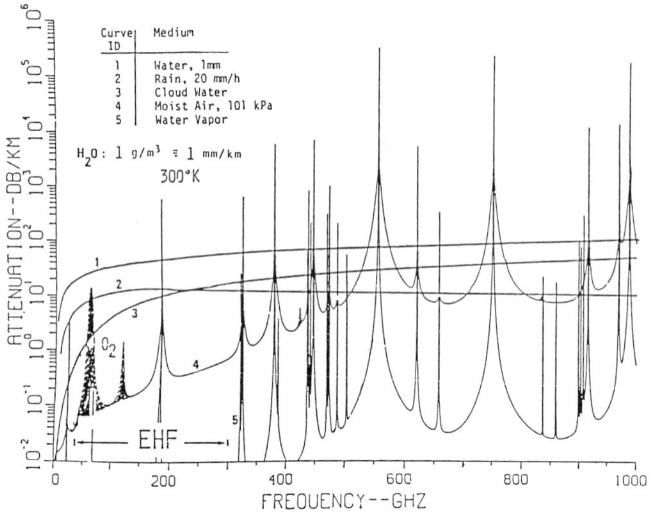

Fig. 1. Normalized (1g/m^3, 300 K) specific H_2O attenuation over the frequency range 5-1000 GHz for five phase states: water 1), rain 2), suspended hydrometeors 3), moist air 4) (numbers shown indicate selected H_2O attenuation peaks), and pure water vapor 5). (Liebe, 1983)

ters in aerosol formation, the influence of small clusters and aerosols on the propagation of electromagnetic radiation, and the importance of small clusters and aerosol surfaces to atmospheric chemical reactions. From a basic point of view, it is evident that the results contribute to a more complete understanding of the forces between ions and neutral molecules. Research in this area is also particularly valuable in the field of interphase physics which is concerned with elucidating the molecular details of the collective effects responsible for phase transitions, nucleation phenomena, the development of surfaces, and ultimately the formation of the condensed state. It is, of course, well known that phase transformations of water vapor to the condensed state have a direct bearing on weather and climate. Furthermore, it is recognized that atmospheric aerosols have a large percentage water content, and water vapor plays a major role in their formation processes. Nevertheless, the mechanisms of these processes are not well known and further work is needed to elucidate the mechanisms of gas-to-particle formation and establish a more rigorous approach to nucleation theory.

The vapor phase clustering of water molecules (both homomolecular and heteromolecular) influences the atmosphere in several ways. Water molecules clustered about ions enhance nucleation, play a role in the formation of noctilucent clouds, and participate in electrical phenomena. Within the infrared windows in the 3 μm to 5 μm and 8 μm to 13 μm regions, simple molecular atmospheric constituents are known not to absorb appreciably. Yet these windows are observed to close up during inclement weather for reasons which are not fully understood. Clusters involving water molecules have been postulated to play a role which has not yet been established. Clustering helps remove gaseous pollutants through gas-to-particle conversion. So far no theory is completely successful in describing clustering phenomena. More work is needed on the basic physics at or near equilibrium under standard conditions.

6.3. *Water in the atmosphere*

Hartmann et al. [1992] supply more details about this topic.

Water is the o n l y substance, that occurs in all three phases - gas, liquid, solid - in the atmosphere; therefore we must consider all three and thus we have to deal with the refractive index not only in its gaseous form but also in its liquid and solid form, as has been partly done by *H. Liebe* [1989]. On top of that we very likely need to consider also a refractive index for H_2O clusters. This has been rarely done until now because so little is known about clusters. While the importance of the solid (e.g. snow and ice) and the liquid form (e.g. rain) of H_2O are well known to a broad public, this is not true for the gaseous phase, water vapor. Compared with other gases in the atmosphere, such as carbondioxide (CO_2) and ozone (O_3), surprisingly little is known about water vapor. Water vapor is, besides clouds, the most

TABLE 3. Resonance frequencies in 10^9 Hz (GHz) for H_2O and O_2

	H_2O	O_2
1	22.235080	53.5957025
2	67.813960	54.1299763
3	119.995940	54.6711450
4	183.310074	55.2213720
5	321.225644	55.7838190
6	325.152919	56.2647660
7	336.187000	56.3633930
8	380.197372	56.9681800
9	390.134508	57.6124893
10	437.346667	58.3238850
11	439.150812	58.4465800
12	443.018295	59.1642150
13	448.001075	59.5909780
14	470.888947	60.3060440
15	474.689127	60.4347760
16	488.491133	61.1505700
17	503.568532	61.8001550
18	504.482692	62.4112230
19	556.936002	62.4862550
20	620.700807	62.9979712
21	658.006500	63.5685200
22	752.033227	64.1277770
23	841.073593	64.6788612
24	859.865000	65.2240045
25	899.407000	65.7647440
26	902.555000	66.3019474
27	906.205524	118.7503430
28	916.171582	
29	970.315022	
30	987.926764	

important contributor to the greenhouse effect and therefore a major regulative component of the earth's climate. Quantitatively there is a very small amount of liquid water and ice in the atmosphere, e.g. between 0.01 g $/\mathrm{m}^3$ and 1g $/\mathrm{m}^3$ in "normal" clouds. But this minute amount plays an important role in the earth's radiation budget. Because there is considerable energy involved in phase transitions and in atmospheric radiative transfer, water vapor plays a central role in the energy budget of the atmosphere. It also plays the major role in many nonlinear, irreversible thermodynamic effects within the atmosphere during stationary and nonstationary conditions which may be described in terms of: clusters, turbulence, and fractals [*Kassner*, 1980; *Lovejoy and Schertzer*, 1986; *Mandelbrot*, 1982; *McElroy and Salawitch*, 1989; *Pilewskie and Twomey*, 1987].

The spatial and temporal distribution of H_2O (all three phases) in the atmosphere is not only important for the climate, weather, biosphere, and atmospheric chemistry - homogenous or heterogeneous - but also for the propagation of electromagnetic waves through the atmosphere. Electromagnetic waves, over a frequency range of more than 10 orders of magnitude, are affected differently by the atmosphere. Because electromagnetic waves are used for communication, navigation, and remote sensing (e.g. photography), it is important to know and/or to try to avoid the disturbances caused by the strongly varying water content in the atmosphere.

Figure 2 shows an example of the H_2O volume mixing ratio (ppmV) [*Hartmann et al.*, 1992]. *Remsberg et al.* [1989] give the latest interim reference model for the variability of middle atmosphere H_2O.

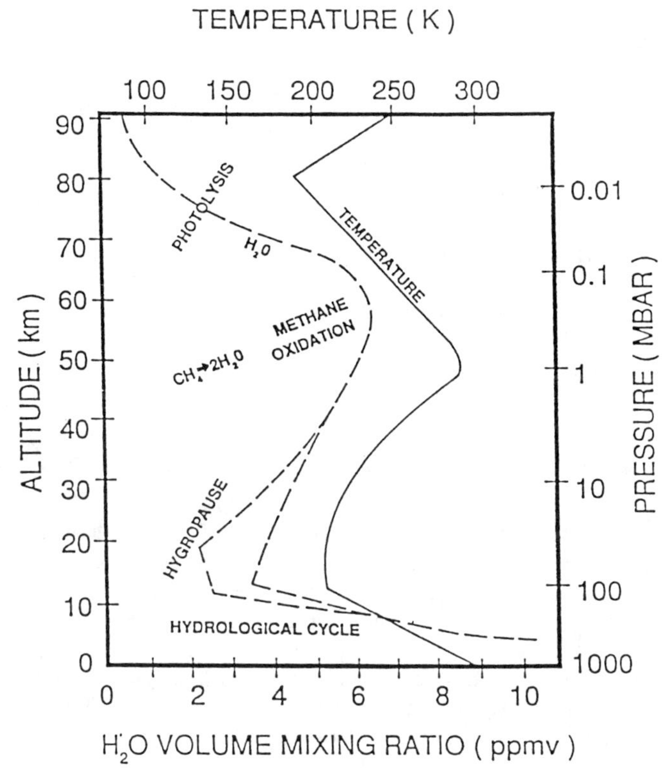

Fig. 2. H_2O volume mixing ratio (ppmV)

7. ERROR CONSIDERATIONS AND CONCLUSIONS

7.1. *General aspects*

Hartmann and Leitinger [1984] supply more details about this topic.

We define as "error" Δ the deviation of the measured length S_M of the optical ray path – between the source S (e.g. the transmitter) and the receiver R – from that of the length S_o of the straight line distance between R and S.

$$\Delta = S_M - S_o = \int_R^S n ds - \int_R^S ds_o \tag{11}$$

where $S_M = S^* + \Delta S_M$. S^* is the electrical length which would be measured by an ideal system free from any technical measuring error. ΔS_M is the technical measuring error, which includes those of the transmitter and the receiver, i.e. the entire system.

For better physical interpretation we split the error Δ into two parts Δ_1 and Δ_2, where

$$\Delta_1 = \int_R^S ds_o - \int_R^S n ds_o \tag{12}$$

and

$$\Delta_2 = \int_R^S n ds_o - \int_R^S n ds \tag{13}$$

Δ_1 contains the first order error and the contributions due to any truncation of the refractive index formulas and Δ_2 is the refraction error.

If we replace the refractive index n by the refractivity N, eq. (6), then we get for the measured range error Δ_{atm} due to the atmosphere [*Leitinger*, 1974; *Leitinger et al.*, 1978; *Hartmann and Leitinger*, 1984; *Ajayi et al.*, 1980]

$$\Delta S_{atm} = 10^{-6} \int_S^R N ds_o + \{ \int_S^R n ds - \int_S^R (1 + 10^{-6} N) ds_o \} + \Delta S_M \tag{14}$$

where ds_o is the straight line element [m] and ds is the ray path element [m]. The term in { } includes the contribution due to the curvature of the ray path in the neutral atmosphere (refraction) and for the ionosphere. Integration has to be carried out from the source (transmitter) to the receiver. The formula holds for all situations (ground station to ground station ranging, space to ground, ground to space, satellite to satellite, etc.). In the special but very important case of trans-ionospheric propagation the transmitter (source S) is located above the ionosphere and the receiver is located on the ground. Then the integrals can be split into two parts, $\int_S^R = \int_S^A + \int_A^R$, A being a point on the ray path below the ionosphere and above the lower stratosphere. Below A the contribution of free electrons to the refractive index is negligible, above A the contribution of the neutral atmosphere is negligible: for trans-ionospheric propagation the splitting of error terms into those from the "neutral atmosphere" ("tropospheric error") and "ionosphere" ("ionospheric error") is not only practical but has a geophysical background as well.

The first term of eq. (14) represents – according to eq. (12) – the first order error. If the refraction error and ΔS_M are small compared to this first term, we get:

$$\Delta S_{atm} \approx 10^{-6} \int_S^R N ds_o \tag{15}$$

This is the most commonly used formula from which ΔS_{atm} is calculated using either eq. (7) or eq. (8) and which in the latter case yields the propagation delay rate β of eq. (10).

To include the refraction error we have to do ray tracing. In context with ionospheric research such computer programs were developed and might be adapted for use in the neutral atmosphere. This is mostly necessary when our transionospheric and transtropospheric observations occur at elevation angles below 15°.

7.2. *Errors due to H_2O*

Considering equations (7), (14), and (15) we have for the first order path length error due to water vapor

$$\Delta_{atm}\ H_2O \approx k \int_0^\infty \frac{\rho_v(s)}{T(s)} ds \approx k \int_0^{T_{RP}} \frac{\rho_v(s)}{T(s)} ds \tag{16}$$

where $k = 1.723 \times 10^{-3}$ K/(g/m^3), s is the propagation path [m], $\rho_v(s)$ is the water vapor density along the propagation path s [g/m^3], $T(s)$ is the temperature along the propagation path s [K] and T_{RP} is the location of the tropopause. Since we deal only with the first order approximation and since the H_2O contributions from above the tropopause are less than or equal to 1 % [*Hartmann*, 1988; *Peter et al.*, 1988; *McCormick et al.*, 1985] we can use T_{RP} as integration limit for equation (16) with sufficient accuracy for all observing elevation angles above 15°.

The recent progress in solid state technology allows the design of sensitive microwave and millimeter wave radiometers that allow passive measurements of water vapor in the atmosphere which can be used to correct the wet path delay. With microwave-spectrometers [*Puliafito and Bürki*, 1991; *Schanda*, 1986; *Ulaby and Carver*, 1983] we can measure the brightness temperature $T_B(f,s)$ of the atmosphere which is proportional to the emitted radiation intensity (thermal noise) of the atmosphere at this frequency. Since water vapor and liquid water in the atmosphere play the most important role for ground based microwave observations two observing frequencies should be selected such that one is predominantly affected by water vapor and the other one by liquid water [*Puliafito*, 1989; *Puliafito and Bürki*, 1991; *Westwater*, 1978, 1980; *Wu*, 1979]. When the measured brightness temperature for the first case is $T_{B,1}$ – the frequency lies around 22 GHz – and for the second case $T_{B,2}$ – the frequency lies around 31 GHz – then the opacity of the atmosphere is small (high transparency) – see eq. (A1.2) – and the linearized expression of T_B, namely $T_B^\star$, is a very good approximation.
Then we get

$$\Delta_{atm}\ H_2O \approx a_0 + a_1 T_{B,1}^\star + a_2 T_{B,2}^\star \tag{17}$$

where $T_{B,1}^\star$ is the linearized brightness temperature for f_1 = 23.8 GHz, $T_{B,2}^\star$ is the linearized brightness temperature for $f_2 = 31.5$ GHz, and a_0, a_1, a_2 is the inversion coefficients [see *Puliafito and Bürki*, 1991; *Westwater*, 1978; *Wu*, 1979]

Puliafito and Bürki used a set of one year radiosonde measurements from Payerne (Switzerland) to calculate the inversion coefficients a_0, a_1, and a_2. The frequencies f_1 and f_2 are selected such that the effect of the annual variation of a_0, a_1, and a_2 is minimal. Puliafito and Bürki also used seasonally calculated inversion coefficients which gave slightly better results than the calculated annual values.

7.3. *Conclusions*

Puliafito and Bürki [1991] presented measured results and comparisons with model calculations showing to which extent the first order residual error in Δ_{atm}, H_2O can be minimized. They got:

1.	Usage of numerical models only implies: Δ_{atm}, H_2O	>	10 cm
2.	P, T, and e measured at the ground	≈	6 cm
3.	Usage of 2 daily nearby radiosonde launches	≈	3 cm
4.	Usage of a 23.8/31.5 GHz radiometer	≈	0,8 cm

In context with the usage of the radiometer *Puliafito and Bürki* [1991] report a limiting residual error of 0,8 cm, where 0,46 cm is due to calibration inaccuracies and 0,30 cm due to inaccuracies in the calculation of the inversion coefficients. This seems optimistic. Due to the inherent systematic errors and due to the fact that statistical measurements are not always available a more conservative figure is adviseable until a final figure is available. However, obtaining a more rigorous estimate is difficult, requiring further measurements – also at several other locations – and further model calculations especially with ray tracing. Even if we take a residual error of 1,5 cm instead of 0,8 cm then we are still a factor of two better in correcting the path delay due to water vapor then it was previously possible with data from 2 daily radiosondes. Thus microwave radiometry can provide a significant improvement over the values given above in 1, 2, 3. Summing up: Right now there seems to be a limiting value for the correction of the water vapor path delay of the atmosphere around 1 cm. To obtain this delay with higher accuracy and for different meteorological conditions – rather than only for an area around Zürich/Switzerland – requires more measurements and more model calculations.

REFERENCES

Ajayi, G.O., A. Hedberg, G. Hamberg: Accurate determination of ionospheric effects on satellite based positioning systems using a versatile 3-dimensional ray tracing program. *Radio Science*, 15, 5, 1009-1-16, 1980.

Bean, B.R.: The Radio Refractive Index of Air. *Proceedings of IRE*, 50, 1962.

Born, M., E. Wolf: *Principles of Optics.* 4th Edition Pergamon Press, Oxford, London, New York, 1970.

Budden, K.G.: *The Propagation of Radio Waves.* Cambridge University Press, 1985.

Davies, K.: *Ionospheric Radio Propagation.* Washington, D.C., Supt. of Documents, U.S. Government Printing Office, 1965.

Davies, K.: *Ionospheric Radio.* Peter Peregrinus Ltd., London, UK, 1990.

Debye, P.: *Handbuch der Radiologie*, Leipzig, 1925.

Gantham, D.D. et al.: Water vapor, precipitation, clouds and fog. In: *Handbook of Geophysics and the Space Environment*, chapter 16, ed. by A.S. Jursa, Airforce Geophysical Laboratory, National Technical Information Service, 5285 Port Royal Road, Springfield, VA 22161, USA, 1985.

Handbook of Chemistry and Physics: ed. by R.C. Weast, p. E-233, 50th ed. 1969–1970. The Chemical Rubber Co., Cleveland, Ohio, 44128, USA.

Hartmann, G.K. and R. Leitinger: Range errors due to ionospheric and tropospheric effects for signal frequencies above 100 MHz. *Bull. Geod.*, **58**, 109–136, 1984.

Hartmann, G.K.: Millimeter Wave Atmospheric Sounder (MAS), in: P.D. Craven, M.R. Torr: ATLAS Mission 1, *NASA Technical Memorandum 4101*, NASA Scientific and Technical Information Division, I-27 to I-31, 1988.

Hartmann, G.K., J.J. Olivero, and G. Haaf: *H_2O in the Atmosphere.* In: *Physics of the upper atmosphere*, ed. by W. Dieminger, G.K. Hartmann, R. Leitinger, Springer-Verlag, Heidelberg, to be published 1992.

Hopfield, H.S.: Tropospheric effect on electromagnetically measured range: prediction from surfach weather data. *Radio Science*, 6, 357–367, 1971.

Hufford, G.: A model for the complex permittivity of ice at frequencies below 1 THz. *Int. Journ. Infrared and Millimeter Waves*, 12, 7, 677–682, 1991.

Kassner, J.L. et al.: Fundamental Studies on Vapor Phase Water Clusters. In: *Atmospheric Water Vapor.* (Ed. Deepak et al.), 613-649, Academic Press, 1980.

Kruspe, G.: Der atmosphärische Brechungsindex und seine Messung mit Mikrowellen–Refraktometern. *Technische Mitteilung Nr. 1*, Inst. für Radiometeorologie und Maritime Meteorologie der Universität Hamburg, 1965.

Landolt–Börnstein: *Zahlenwerte und Funktionen* aus vol. II, part 6, pp. 874–908, Springer-Verlag, Berlin, Göttingen, Heidelberg, 1959.

Landolt–Börnstein: ed. by K.H. and A.M. Hellwege, Part 8, *Optical Constants*, 871–901 (878), Springer-Verlag, Berlin, Göttingen, Heidelberg, 1962.

Landolt–Börnstein: ed. by K.H. Hellwege and H.H. Voigt, vol. I, *Astronomy and Astrophyscis*, "Influence of the Earth's atmosphere", 48–61, Springer-Verlag, Berlin, Heidelberg, New York, 1965.

Leitinger, R.: Der Einfluss ionosphärischer Ausbreitungsfehler bei der geodätischen Anwendung von Navigationssatelliaten. *Kleinheubacher Berichte*, **17**, 321–332, 1974.

Leitinger, R., R.S. Allen, D.E. Donatelli, and G.K. Hartmann: Adaptive mapping of ionospheric features. In: *Proceedings of the Symposium on the Effect of the Ionosphere on Space and Terrestrial Systems*, edited by J.M. Goodman, 530–537, U.S. Government Printing Office, Washington, D.C., 1978.

Liebe, H.J.: Atmospheric EHF window transparencies near 35, 90, 140, and 220 GHz. *IEEE Trans. Ant. Prop.*, AP-31, 1, 1983.

Liebe, H.J.: An updated model for millimeter wave propagation in moist air. *Radio Science*, **20**, 5, 1069–1089, 1985.

Liebe, H.J.: MPM – An Atmospheric Millimeter Wave Propagation model. *Int. Journ. Infrared and Millimeter Wave*, **10**, 6, 631–650, 1989.

Liebe, H.J., G. Hufford, and T. Manabe: A model for the complex permittivity of water at frequencies below 1 THz. *Int. Journ. Infrared and Millimeter Waves*, **12**, 7, 659–675, 1991.

Lovejoy, S. and D. Schertzer: Scale Invariance, symmetries, Fractals and Stochastic Simulations of Atmospheric Phenomena. *Bulletin American Meteorological Society*, **67**, 1, 21-32, 1986.

Mandelbrot, B.B.: *The fractal geometry of nature*, Freeman and Co., New York, 1982.

McCormick, M.P., P. Hamill, and U.O. Farrukh: Characteristics of Polar Stratospheric Clouds as observed by SAM II, SAGE, and lidar. *J. Meteorol. Soc. Japan*, **63**, 267-276, 1985.

McElroy, M.B. and R.J. Salawitch: Changing composition of the global stratosphere. *Sciencc*, **243**, 763-770, 1989.

Peter, R., K. Künzi, and G.K. Hartmann: Latitudinal Survey of Water Vapor in the Middle Atmosphere using an Airborne Millimeter Wave Sensor, *Geophys. Res. Lett.*, **15**, 11, 1173-1176, 1988.

Pilewskie, P. and S. Twomey: Discrimination of Ice from Water in Clouds by Optical Remote Sensing. *Atmospheric Research*, **21**, 113–122, 1987.

Puliafito, E.: Bestimmung von Wasserdampf- und Ozonprofilen in der mittleren Atmosphäre durch Millimeterwellenspektroskopie. *Ph.D. Thesis*, University of Braunschweig (FRG), 1989.

Puliafito, E. and B. Bürki: Tropospheric path length correction using a dual frequency transportable radiometer. *Proceedings of IGARS'91*, vol. III, 1337–1340, Helsinki University of Technology, June 1991.

Ratcliffe, J.A.: *The Magneto-Ionic Theory and its Application to the Ionosphere.* Cambridge at the University Press, 1959.

Rawer, K. and K. Suchy: *Radio observations of the ionosphere.* In: Encyclopedia of Physics, ed. by S. Flügge, vol. XLIX/2 – geophysics II/2 –, pp. 1–537, Springer-Verlag, Berlin, Heidelberg, New York, 1967.

Remsberg, E.E., J.M. Russell III, and C.Y. Wu: An interim Reference Model for the variability of the middle atmosphere H_2O vapor distribution. In: *Handbook for MAP*, **31**, 50–66, ed. G.M. Keating, SCOSTEP Secretary, Urbana, Illinois 61801, USA, 1989.

Schanda, E.: *Physcial Fundamentals of Remote Sensing*, Springer Verlag, Berlin, Heidelberg, New York, Tokyo, 1986.

Sommerfeld, A.: *Optik*, 2. Auflage, Akademische Verlagsgesellschaft, Geest & Portig KG, Leipzig, 1959.

Tatarski, V.E.: *Wave Propagation in a Turbulent Medium.* New York, McGraw–Hill, 1961.

Tatarski, V.E.: *The Effects of the Turbulent Atmosphere on Wave propagation.* (translated from Russian) Springfield, V.A., National Technical Information Service, U.S. Dept. of Commerce, 1971.

Ulaby, F.T. and K.R. Carver: Passive Microwave Radiometry, in: R. N. Colwell (Ed.): *Manual of Remote Sensing*, Vol. 1, American Society of Photogrammetry, Falls Church, VA 22046, USA, 475-516, 1983.

Waters, J.W.: Absorption and Emission by Atmospheric Gases, in: L. Marton (Ed.): *Methods of Experimental Physics*, Vol. 12, Astrophysics, Academic Press, New York, San Francisco, London, 142-176, 1976.

Westwater, E.R.: The accuracy of water vapour and cloud liquid determination by dual frequency ground-based microwave radiometry. *Radio Science* **13**, 4, 677–685, 1978.

Westwater, E.R. and N.C. Grody: Combined Surface and Satellite Temperature Profile Retrieval. *Journal of Applied Meteorologie*, **19**, 12, 1438–1444, 1980.

Wu, S.: Optimum Frequencies of passive Microwave Radiometer for Tropospheric Path-Length Correction. *IEEE Trans. Ant. and Prop.*, AP-27, 2, 1979.

G. K. Hartmann, Max-Planck-Institut für Aeronomie, Postfach 20, D - W 3411 Katlenburg-Lindau, Germany

Modeling and Prediction of Density Changes and Winds Affecting Spacecraft Trajectories

T. L. KILLEEN, A. G. BURNS, R. M. JOHNSON

Space Physics Research Laboratory, Department of Atmospheric, Oceanic and Space Sciences, University of Michigan, Ann Arbor, Michigan

F. A. MARCOS

Phillips Laboratory/LIS, Hanscom Air Force Base, Massachusetts

The various thermospheric processes that are capable of perturbing the orbital characteristics of low-earth-orbiting satellites are reviewed and their importance assessed. The drag force acting on orbiting vehicles is linearly dependent on atmospheric density and quadratically dependent on the difference in speed between the satellite velocity and the in-track atmospheric wind. Significant perturbations in ambient densities and winds are common at thermospheric altitudes and are manifested by waves and tidal structures which propagate upwards from below, and by changes to the global circulation pattern due to solar and magnetospheric forcing processes. Variations in thermospheric density and wind of up to an order of magnitude are commonplace, depending on the specific geophysical conditions and spacecraft altitude. Such variations can lead to loss of spacecraft acquisition in orbital tracking programs and even to premature re-entry. The dependencies of thermospheric densities and winds on season, solar-cycle, altitude, latitude and local time are relatively well understood and modeled. Less well understood are the variations driven by geomagnetic storms and tidal and stochastic wave forcing processes. Uncertainties also remain in the understanding of the physics of the drag process itself, as evidenced by the lack of reliable individual satellite drag coefficients in many cases. This paper reviews the status of our understanding of thermospheric density and wind perturbations responsible for orbital variations and describes some of the current modeling efforts designed to improve our ability to specify these perturbations.

1. Introduction

The stability of satellite trajectories in low earth orbit (LEO) is a subject of considerable practical and theoretical interest in the current era of increasing orbital activities and manned space flight. Aspects of concern include spacecraft lifetime prediction, control, tracking, on-board fuel requirements and propulsion, and reentry. This interest is perhaps most clearly brought to the attention of the public when premature re-entry of orbiting vehicles occurs due to unexpected increases in atmospheric density associated with high solar activity. While such drastic events are fairly rare, the precision of routine tracking and orbit prediction are of great and continuing importance for many operational systems.

Environmental Effects on Spacecraft Positioning and Trajectories
Geophysical Monograph 73, IUGG Volume 13

In this paper, we review the various changes in thermospheric density and wind that can affect satellite trajectories and describe the current ability of models to provide accurate specifications of the thermospheric state variables for use in orbital tracking programs. The organization of the paper is as follows. Section 2 presents the satellite drag equation and describes the uncertainties that result from the estimation of satellite drag accommodation coefficients. Sections 3 and 4 present discussions of thermospheric density and wind perturbations, respectively, and assess their importance for satellite drag calculations. Section 5 provides a discussion of current modeling efforts to improve density and wind specification, with an emphasis on the new Vector Spherical Harmonic (VSH) model. Section 6 presents a short summary.

2. Satellite Drag

The Drag Equation

The orbital motion of spacecraft in low earth orbit, described mathematically by the Lagrange planetary equations

[Fitzpatrick, 1970], is continually influenced by three primary perturbing influences. These are: i) the non-spheroidicity of the Earth; ii) the third-body gravitational interactions with the Moon and the Sun; and iii) the drag experienced by the interaction with the extended upper atmosphere of the Earth. This paper focuses on the last of these three perturbing influences, namely satellite drag. The reader is referred to the review by Liu [1983] and references therein for a more complete discussion of orbit theory.

The drag force on a satellite in low earth orbit is given by:

$$F_D = \frac{1}{2}\rho\ C_D(V_s - U)^2\ A_r \tag{1}$$

where ρ is the atmospheric density, V_S is the satellite velocity and U is the component of the neutral wind along the satellite velocity vector, C_D is the drag coefficient and A_r is the effective area of the spacecraft which interacts with the ramming ambient atoms and molecules.

The drag equation has been used for various purposes over the past three decades. For example, spacecraft orbital perturbations have been routinely monitored and the calculated orbitally-averaged drag force has been used to provide estimates of the total atmospheric density, on the assumption of zero wind. Such density results were used in the development of the early empirical models of thermospheric density structure [Harris and Jastrow, 1959; King-Hele, 1959; Jacchia, 1965]. The Jacchia model was particularly successful and was subsequently further developed to become a standard against which later models and data sets were compared [Jacchia, 1970; 1971; 1977]. Clearly, the accuracy of such models depends, in part, on the degree to which the zero wind assumption is true and, in part, on the extent to which the product of the drag coefficient and the effective area is well known [Herrero, 1987].

A second use of the drag equation is for the purposes of specifying and predicting orbital perturbations, either using measurements or model predictions of densities. This approach provides the basis for operational satellite tracking programs. The performance of the orbital tracking and prediction codes is obviously strongly dependent on the accuracy of the density inputs used [Marcos, 1985]. In later sections of this paper, we review the status and accuracy of existing density specification models.

The Drag Coefficient

The drag coefficient is a key parameter needed for the calculation of the rate of change of orbital velocity of a spacecraft due to the influence of the drag force. In the literature, the drag parameter can take several forms, e.g., the drag coefficient (C_D), ballistic coefficient ($B = (C_D/M)$, where M is the spacecraft mass), rate of change of mean motion ($\dot{S}$), and others [Snow and Liu, 1991]. In principle, the drag coefficient is a constant which can be determined either theoretically or from tracking data. In practice, however, spacecraft tracking programs allow the drag coefficient to vary with time to account for errors in density models, errors in the specification of the geopotential and changes in the satellite area-to-mass ratio due to expulsion of expendibles, etc.

Accurate drag coefficient specification is needed for density measurements using satellite orbital variations. Herrero [1987] has discussed the precision of drag coefficient estimates for spacecraft in low earth orbit. We review here some of the main issues. There are indications that the estimated drag coefficients, even for spacecraft with very simple shapes, may be in error by several 10's of percent. The central problem with the estimation of $C_D$is related to the lack of quantitative understanding of the microscopic physical process of atom-surface momentum transfer. Generally, satellites flown or used to provide density data have simple (often spherical or cylindrical) shapes and the drag coefficient is often given the nominal value of ~2.2 (Equations 6 and 7, below). Furthermore, it is often assumed that, for the sphere, $C_{D,}$ may never be less than 2.0.

The numerical value of C_D is given by integrating the momentum imparted to each element of the spacecraft surface by the impinging atmospheric atoms and molecules. The momentum exchange is a function the angle of attack, α, of the air-surface interaction. Figure 1, adapted from Herrero

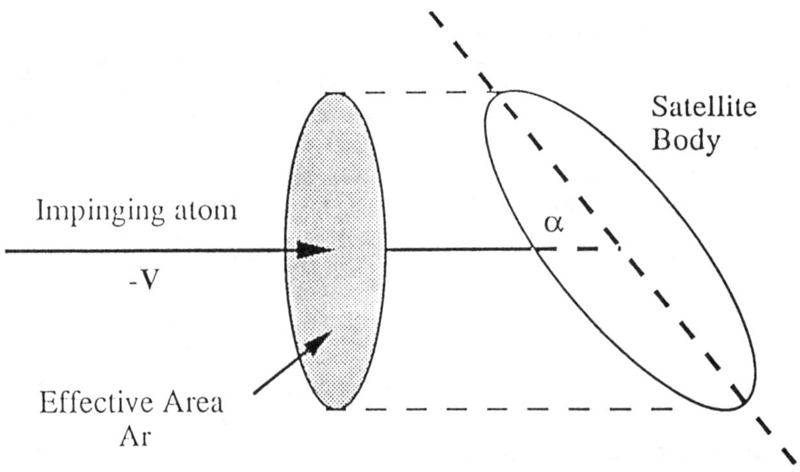

Fig. 1. Motion of impinging atmospheric atoms with respect to the orbiting vehicle.

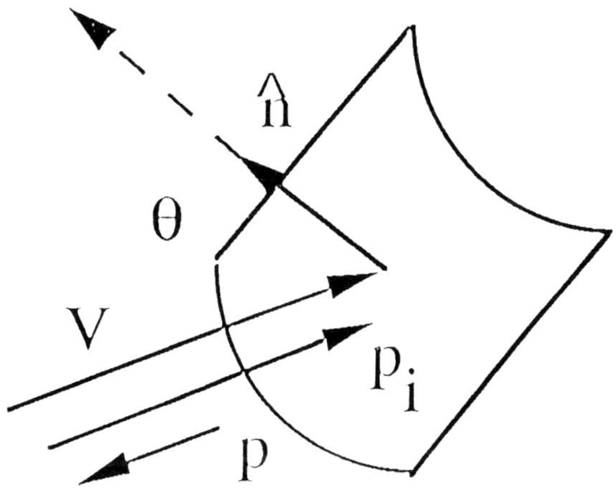

Fig. 2. Component of reflected momentum along the line of incidence.

[1987], shows the simple geometry of this interaction in the reference frame of the spacecraft. Laboratory measurements of the momentum exchange often use the momentum transfer ratio, $f(\theta)$, which represents the fraction of momentum reflected backwards along the line of incidence of the impinging atom. The angle θ here represents the angle between the incident direction of the impinging atom and its recoil direction (Figure 2). The momentum transfer ratio is related to the normal and tangential momentum accommodation coefficients, σ_n and σ_t, respectively. The accommodation coefficients provide a quantitative indication of the fraction of incident momentum converted to drag for each impacting atom or molecule. The momentum transfer ratio, $f(\theta)$, and the accomodation coefficients, σ_n and σ_t, are given by:

$$f(\theta)=\frac{p_r}{p_i}=(1-\sigma_n)\cos^2\theta\;-\;(1-\sigma_t)\sin^2\theta\;+K\sigma_n\cos\theta \quad (2)$$

where

$$\sigma_n=\frac{(p_{ni}-p_{nr})}{p_{ni}};\qquad \sigma_t=\frac{(p_{ti}-p_{tr})}{p_{ti}} \quad (3)$$

The p's represent components of momentum, with subscript i denoting the incident component and subscript r the reflected component. K is a constant depending on the surface temperature. Herrero [1987] quotes a value for K of 0.038 for 300 K. Real surfaces have properties that are intermediate between the two extremes of "full" and "zero" accommodation of the impinging atom at the spacecraft surface. Full accommodation is equivalent to diffuse reflection at the surface (i.e., the assumption is made that the impinging atom is fully absorbed by the surface and is later re-emitted at an arbitrary isotropic direction). Zero accommodation is equivalent to specular reflection, with the angle of incidence equalling the angle of reflection. The two extreme situations give the following:

a) Full Accommodation:

$$\sigma_t=\sigma_n=1.0;\qquad f(\theta)=0.038\cos(\theta) \quad (4)$$

b) No Accommodation:

$$\sigma_t=\sigma_n=0.0;\qquad f(\theta)=\cos(2\theta) \quad (5)$$

Figure 3 shows some typical laboratory measurements of $f(\theta)$ from i) Seidl and Steinheil [1974] for ~2eV helium atoms impacting a smooth sapphire surface, ii) Knechtel and Pitts [1973] and iii) Herrero [1987] for ~10eV nitrogen molecules and ions on Aluminum. For comparison, the full accommodation and the zero accommodation curves are shown. A negative $f(\theta)$ indicates that the reflected momentum has a component, p_r, along the original direction of the incident momentum, p_i. As can be seen, the more realistic situation is close to neither of the two extremes of full and zero accommodation. Thus, partial accommodation of impinging atoms and molecules occurs in practice and can vary with time as surfaces age.

To calculate the numerical value of the drag coefficient, C_D, for a spacecraft body of surface Σ, the following formula can be used [*Herrero*, 1987]:

$$C_D=2(1+\frac{1}{A_r}\int_\Sigma f(\theta)\cos\theta\;dA\;) \quad (6)$$

which, for a spherical satellite reduces to:

$$C_D=2\;(\;1+\int_\Sigma f(\theta)\cos\theta\;\sin\theta\;d\theta\;) \quad (7)$$

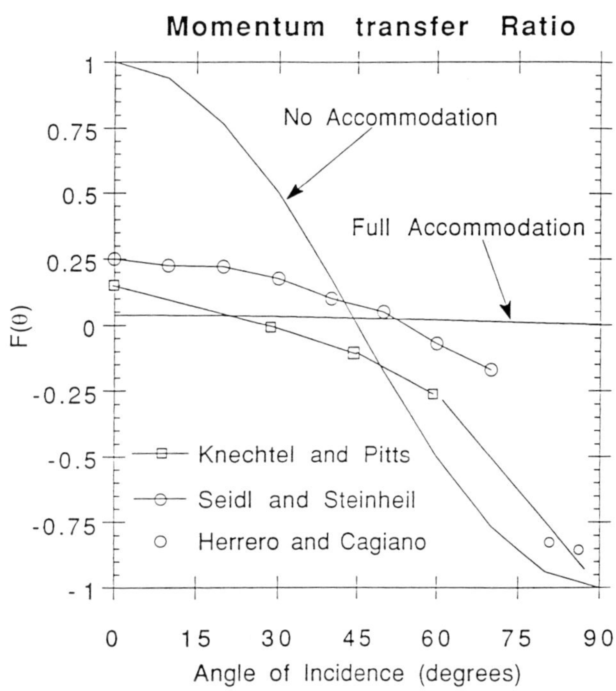

Fig. 3. Measured variation of momentum transfer ratio as a function of angle of incidence. The referenced measurements are shown together with calculations for no accommodation and for full accommodation of impinging atoms at the spacecraft surface.

The weighting term $2\cos\theta \sin\theta$ maximizes at 45°, giving greatest weight to the momentum transfer ratio values near this angle of incidence. From an examination of this dependence, and of the data of Figure 3, it can be seen why typical values of C_D are slightly greater than 2.0. Values smaller than 2.0 are not unknown, however, for situations where the momentum transfer ratio becomes negative at smaller angles than 45° [Knechtel, 1973].

Herrero [1987] has shown that values of C_D for a spherical spacecraft can vary between ~1.6 - 2.7, depending on both the nature of the surface and the degree of accommodation of the impinging molecules. The variability arises from the need to measure the accomodation properties of differing exposed surfaces. Since measurements which replicate the high energies of the incoming atoms are difficult to perform in the laboratory and require detailed study of the angular dependence of the recoil, most spacecraft are given estimated values for C_D, based on relatively simple theoretical arguments. These estimates are probably accurate only to within a few 10's of percent. Furthermore, the drag coefficient can vary in orbit due to spacecraft orientation changes and surface aging, as mentioned above.

Because of the uncertainties surrounding individual drag coefficients, absolute density estimates from orbital drag determination are also subject to significant inaccuracies. For the purposes of satellite orbit determination, it is the product of ρ and C_D that is important, however. In general, it is the first of these two parameters that is most variable in real situations, although in many tracking codes, the latter parameter is often varied to allow for the effects of atmospheric density perturbations.

For improved specification and forecast of satellite orbital perturbations due to time-varying atmospheric drag, it is essential to obtain accurate absolute values for atmospheric density and to use an accurate drag coefficient which is allowed to vary only slightly in orbit. This is particularly true for forecasting requirements where the spatial and temporal variations in drag are likely to be primarily due to geophysical variations in ρ.

3. Atmospheric Density and Composition Variations

In this section, the major physical processes responsible for thermospheric composition and density changes at altitudes of relevance to spacecraft in low earth orbit are reviewed. Emphasis will be placed on the altitude range ~180-400 km.

Theoretical Development

The governing (or "primitive") equations of the thermosphere are given by appropriate expressions for the conservation of mass, momentum, and energy for the gas as a whole, as well as for the individual species within the gas. These equations may be derived by taking the first three moments of the Boltzmann transfer equation [Schunk, 1977], where certain assumptions are generally made regarding the form of the viscous stress tensor and the heat flux vector (from higher order moments) in order to close the set of equations. The conservation equations are coupled and require appropriate boundary conditions and information on a number of transport parameters (e.g., coefficients of viscosity, thermal conduction, eddy diffusion, etc.) for their solution. The transport parameters are obtained either experimentally or from considerations of gas kinetic theory. This theoretical formulation is considered to be sufficiently accurate for all practical purposes involving neutral densities and composition, although some of the assumptions are questionable in certain domains (generally high altitudes) involving ionospheric and magnetospheric plasma populations.

The basic mechanisms that control the composition and thereby the total density of the thermosphere are perhaps best appreciated by considering the individual species continuity equation. Although we are mainly interested here with total density, a knowledge of thermospheric composition is required for a complete specification of the region.

In the altitude range of interest, the three major neutral species are O, O_2, and N_2, with O being lighter than the mean mass and O_2 and N_2 being heavier than the mean mass. Above the turbopause (near 105-110 km altitude), the process of molecular diffusion tends to return the major constituents towards individual "diffusive equilibrium" profiles given by their respective masses. Since molecular nitrogen is largely chemically inert, it maintains a profile that resembles one of diffusive equilibrium, but one that can also be perturbed significantly by dynamical processes. Atomic and molecular oxygen, on the other hand, are controlled by both chemical production and loss (photodissociation and recombination) as well as by dynamical and diffusive processes. The assumption of diffusive equilibrium (i.e., zero diffusive flux) may not always be justified [Mayr, 1984; Rees et al., 1988; Burns et al., 1989], even in a relatively quiet atmosphere and for inert species, such as molecular nitrogen.

The continuity equation for a single thermospheric species may be written in terms of the partial derivative with time of the mass mixing ratio, Ψ_i, of the ith species. Dickinson et al. [1984] used horizontal spherical coordinates and a log pressure vertical coordinate (λ, ϕ, z), where λ = longitude, ϕ = latitude and $z = \ln(P_o/P)$, P is the pressure, and P_o (= 50 μPa) is a reference pressure, to write the following vector equation for the mass mixing ratio of each species:

$$\begin{aligned}\frac{\partial\tilde{\Psi}}{\partial t} &= -\frac{e^z}{\tau}\frac{\partial}{\partial z}\left[\frac{\overline{m}}{m_{N_2}}\left(\frac{T_o}{T}\right)^{0.25}\frac{\mathbf{L}}{\tilde{\alpha}}\tilde{\Psi}\right] \\ &+ e^z\frac{\partial}{\partial z}\left[K(z)e^{-z}\left(\frac{\partial}{\partial z}+\frac{1}{\overline{m}}\frac{\partial\overline{m}}{\partial z}\right)\tilde{\Psi}\right] \\ &- \left[\vec{V}\bullet\nabla\tilde{\Psi} + w\frac{\partial}{\partial z}\tilde{\Psi}\right] + \tilde{S} - \tilde{R}\end{aligned} \tag{8}$$

Where $\tilde{\Psi}$ is the vector mass-mixing ratio, given by (Ψ_{O_2}, Ψ_O) and Ψ_i is defined by:

$$\Psi_i = \frac{n_i m_i}{\sum n_j m_j} \tag{9}$$

and the mixing ratio of N_2 is defined as:

$$\Psi_{N_2} = 1 - \Psi_{O_2} - \Psi_O \quad (10)$$

Also, $\vec{V}$ = the horizontal velocity vector; t = time, z = vertical coordinate,u = the eastward neutral velocity; v = the northward neutral velocity; w = the vertical neutral velocity; T = temperature; n_i is the number density of the ith species, and m_i is the mass of the ith species. $K(z)$ is the eddy diffusion coefficient, $\overline{m}$ is the mean mass, P_o is the atmospheric pressure at the ground (= 10^5 Pa), T_o = 273K, τ is the diffusion time scale.

The matrix operator **L** has elements:

$$L_{ij} = \delta_{ij}\left(\frac{\partial}{\partial z} - \varepsilon_{ij}\right) \quad (11)$$

where δ_{ij} is the delta function and ε_{ij} is defined by

$$\varepsilon_{ij} = 1 - \frac{m_i}{\overline{m}} - \frac{1}{\overline{m}}\left[\frac{\partial \overline{m}}{\partial z}\right] \quad (12)$$

The **L** matrix defines diffusive equilibrium solutions through the equation **L** $\tilde{\Psi}$=0.

The variables $\tilde{S}$ and $\tilde{R}$ represent chemical source and sink terms for the individual species, respectively. Photodissociation of O_2 provides a source of O given by:

$$\tilde{S} = J_{O_2}\begin{bmatrix} -\Psi_{O_2} \\ +\Psi_{O_2} \end{bmatrix} \quad (13)$$

where J_{O_2} is the molecular oxygen photodissociation rate. Three-body recombination provides a source of O_2 and a sink of O via the reaction (O + O + M $\rightarrow$ O_2 + M). The rate coefficient for this reaction is given by:

k = 3.8 x 10^{-30} exp(-170/T)/T for M = O_2 [*Johnson*, 1968] and

k = 4.8 x 10^{-33} for M = N_2 [*Campbell*, 1973].

The components of the vector $\tilde{\alpha}$, which appears in the first term on the right hand side of Equation 8 are given by:

$$\begin{aligned} \alpha_{11} &= -[\phi_{13} + (\phi_{12} - \phi_{13})\Psi_2] \\ \alpha_{22} &= -[\phi_{23} + (\phi_{21} - \phi_{23})\Psi_1] \\ \alpha_{12} &= (\phi_{12} - \phi_{13})\Psi_1 \\ \alpha_{13} &= (\phi_{21} - \phi_{23})\Psi_2 \end{aligned} \quad (14)$$

where the subscript 1 denotes O_2; 2 denotes O; and 3 denotes N_2. The ϕ_{ij} are defined by:

$$\phi_{ij} = \left(\frac{D}{D_{ij}}\right)\left(\frac{m_3}{m_j}\right) \quad (15)$$

where D_{ij} is the mutual diffusion coefficient given by *Colegrove* [1966] and D is the molecular diffusion coefficient given by:

$$D = D_o\left[\frac{P_o}{P}\right]\left[\frac{T}{T_o}\right]^{1.75} \quad (16)$$

and D_o is a characteristic diffusion coefficient (= 2 x 10-9 cm2 s-1) at pressure P_o and temperature T_o.

The terms on the right hand side of equation 8 represent, respectively, the changes in the composition due to molecular diffusion, eddy diffusion, horizontal and vertical advection, and chemical production and loss.

Departures from diffusive equilibrium are driven by the hydrodynamic transport and the chemical terms in Equation 8. In this paper, we will use calculations of the individual terms from the National Center for Atmospheric research (NCAR) Thermospheric general circulation model (TGCM) to illustrate the importance of vertical and horizontal winds for the advective control of thermospheric density.

Mean Thermospheric Density and Forcing Mechanisms

The specification of thermospheric density may be treated conveniently as separable into two parts: 1) the mean density structure that would exist if there were no transport processes operating (i.e. the zero flux or diffusive equilibrium solution to the coupled continuity and momentum equations), and 2) perturbations from the mean due to transport processes. In this section we briefly review the mean density fields at satellite altitudes. The subsequent section describes density perturbations.

Thermospheric density and temperature measurements made by a variety of techniques over the years have led to the development of a good first-order description of the principal morphological characteristics for the unperturbed case. Table 1, taken from Hedin [1988] provides a list of the key data sources used to develop the Cospar International Reference Atmosphere (CIRA-86). The key to the references may be found in the original paper by Hedin. Figure 4, also taken from Hedin [1988], provides a schematic of the historical development of empirical thermosphere models up to ~1989. The CIRA-86 model is a version of the MSIS series of empirical models [Hedin, 1987], based primarily on data from satellite-borne neutral mass spectrometers and ground-based incoherent scatter radars. The most recent member of this family also incorporates a description of global thermospheric winds [Hedin et al., 1991].

Marcos et al. [1989] have carried out a comprehensive assessment of the capability of several models to predict density and drag using satellite accelerometer (SETA) density measurements obtained during the period 1974-84. Figure 5 shows the results of this statistical validation study. Model accuracies were determined by comparing each model prediction with the appropriate measurement and calculating the standard deviation for the entire data set. The quantity statistically analyzed is the ratio between modeled and measured density. The models of Jacchia are denoted by "J" and the MSIS models are denoted by "MS". The plot shows the evolution of standard deviations with time (in years) for data comparisons using four satellites. More recent models are seen to provide about the same level of accuracy as older models, indicating the difficulty of making improvements in density specification with incremental approaches based on adding more and more data to the representation. The scatter of the

TABLE 1. Data Sources for CIRA 1986

Data Set	Measurement	Nominal Altitudes
OGO-6 (MS)	N_2,O,He	400-700 km
San Marco-3 (MS)	N_2O, He,Ar	160-250 km
Aeros-A NATE (MS)	N_2O,He,Ar	200-500 km
AE-C NATE (MS)	N_2,O,He,Ar	250-400 km
AE-C NATE (MS)	Tp	135-300 km
AE-C OSS (MS)	N_2,O,He,Ar,O_2,N	140-500 km
AE-C BIMS (IMS)	H	200-500 km
AE-C MIMS (IMS)	H	160-300 km
AE-C EUVS (absorp)	O_2	100-150 km
AE-D OSS (MS)	N_2,O,He,Ar,O_2,N	140-450 km
AE-E NATE (MS)	Tp	140-400 km
AE-E NACE (MS)	N_2,O,He,Ar	140-450 km
AE-E NATE (MS)	Tp	140-400 km
AE-E OSS (MS)	O_2,N	140-400 km
AE-E-BIMS (IMS)	H	200-500 km
AE-E EUVS (absorp)	O_2	100-150 km
ESRO-4 (MS)	N_2,O,He,Ar	200-350 km
DE-B NACS (MS)	N_2,O,He,Ar,N	200-900 km
DE-B WATS (MS)	Tp	200-500 km
Rockets (MS)	N2,Ar,O_2,Tp	100-300 km
Rockets (absorp)	O_2	120-150 km
Rockets (gauge)	ρ,Tp	85-130 km
Rockets (falling sphere)	ρ,Tp	85-120 km
Rockets (grenade)	ρ,Tp	85-110 km
Millstone Hill (IS)	Tp	100-130 km
Millstone Hill (IS)	Tp	Exospheric
St. Santin (IS)	Tp	Exospheric
Arecibo (IS)	Tp	Exospheric
Arecibo (IS)	Tp,N_2	100-135 km
Jicamarca (IS)	Tp	Exospheric
Malvern (IS)	Tp	240-400 km

Tp: neutral temperature either measured directly or indirectly from ion temperature; ρ: total mass density; MS: neutral mass spectrometer; IMS: ion mass spectrometer; IS: incoherent scatter.

points for individual models shown in Figure 5 is, at least in part, a consequence of the differing accuracies and precisions of the experimental data. Additional model-data validation results are presented in a later section of this paper. To illustrate basic thermospheric density variations in the following discussion, we will present output from the Vector Spherical Harmonic (VSH) model of Killeen et al. [1987]. This is a hybrid model, based in part on the calculations of the NCAR-TGCM and in part on empirical observations and is described in greater detail below.

The primary variation observed in upper thermospheric density at any given altitude is controlled by the heating due to solar EUV absorption. A basic diurnal characteristic is observed, as seen in Figure 6a, which illustrates the Spring Equinox density field for moderate geomagnetic and solar activity levels (Kp=3; $F10.7$=180) at 400 km altitude. As can be seen from Figure 6a, the upper thermospheric density structure at mid- and low latitudes is characterized by a diurnal peak in the early afternoon and a minimum in the early morning hours. This diurnal signature is modified at high latitudes due the important sources of energy associated with magnetospheric-ionospheric coupling processes (Joule and particle heating). The day-to-night density variation is typically a factor of two or more.

At lower thermospheric altitudes, the available data base is far more sparse than at higher altitudes. It is known, however, from ground-based optical and radar techniques, that the situation is typically dominated by semi-diurnal tidal structures. Figure 6b shows the VSH-calculated global density structure for an altitude of 120 km at equinox. Since the global tidal climatology is not as yet well specified or understood, these calculations are much less reliable than those at higher altitudes. They do depict, however, the semi-diurnal structures that are known to have complex latitudinal characteristics. Semi-diurnal variations of several 10's of percent in total density are predicted, primarily at low and middle latitudes.

Figure 7 shows the corresponding model density outputs at high latitudes in a polar dial format. Figures 7a and 7b show, respectively, high-latitude thermospheric density calculations at the 400 and 120 km altitude levels for the northern hemisphere. The density contours at the higher altitude (Figure 7a) illustrate the effects of the day-to-night pressure and temperature gradients in the polar upper thermosphere. The contours are approximately orthogonal to the expected flow direction for the solar-driven winds, which are forced to flow away from the mid-afternoon mid-latitude pressure bulge caused by solar EUV heating. The density field at the lower altitude depicts a localized high density region associated with vortical motion in the lower thermosphere [Killeen et al., 1992].

Figure 8 illustrates the relationship between the neutral wind pattern and the global thermospheric density field. This figure presents VSH model calculations of the density and neutral wind field at 300 km altitude for moderately active solar and geomagnetic conditions during the December solstice period (12:00 UT). The three panels depict the winter and summer polar regions and the corresponding global fields in a cartesian representation. It can be seen that the winds of greatest magnitude (in excess of 400 m/sec) occur at high latitudes, where the characteristic cellular vortices are driven by magnetospheric forcing processes [Killeen and Roble, 1988]. The neutral wind morphology has a significant effect on the density morphology, with localized regions of low density being approximately co-located with the region of cyclonic vorticity (counter-clockwise).

As can be inferred from an examination of Equation 1 and the information shown in Figures 6-8, spacecraft in low Earth orbit (even those in circular orbits) are likely to be subject to significant drag variations during individual orbits due to these density variations, even for the mean situation as calculated using "steady-state" or diurnally-reproducible forcings within the TGCM. The specific amount of drag experienced will be dependent on the exact nature of the orbit and its geometrical relationship to global density structures.

In order to arrive at an improved understanding of the causes of density variations such as those described briefly above, recent work has focused on a quantitative examination of the responsible forcing mechanisms. As discussed in section 3.1, thermospheric density is largely controlled by a combination

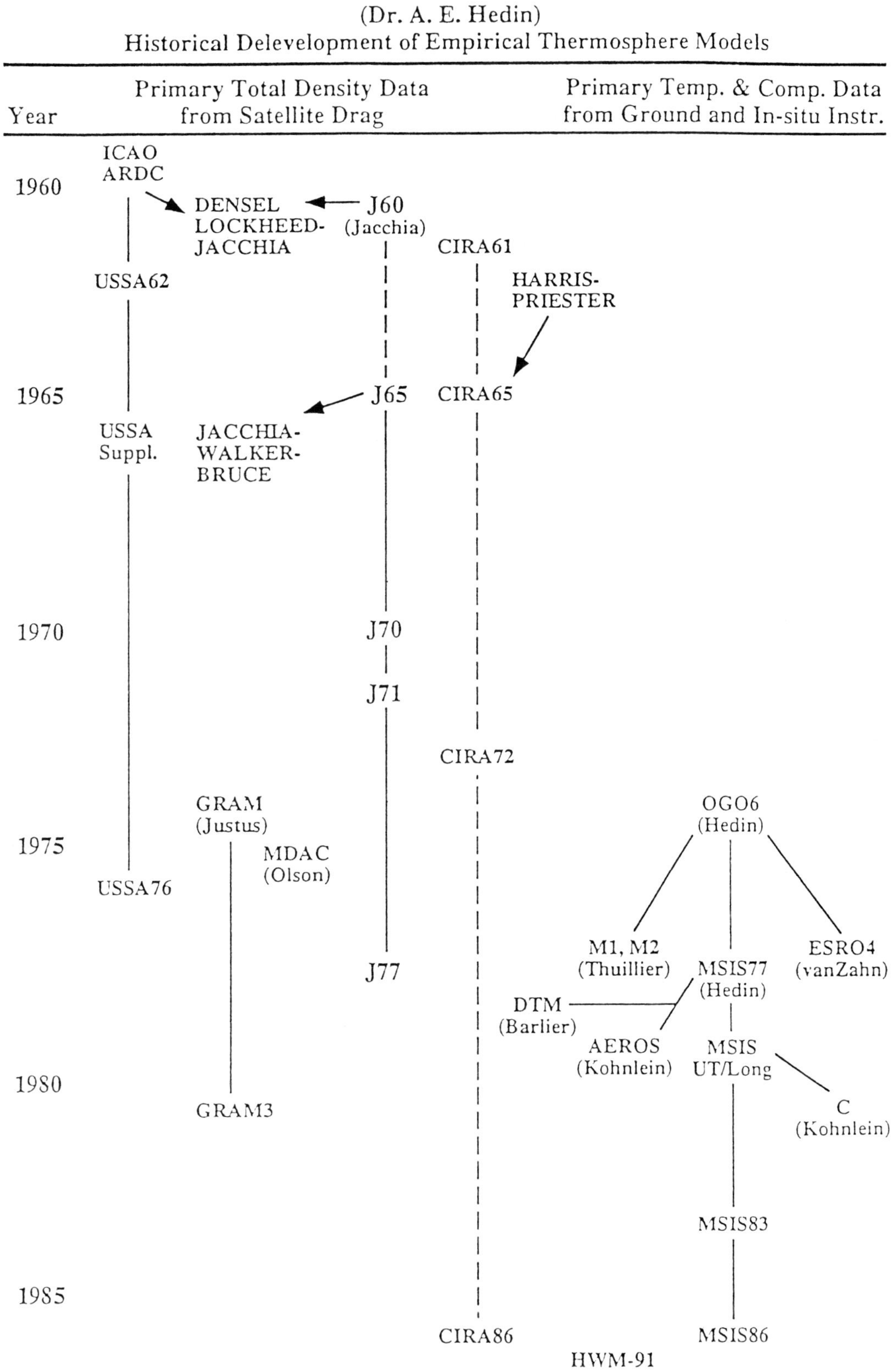

Fig. 4. Schematic illustrating historical development of empirical thermospheric models. Taken from Hedin [1987].

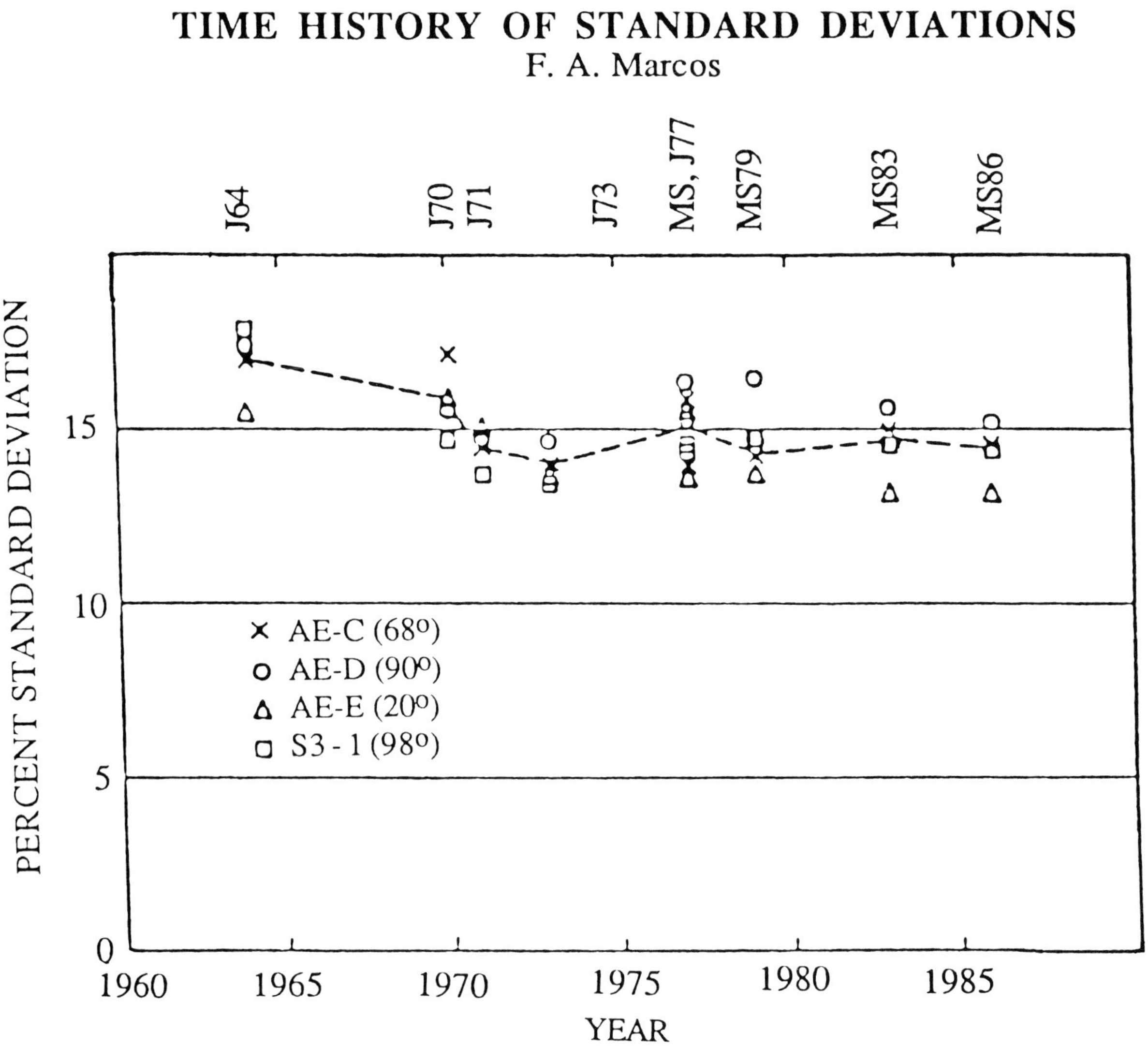

Fig. 5. Statistical assessment of accuracy for thermospheric density specification using several empirical models. Taken from Marcos et al. [1989]. Figure reproduced with the permission of the American Astronautical Society.

of advective and diffusive processes. To illustrate the characteristics and spatial morphology of the "compositional adjustment forcing terms", Burns et al. [1989] have used runs of the NCAR-TGCM to calculate the magnitudes of the respective terms. Figure 9 shows results of the z=1 constant-pressure-level (~320 km) analysis for the molecular nitrogen mass mixing ratio, Ψ_{N_2}, as a function of location (latitude and longitude) at 03:00 hrs UT. Total density variations will have similar dependencies to those calculated for the molecular nitrogen single-constituent mixing ratio.

The individual compositional adjustment forcing terms due to horizontal advection, vertical advection and molecular diffusion terms are shown in Figures 9a, b and c, respectively, in units of sec^{-1}, while Figure 9d illustrates the instantaneous total rate of change of the mass mixing ratio for N_2, given essentially by the sum of these three terms. For comparison, the calculated Ψ_{N_2} is given in Figure 9f and the neutral horizontal wind (arrows, with scale at lower right in m/sec) and vertical wind (contours, with units of 10^6 sec^{-1}) calculated by the TGCM at the same time are shown in Figure 9e. The vertical "winds" are normalized to the local scale height in order to simplify the relationship with vertical advection on a constant pressure surface and hence the units are sec^{-1}. The relative significance of the contributions of winds and horizontal gradients in mass mixing ratio to the advective terms can be assessed by reference to these last two figures.

The three major upper thermospheric compositional terms are shown in Figure 9. Vertical and horizontal advection terms are of the same order, with molecular diffusion typically being ~25% of the other two terms. The molecular diffusion term has

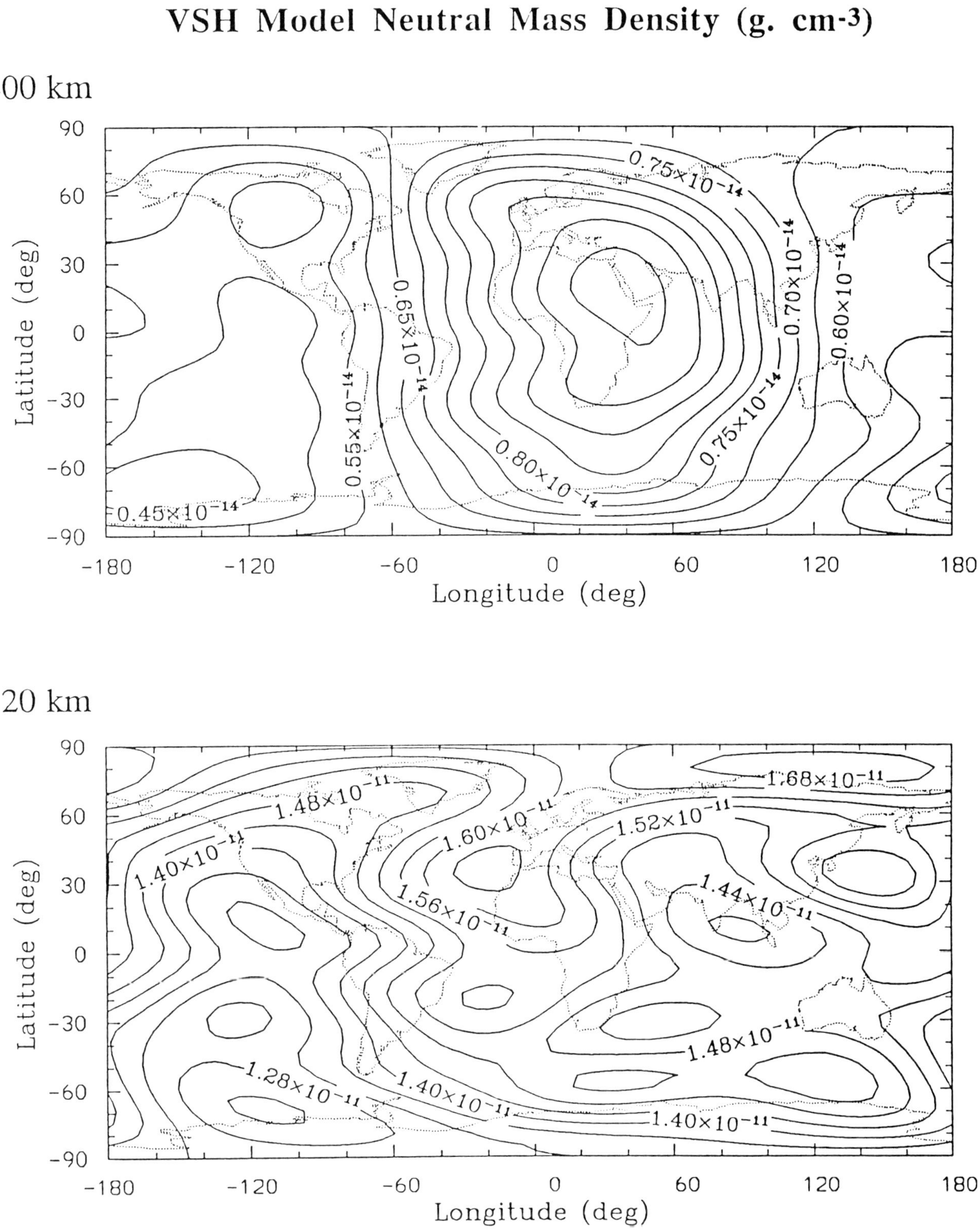

Fig. 6. Modeled total neutral mass density at a) 400 km and b) 120 km altitude. The VSH model was used to generate these plots for 12:00 hrs Universal Time and geophysical conditions corresponding to equinox, moderate geomagnetic activity (*Ap* =12) and moderate solar activity (*F10.7* = 180).

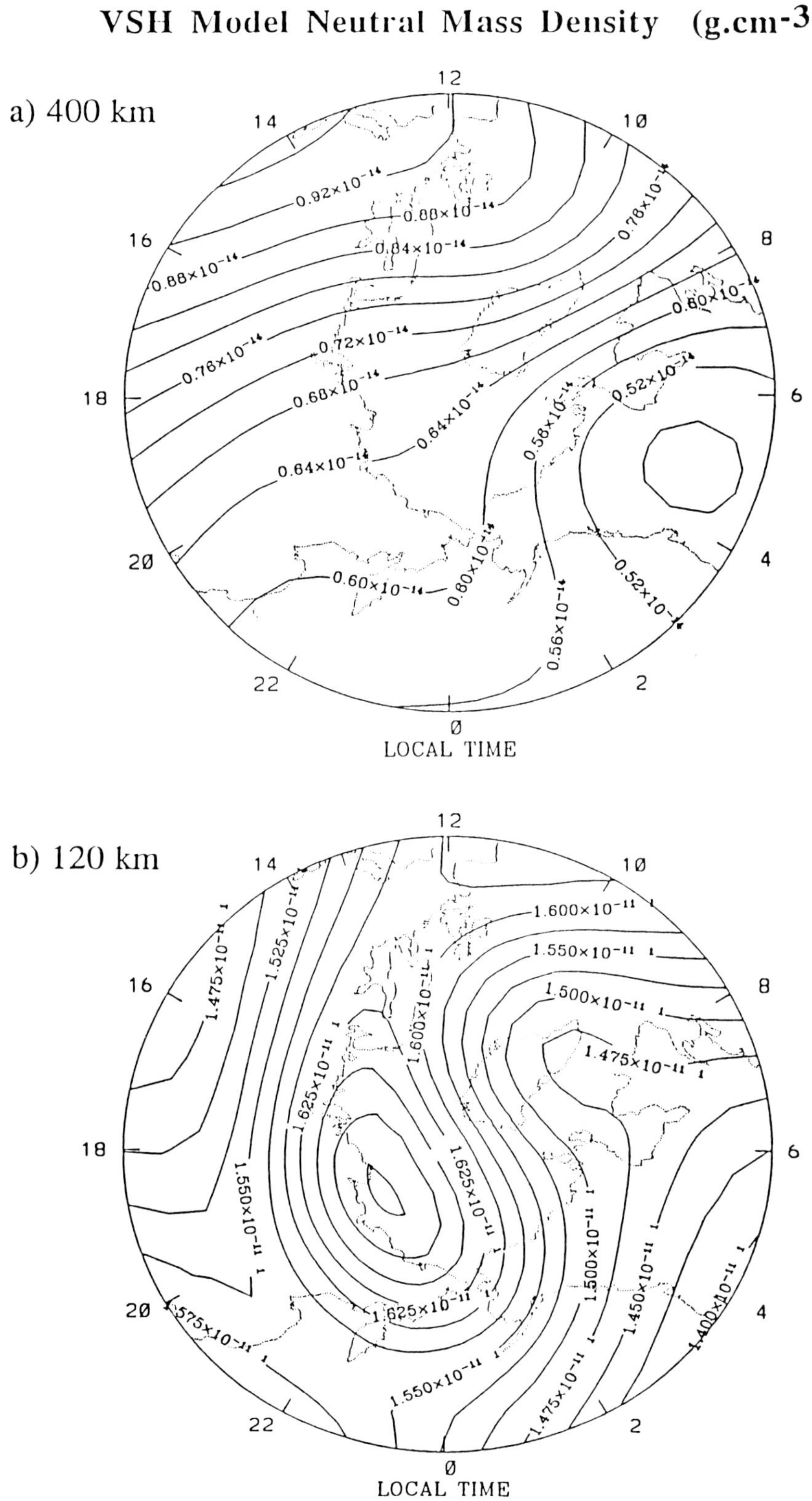

Fig. 7. As for Figure 4 except that the densities are plotted in polar coordinates (geographic latitude and solar local time) for the Northern hemisphere (outer circle at 40° N latitude).

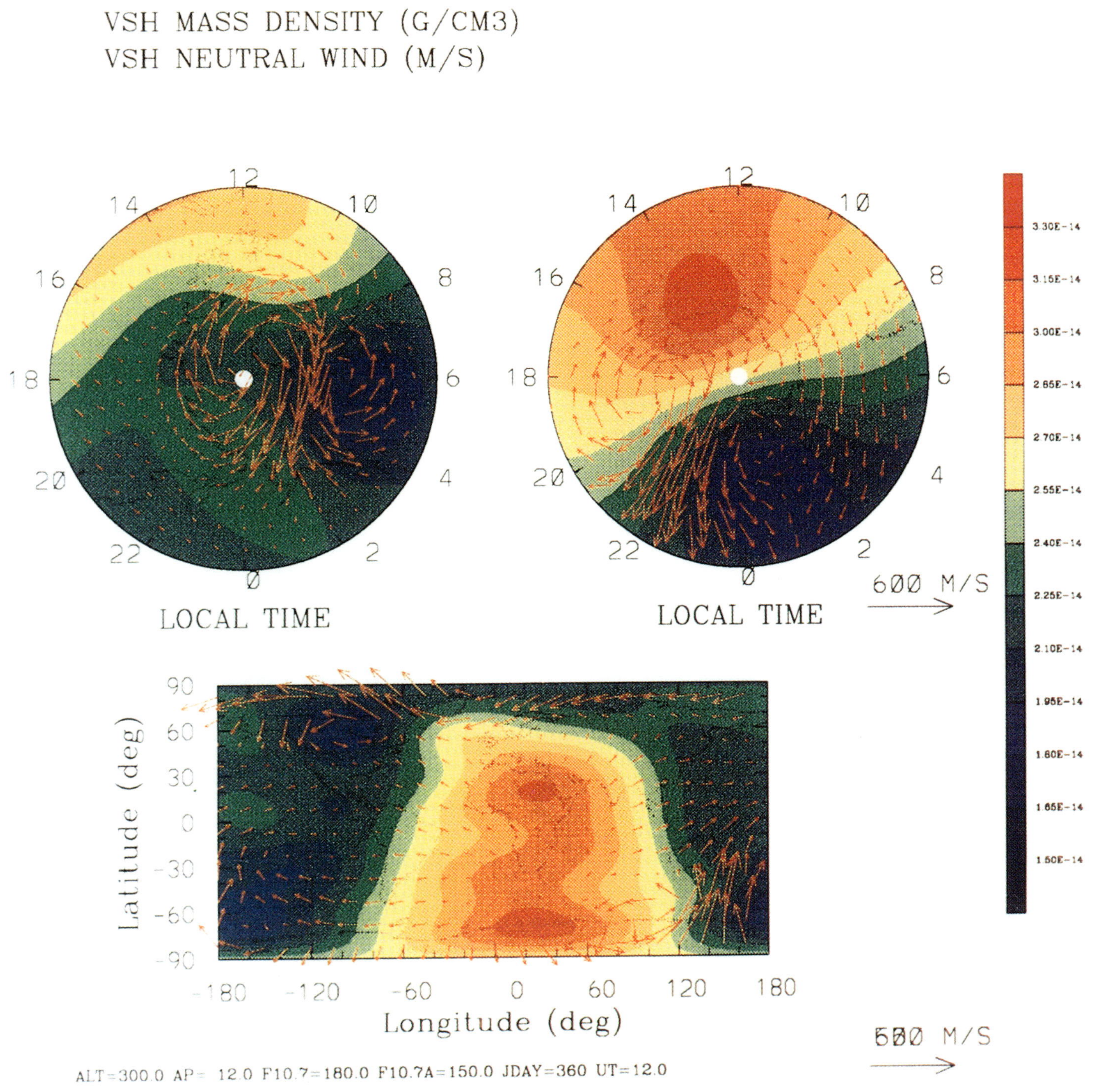

Fig. 8. Neutral density (g.cm^{-3}) and neutral wind fields (m.s^{-1}) at 300 km altitude for 12:00 UT, December solstice condition, with Ap=12 and $F10.7$=180, as calculated by the VSH model. Panels (a) and (b) show the polar projections from 40° latitude to the north and south geographic pole, respectively. Panel (c) shows the corresponding global fields in a cartesian projection.

a generally smaller absolute magnitude in the northern (winter) hemisphere, implying that more significant departures from diffusive equilibrium occur in the summer hemisphere (Figure 9c). Large local compositional effects related to dynamical features can be seen. For example, large positive rates of change of Ψ_{N_2} occur in the post-midnight region of the mid- to high-latitude summer hemisphere as a result of the advection of nitrogen-rich air by the strong horizontal winds (Figure 9e)

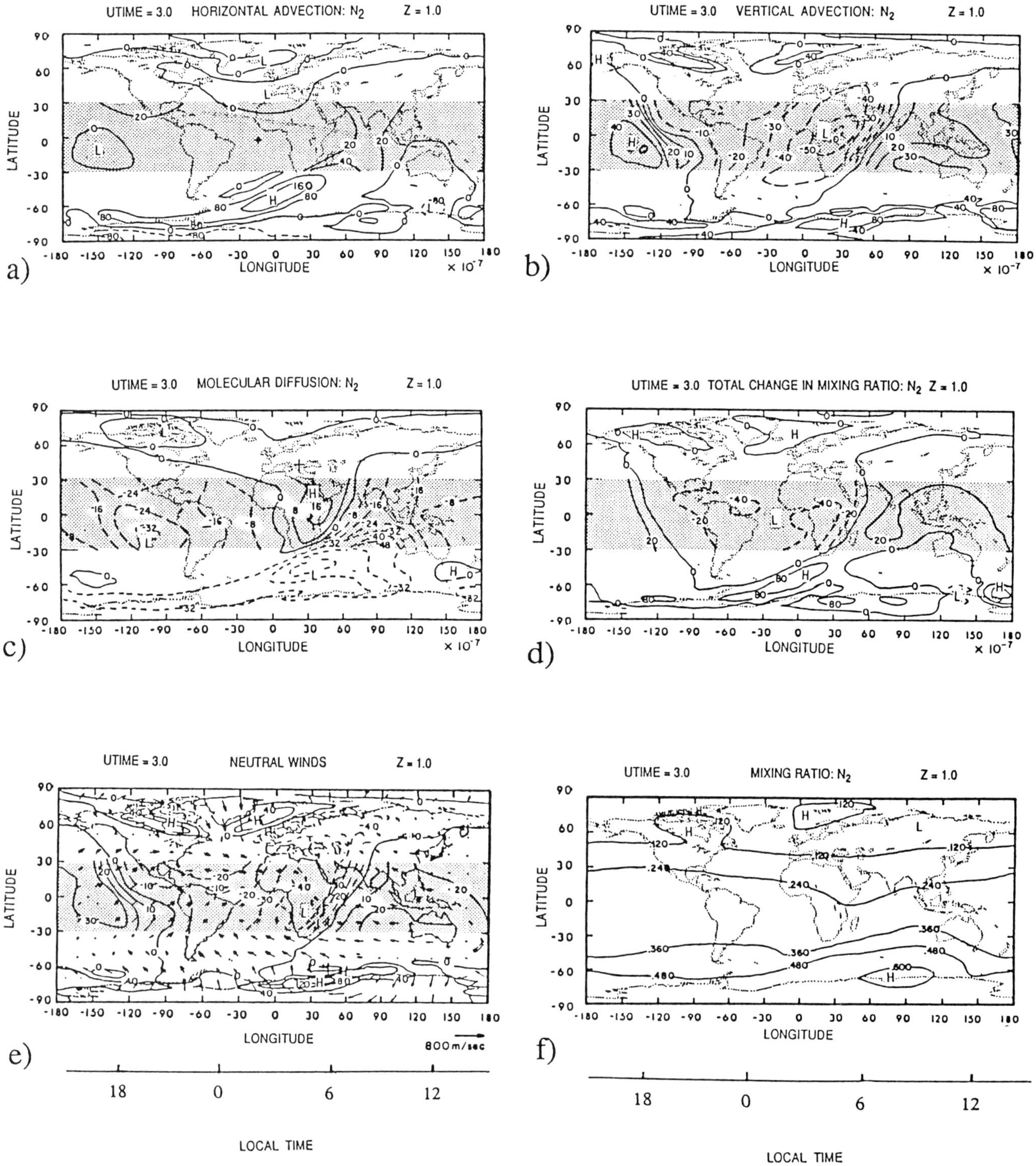

Fig. 9. Calculated individual compositional adjustment forcing terms due to (a) horizontal advection, (b) vertical advection and (c) molecular diffusion terms in units of sec^{-1}. The instantaneous total rate of change of the mass mixing ratio for N_2, given essentially by the sum of these three terms is shown in panel (d). For comparison, the calculated Ψ_{N2} is given in panel (f) and the neutral horizontal wind (arrows, with scale at lower right in m/sec) and vertical wind (contours, with units of 10^6 sec^{-1}) calculated by the TGCM at the same time are shown in panel (e). The vertical "winds" are normalized to the local scale height in order to simplify the relationship with vertical advection on a constant pressure surface and hence the units are sec^{-1}. Figure taken from Burns et al. [1989].

blowing equatorward over the magnetic pole (i.e., near -50 to +30° longitude and -50° to -80° latitude). Such strong horizontal compositional advection is not apparent at corresponding locations (e.g., near +60°latitude) in the mid- to high-latitude winter hemisphere, even though the horizontal neutral winds there are of comparable magnitude. This seasonal asymmetry occurs because the meridional Ψ_{N_2} gradients in the winter hemisphere are weaker than in the summer hemisphere.

Compositional adjustment due to horizontal advection is a less important process in the mid- and low-latitude region, where the horizontal winds are weaker in magnitude and the horizontal gradients of mixing ratio are also small. Changes in the mixing ratio are dominated by vertical advection at these latitudes; downward winds lead to decreases in the mixing ratio of N_2 at night, while the upward winds in daytime advect nitrogen-rich air upwards (Figure 9b). Vertical advection also dominates the changes in the Ψ_{N_2} in the auroral zones, where Joule heating causes strong vertical winds that advect nitrogen-rich air upwards (Figure 9e). Molecular diffusion acts to compensate for departures from diffusive equilibrium caused by dynamical effects over the whole globe.

The net result of these individual compositional forcing terms at mid- and low-latitudes (Figure 9d) results in a relatively simple diurnal variation of mixing ratio, driven primarily by the vertical winds (Figure 9f), associated with the diurnal "breathing" of the upper thermosphere. Cooling at night leads to downward vertical winds and hence decreases in Ψ_{N_2} , while heating during the day leads to upward vertical winds and hence increases in Ψ_{N_2} . At equatorial latitudes the minimum values for Ψ_{N_2} are found near dawn, and the maximum values are found near dusk.

Studies such as these aid in reaching an improved understanding of thermospheric density variations by assessing the relative importance of the various forcing mechanisms. It is to be expected that the improved understanding of the causal physics will aid in the development of improved empirical and semi-empirical models of the region. A primary conclusion is that the density field is controlled through a combination of three major forcing terms which have roughly commensurate magnitudes. These terms are horizontal advection, vertical advection and molecular diffusion.

Density Perturbations

This section discusses some of the primary causes of thermospheric density perturbations. Variations in density of up to an order of magnitude are commonplace, depending on the specific geophysical conditions. Percentage variations tend to grow with altitude, with the greatest perturbations seen at upper thermospheric altitudes.

Gravity waves. Atmospheric gravity waves (waves with gravity acting as the restoring force) are a ubiquitous phenomenon in the upper atmosphere at all altitudes, but particularly at mesospheric and lower thermospheric altitudes where they can dominate the dynamics. Mesospheric gravity waves are primarily of tropospheric and stratospheric origin. Processes, such as flow over mountainous topography, jet streams, earthquakes, volcanoes and strong frontal passages are all capable of generating upward-propagating gravity waves. As the waves move upwards they grow in amplitude in order to conserve energy density, and are filtered by the mean flow. The upward-propagating waves "break" at mesopause altitudes and below, depositing their energy and momentum locally and affecting global energy and momentum budgets, as well as producing turbulence and diffusion. At thermospheric altitudes, the largest gravity waves are primarily of auroral origin, due to the episodic and impulsive insertions of energy and momentum from the magnetosphere.

Gravity waves have typical periods ranging from ~4 minutes to hours, horizontal wavelengths of 10's to 1000's km, and vertical wavelengths of 10's km. Wave production is essentially a stochastic process and, as such, non predictable. Figure 10 shows a decomposition of density wave data obtained from a single orbital pass of the satellite housing the SETA accelerometer experiment of Marcos [1985]. The density data have been filtered using a Behannon and Ness least squares filter to illustrate the variations in specific spatial scales. As can be seen, while the primary variations at altitudes of 170-250km occur on "planetary" spatial scales (>20°), significant variations also occur on smaller scales. Figure 11 illustrates an integrated power spectrum for a full day of SETA data. This graph shows the percentage power occurring at scales smaller than the abscissa value, with spatial scale units of degrees latitude. The vertical dotted lines represent the spatial resolution of the MSIS and TGCM models. As can be seen, the models are only capable of accounting for ~55% and ~70% of the power seen on all orbits occurring on this day, respectively.

A simple scale analysis indicates that the density perturbations due to gravity waves may be approximated by the following:

$$\left|\frac{\rho'}{\overline{\rho}}\right| \sim \frac{\overline{N}}{g}|u'| \sim \frac{|u'|}{500ms^{-1}} \tag{17}$$

where the ρ' represents the density perturbation about the mean value, $\overline{\rho}$, due to the gravity wave passage, $\overline{N}$ represents the Brunt-Vaisala frequency, u' is the perturbation wind velocity and g is the gravitational acceleration. Observations show that the perturbation wind velocities associated with gravity wave passage are typically in the range 20-40 ms^{-1}. Thus density perturbations due to gravity wave passage are normally in the range of 5-20%.

Since the spectrum of gravity waves is unpredictable, it is of importance to determine experimentally their statistical occurrence frequency as a function of space and time. Preliminary studies of this type have suggested that the greatest activity occurs during nighttime hours, in winter, and at high and mid latitudes. The density variations caused by gravity waves are often considered to provide the accuracy "floor" for thermospheric density specification. This is due to the difficulty of assessing their likelihood of occurrence and

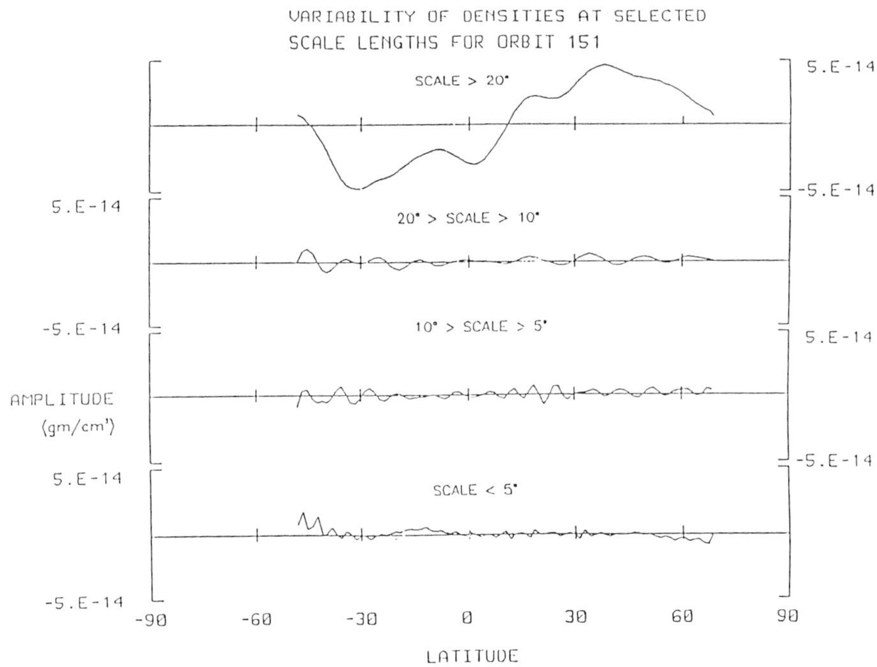

Fig. 10. Measured variations in specific spatial scales as measured using the SETA instrument for orbit 151. Density variations in the specified spatial scale intervals were obtained by normalizing the data to 200 km and passing a Behannon and Ness least squares filter through the spatial series.

the lack of statistical models. It is likely, however, that gravity-wave-induced changes to satellite orbital parameters (which respond to orbit-integrated drag rather than local effects) are relatively small, due to the tendency for the effects of the waves to average out when integrated over large fractions of full orbits.

Tides. Atmospheric tides are coherent oscillations in atmospheric state variables that occur at subharmonics of a solar day. Since the tidal variations are very changeable (i.e. they vary in phase and amplitude greatly on a day-to-day basis) and are overlaid by the rich spectrum of gravity waves, it is usually necessary to derive multi-day averages of experimental

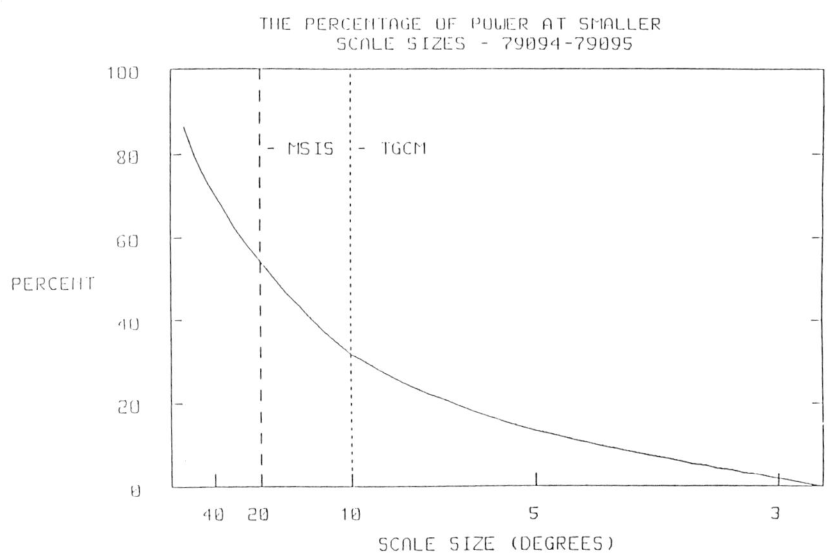

Fig. 11. Integral power spectrum, illustrating the percentage power remaining at scales smaller than the abscissa for the SETA measurements from Julian day 79094-5. Units are degrees of latitude. The dashed and dotted lines represent the MSIS-86 and TGCM spatial resolutions, respectively.

measurements to determine with accuracy the characteristic phases and amplitudes of the various tidal components. Tidal activity at the base of the thermosphere is usually dominated by semi-diurnal (12-hour) waves that propagate upwards and are dissipated by ion drag and viscosity at higher altitudes. The density calculations shown in Figure 6b illustrate tidal variations at 120 km altitude that are primarily semi-diurnal. These tides have complex latitudinal and altitudinal dependences that are strongly seasonally dependent. The calculations at 300 km altitude (Figure 6a) indicate that the upper thermospheric structure is primarily diurnal in nature and is forced in-situ.

Most atmospheric tides are thought to be generated by solar insolation at stratospheric altitudes. The generation of semi-diurnal (12-hour) tides is favored due to the matching of their tropical latitudinal/local time characteristics to the shape of the causative solar heating profiles. The study of tidal climatology is in its infancy [Vial and Forbes, 1989], with data available only from a limited set of ground radar locations [e.g., Manson et al., 1989; Johnson and Martin, 1990; Reese et al., 1990]. The distribution of ground-based tidal data in the 105-200 km region is particularly sparse at present, though planned satellite programs will remedy this deficiency over the next decade.

Since direct measurements of tidal variations in density are not available, we use neutral wind measurements here to illustrate tidal variability as seen from the ground. Figure 12 shows experimental measurements of tidal wind variations observed from the incoherent scatter radar at Sondrestrom for two representative days. The upper panels correspond to geomagnetic quiet conditions. For this case, strong semidiurnal tidal structures can be seen to propagate (diagonally) upwards and dissipate in the lower thermosphere. In the lower two panels, corresponding to geomagnetically active conditions, the tidal forms are swamped by auroral forcings.

Tides have a significant effect on thermospheric density structures at satellite altitudes, as can be seen from a simple scaling analysis. If we assume that tidal temperature perturbations are small compared to the mean background temperature, we may derive the following approximate expression for the density perturbation:

$$\frac{\rho'}{\bar{\rho}} \approx \left\{1-\frac{T'}{\bar{T}}\right\}\exp\left\{-x_m\right\}-1 \tag{18}$$

$$x_m = \frac{-g\bar{m}}{R}\int_0^z \frac{T'}{\bar{T}^2}\,dz' \tag{19}$$

Evaluation of this expression indicates that density perturbations of the order of 10-20% at 120 km may occur due to tides, consistent with the model calculations shown in Figure 6b. At higher altitudes, the tidal forcing processes are replaced by diurnal in-situ forcings due to the direct absorption of solar UV and EUV (see Figure 6a).

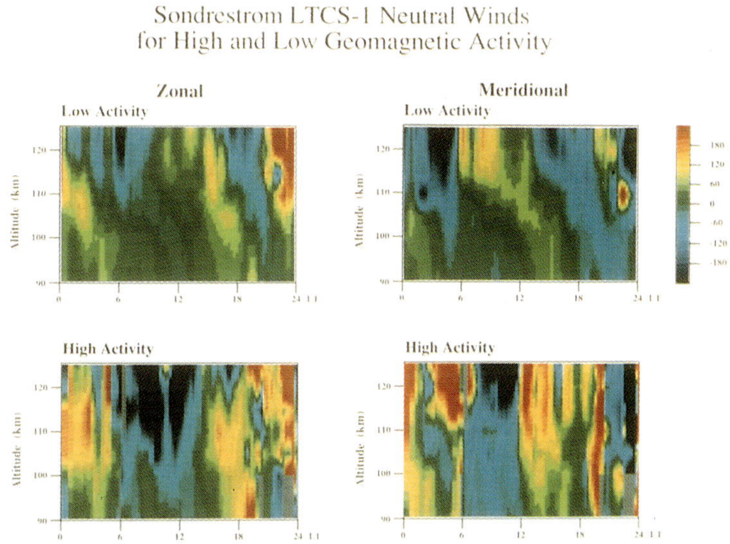

Fig. 12. Experimental measurements of tidal wind variations (zonal and meridional components) observed from the incoherent scatter radar at Sondrestrom for two representative days. The upper panels correspond to geomagnetic quiet conditions. For this case, strong semidiurnal tidal structures can be seen to propagate (diagonally) upwards and dissipate in the lower thermosphere. In the lower two panels, corresponding to geomagnetically active conditions, the tidal forms are strongly perturbed by auroral forcings.

Other density variations . Oscillatory density perturbations of longer periods than those associated with gravity waves and tides are also known to occur. Chief among these, of course, are variations associated with the solar cycle. Figure 13 shows output from the VSH model for an altitude of 300 km at equinox for (a) solar maximum and (b) solar minimum conditions. Density changes of up to an greater than a factor of 5 are seen at this fixed altitude. The solar cycle density ratios are greater at higher altitudes and smaller at lower altitudes, with factors of ~1.7 at 200 km and ~20 at 800 km [Walterscheid, 1989]. The solar cycle takes approximately 11 years, with ~5.5 years between successive maxima and minima. These solar cycle variations are associated with changes in the radiative heating from absorption of the most variable component of the solar spectrum, that is, in the extreme ultraviolet (EUV). Changes of similar magnitude are also observed over much shorter time scales whenever large short-term changes in solar EUV irradiances occur (e.g., in association with solar flares, etc.).

In addition to the fairly well modeled seasonal variation in density, there is a semi-annual variation which is, as yet, poorly understood. This variation is seen in empirical models such as MSIS, and is firmly established by experiment throughout the thermosphere. The NCAR-TGCM and other numerical models, however, have not yet been successful in predicting the observed variation, which is manifested by an October maximum and a July minimum. Figure 14 illustrates some of the relevant semi-annual maximum-to-minimum density ratio measurements that were summarized by Moe and

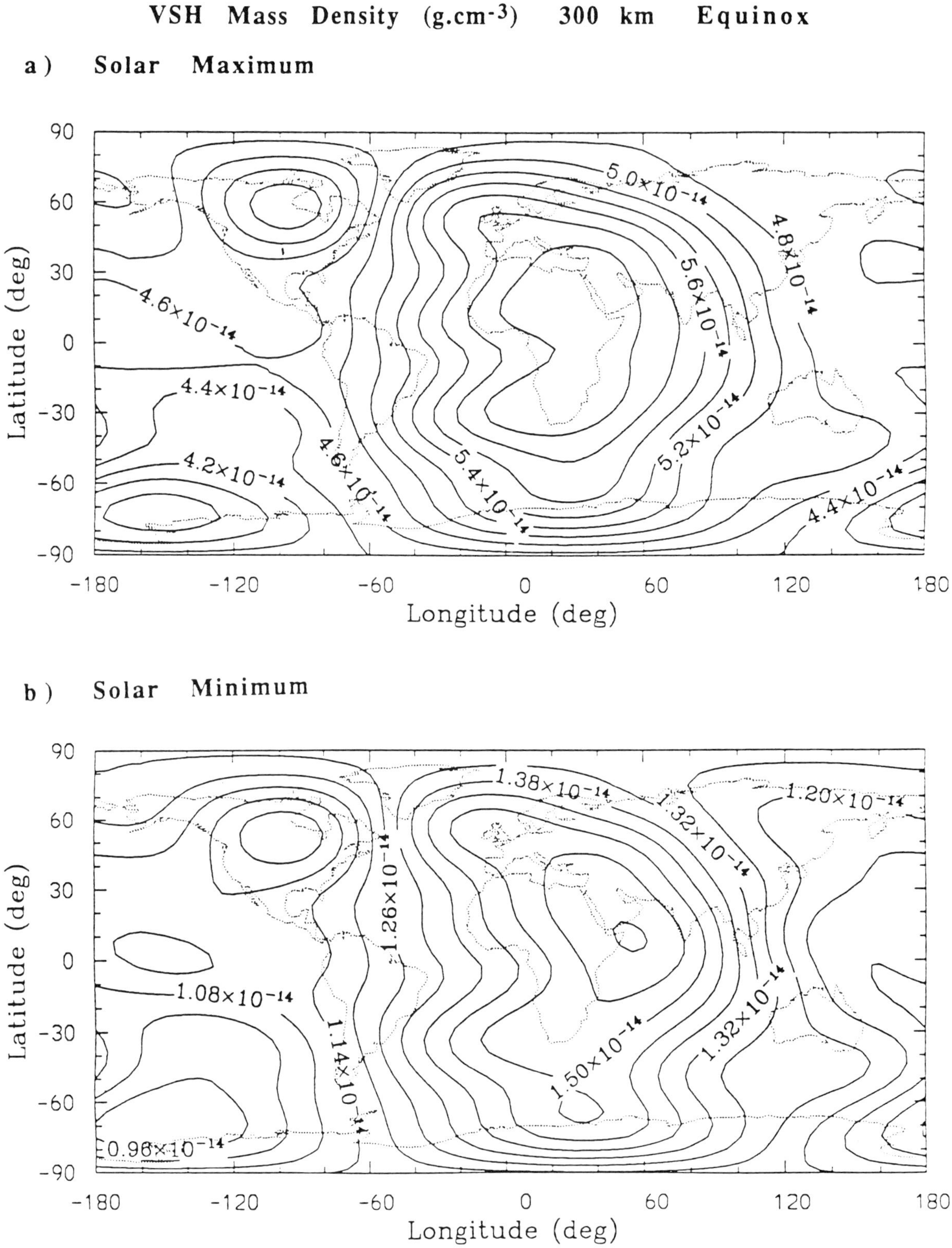

Fig. 13. Solar cycle variations in total thermospheric density at 300 km altitude illustrated by calculations of a) density during solar cycle maximum (F10.7=220) and b) density during solar cycle minimum (F10.7=80). The VSH calculations are for equinox conditions and moderate geomagnetic activity.

Moe [1975]. Figure 14a depicts the results of various experimental determinations of the October-to-July density ratio, $R(z)$, as a function of altitude (see Moe and Moe [1975] and Volland et al., [1972] for the full set of experimental references). Figure 14b shows the results of a simple fit to the semi-annual variation given by (1+ $G(R(z)$ -1)) (see Moe and Moe [1975] for details). As can be seen, the semi-annual variation in density may lead to density variations of ~±20% at 300 km over the period from minimum to maximum. The magnitude of the semi-annual variation is significant for satellite tracking purposes and needs to be taken into consideration in density specification models.

Storm-time variations. Whereas the density variations due to tides and gravity waves are of the order of 10-20%, variations exceeding these values can occur due to impulsive thermospheric forcings associated with geomagnetic storms. Burns et al. [1991] have used the NCAR-TIGCM (a more advanced version of the TGCM) to study the changes in thermospheric composition that result from geomagnetic storms. Storm-time density variations have a complex latitudinal and local time morphology that is dependent on the details of the time history of the storm forcing. Perturbations tend to propagate away from the region of energy insertion, that is from high-to-low latitudes.

Figure 15, adapted from the Killeen et al. [1991] study, shows NCAR-TIGCM calculations of the terms responsible for changes in the mass mixing ratio of N_2 (Ψ_{N_2}) during the large geomagnetic storm that occurred on the 24th of November 1982. The terms are referred to an altitude of ~280 km. The plots show the differences between storm and quiet times for the following terms: a) vertical advective forcing of Ψ_{N_2} ; b) horizontal advective forcing of Ψ_{N_2} ; c) molecular diffusion effects and d) Ψ_{N_2} itself. The first three plots (15a, b and c), showing compositional adjustment forcing terms, are for 14:00 hrs UT, some 3 hours after the commencement of the storm, and the plot showing the mass mixing ratio (15d) is for 19:00 hrs UT. Thus, these plots illustrate some of the physical forcing processes that give rise to the changes in mass mixing ratio at ~280 km a few hours after the commencement of a large geomagnetic storm.

The pattern of compositional forcing that is seen at high latitudes for this storm-time case results from dynamical forcings, both horizontal and vertical advection, where the horizontal and vertical winds are due primarily to ion drag and Joule heating, respectively. Figure 15a illustrates the rate of change of mass mixing ratio due to the advective effects of vertical winds. A region of positive values (associated with upwelling near regions of enhanced heating) can be seen to be roughly coincident with the location of the auroral zone and regions of negative forcing (associated with subsidence) lie equatorward of the auroral zone and within the geomagnetic polar cap.

Figure 15b illustrates the rate of change of Ψ_{N_2} due to the advective effects of horizontal winds. The region of large positive values in the post midnight sub-auroral sector is due to the existence of the postmidnight increase in the meridional winds which tends to increase the mass mixing ratio of nitrogen downstream from the polar regions. Figure 15c illustrates the rate of change of Ψ_{N_2} due to molecular diffusion. This term tends to be negative throughout the high-latitude region as it attempts to restore the thermosphere to the diffusive equilibrium that has been interrupted by the high-latitude dynamical processes. Maximum values of the perturbed Ψ_{N_2} (Figure 15d) exceed ~30% and are found in the auroral and sub-auroral morning sector. The magnitudes of the three terms shown in Figure 15a, b and c are roughly commensurate, indicating that all three processes (vertical and horizontal advection and molecular diffusion) are of roughly equal importance in the establishment of the storm-time compositional state of the high-latitude thermosphere. This analysis illustrates how geomagnetic storm-time variations are primarily forced by the large scale winds, both vertical and horizontal, that are established by the high-latitude forcing processes of ion drag and Joule heating.

The storm-time density perturbations may reach values as large as 100% or larger at altitudes associated with low-Earth orbiting vehicles, and their effects can propagate globally with varying time constants. A recent example of equatorward propagation of storm-time density perturbations was presented by Burns and Killeen [1992]. They conducted a superposed-epoch study of several geomagnetic storms, using data from Dynamics Explorer 2 (DE-2). Figure 16 shows the averaged low-latitude density changes (storm-to-quiet ratios) at 350 km in the evening sector (18:00-20:00 hrs LT) as a function of "storm time" and latitude from the Burns and Killeen study. The most noticeable feature in these density plots is the large enhancements of density that are seen 3 to 6 hours after the onset of the storm. These calculated perturbations are associated with large scale gravity waves (sometimes known as traveling ionospheric disturbances or TIDs) that travel at the local speed of sound away from high-latitudes (in both hemispheres) and towards low latitudes where they can meet and interfere with each other. The enhancements are followed by density depletions 6-12 hours after the onset of the storm for N_2 and 6-9 hours after the onset of the storm for O, indicating a large-scale wavelike response at equatorial latitudes. The magnitude of the equatorial post-storm-onset density perturbations was about 40% for N_2 densities and about 25% for O densities. The perturbations observed by Burns and Killeen were seen to return to their quiet time values some 12 hours after the onset of the storm.

These results show that density perturbations of several 10's of percent can occur in response to geomagnetic storm-time forcings, even at equatorial latitudes and for up to 12 hours after the storm onset. Density specification models have to cope with the difficult problems involved with phase and amplitude prediction for such traveling perturbations. The magnitudes of the variations are significant, particularly at high latitudes.

4. Atmospheric Wind Variations

We have discussed how variations in thermospheric winds play a major role in density variability, through their

Semi-Annual Thermospheric Density Oscillation:
Ratio of October maximum to July Minimum

a)

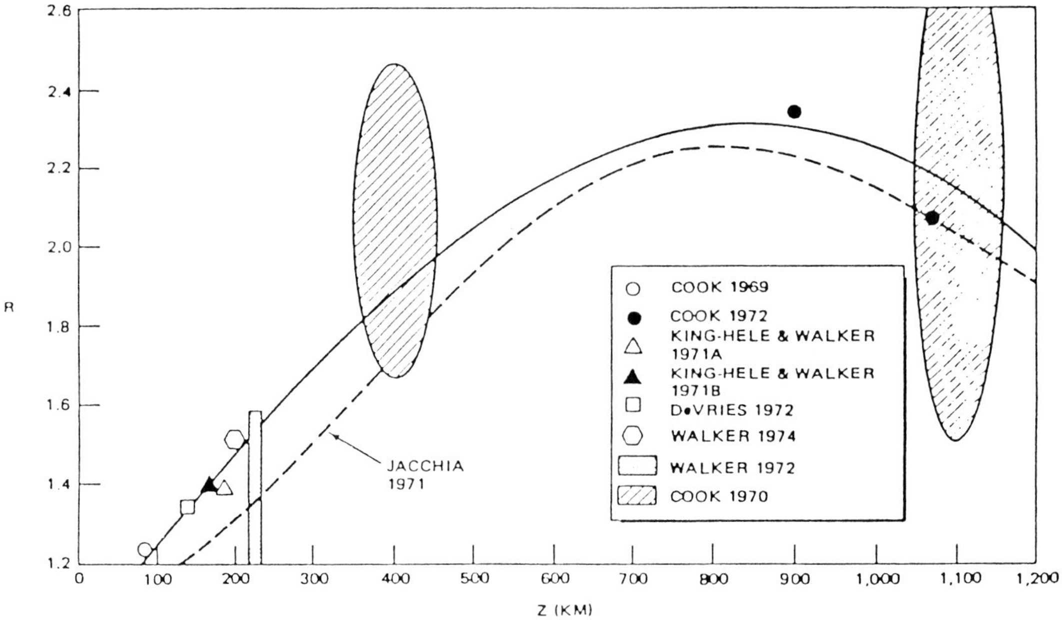

Density multiplicative factor as a function of Z (altitude):
[1 + (R(Z) - 1) G(D)]

b)

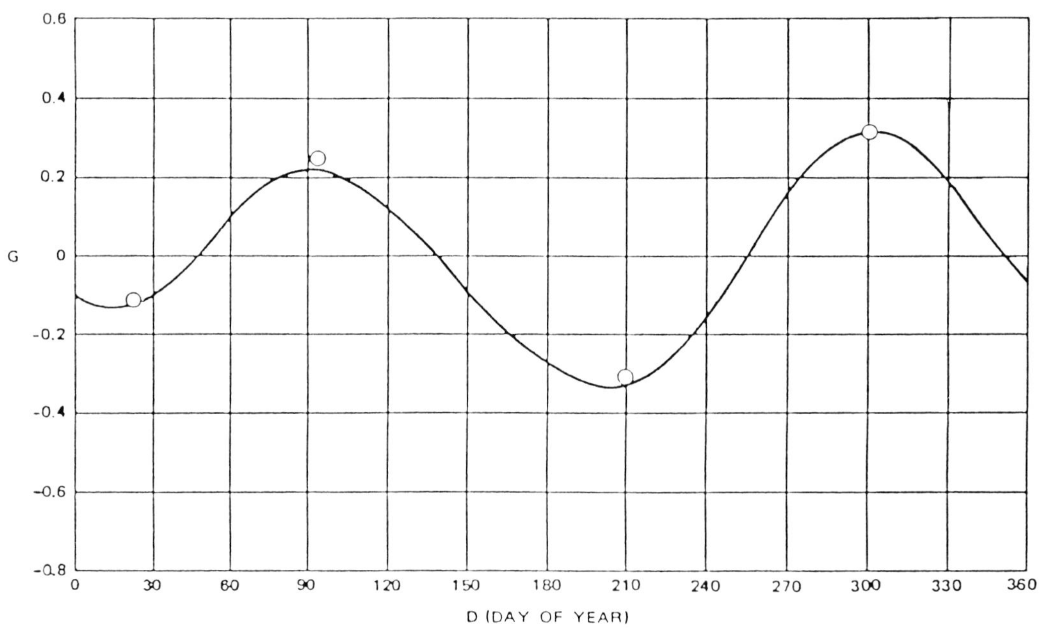

(Moe, Jacchia)

Fig. 14. (a) Experimental estimates of the ratio between the October maximum and the July minimum of the semi-annual density variation as a function of altitude. The original references may by found in the report by Moe and Moe [1975]. (b) Fit for the temporal variation factor, G, as a function of month. Figure reproduced with permission from McDonnel Douglas.

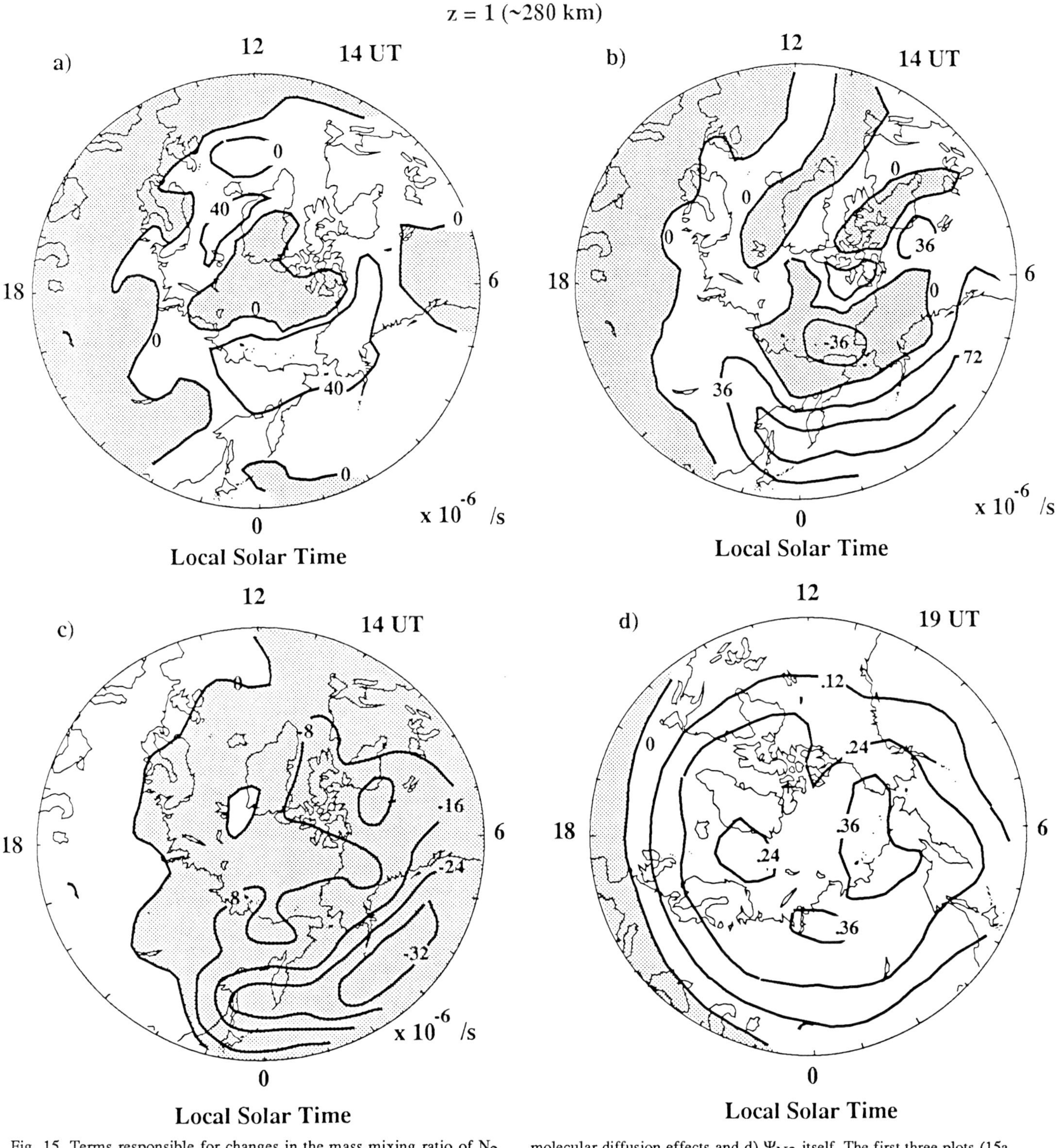

Fig. 15. Terms responsible for changes in the mass mixing ratio of N_2 (Ψ_{N2}) during the large geomagnetic storm that occurred on the 24th of November 1982 as calculated using the NCAR TGCM. The terms are referred to an altitude of ~280 km. The plots show the differences between storm and quiet times for the following terms: a) vertical advective forcing of Ψ_{N2}; b) horizontal advective forcing of Ψ_{N2}; c) molecular diffusion effects and d) Ψ_{N2} itself. The first three plots (15a, b and c), showing compositional adjustment forcing terms, are for 14 UT, some 3 hours after the commencement of the storm, and the plot showing the mass mixing ratio (15d) is for 19 UT or two hours into the storm simulation.

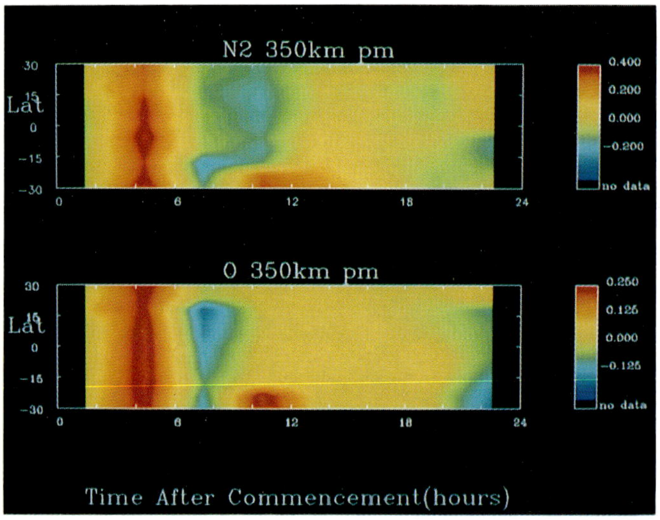

Fig. 16. This figure shows N_2 (top) and O (bottom) density changes that result from a storm in the evening local time sector, using a superposed epoch analysis based on DE-2 data from 8 storms. The vertical axis is the geographic latitude, with the summer hemisphere on the bottom, while the horizontal axis is the time after the onset of the storm. The color scale is given in the fractional change between storm time and the control case.

advective effects on density and composition. Thus, any real understanding of the causes of satellite drag variability requires an associated understanding and description of global winds. The neutral winds, however, also have a *direct* effect on satellite drag through the quadratic dependence indicated in Equation (1) above. This effect is generally small since the in-track wind components are normally a small fraction of satellite velocities. When these winds approach values in excess of 500 m s^{-1}, however, their direct effect on drag quickly becomes significant. A wind of 1000 m s^{-1}, for example, causes a 23% or 27% change in drag for a satellite traveling at 8000 m s^{-1}, depending on whether the wind is blowing along or against the satellite velocity vector. High speed winds such as these, though not frequent, have been measured at high geomagnetic latitudes during geomagnetic storms. The reader is referred to recent reviews on global thermosphere dynamics by Killeen and Roble [1988] and Crowley [1991] for detailed discussions of our current understanding, and for references of work on thermosphere dynamics.

To demonstrate extreme wind variability, Figure 17 shows results from a study of the neutral wind response to the geomagnetic storm of November 28, 1982 [Killeen and Roble, 1988]. Neutral wind measurements from seven orbits of the DE-2 spacecraft over the northern-hemisphere polar region are plotted as a function of the UT of the satellite passage over the north pole and angle along the satellite track (90° AAT at the north geographic pole). These orbits occurred just prior to and following the onset of a major geomagnetic storm. Following the storm sudden commencement, near 09:00 hrs UT, the neutral wind field responded dramatically to the strong magnetospheric forcings associated with the storm (enhanced precipitation, large convection electric fields, expanded auroral oval, etc.). Wind speeds reached magnitudes in excess of 1000 m.s^{-1} for orbit 7217. This example demonstrates how large and rapid perturbations of the high-latitude thermosphere can occur as a function of enhanced and varying levels of geomagnetic activity.

Neutral wind perturbations, such as those shown in Figure 17, clearly influence satellite drag at the 20-30% level for orbits where the component of the wind in the direction of the spacecraft velocity vector is large. Since the enhanced wind speeds are relatively localized (perhaps ~20 degrees in latitude) and normal wind speeds are typically much less than for the storm situation depicted in Figure 17 (cf. Figure 8), the effects on satellite drag are usually small. Figure 18 shows the orbital drag perturbations for a DE-2 orbital pass during a geomagnetic storm whose local time plane was selected so as to have the greatest effect (02:00 - 14:00 LT plane). Drag excursions of up to 20-30% are evident that are localized to the high-latitude regions where neutral wind magnitudes are the greatest. From these and other studies it may be concluded that, while in-track winds can lead to localized drag perturbations of up to 25-30% at times, these excursions occur over relatively small regions at high latitudes, and their effects on orbital-averaged drag are likely to be small.

5. Atmospheric Density and Wind Specification Models

Rapid progress in understanding and theoretically modeling the state variables of the Earth's upper atmosphere (temperature, density, wind, and composition) has been sustained over the past decade, with several coordinated experimental programs, such as the NASA Atmosphere Explorer (AE) and Dynamics Explorer (DE) missions and the U.S. Air Force Space Test Program, providing important global-scale experimental data sets in various altitude regions. The fundamental theoretical understanding of the upper atmosphere has greatly benefitted from these programs, and a new generation of sophisticated numerical models (thermospheric general circulation models, TGCMs) has been tested and constrained by the global data sets thereby obtained.

The morphological description of the structure of the upper atmosphere has also improved due to the continued development and enhancement of semi-empirical models, such as the Mass Spectrometer-Incoherent Scatter (MSIS) model [Hedin, 1983; 1987; Hedin et al., 1991], and the earlier Jacchia models based on satellite drag measurements [Jacchia, 1965; 1970; 1971; 1977]. Operational models of thermospheric density and temperature are currently based on such semi-empirical models. These models reflect the mean behavior of the thermosphere as described by the large bodies of data from the various previous experimental programs. Semi-empirical models use analytical functions to fit to randomly-selected subsets of the available data, generating spectral coefficients that can be used conveniently to reconstitute thermospheric densities and temperatures as a

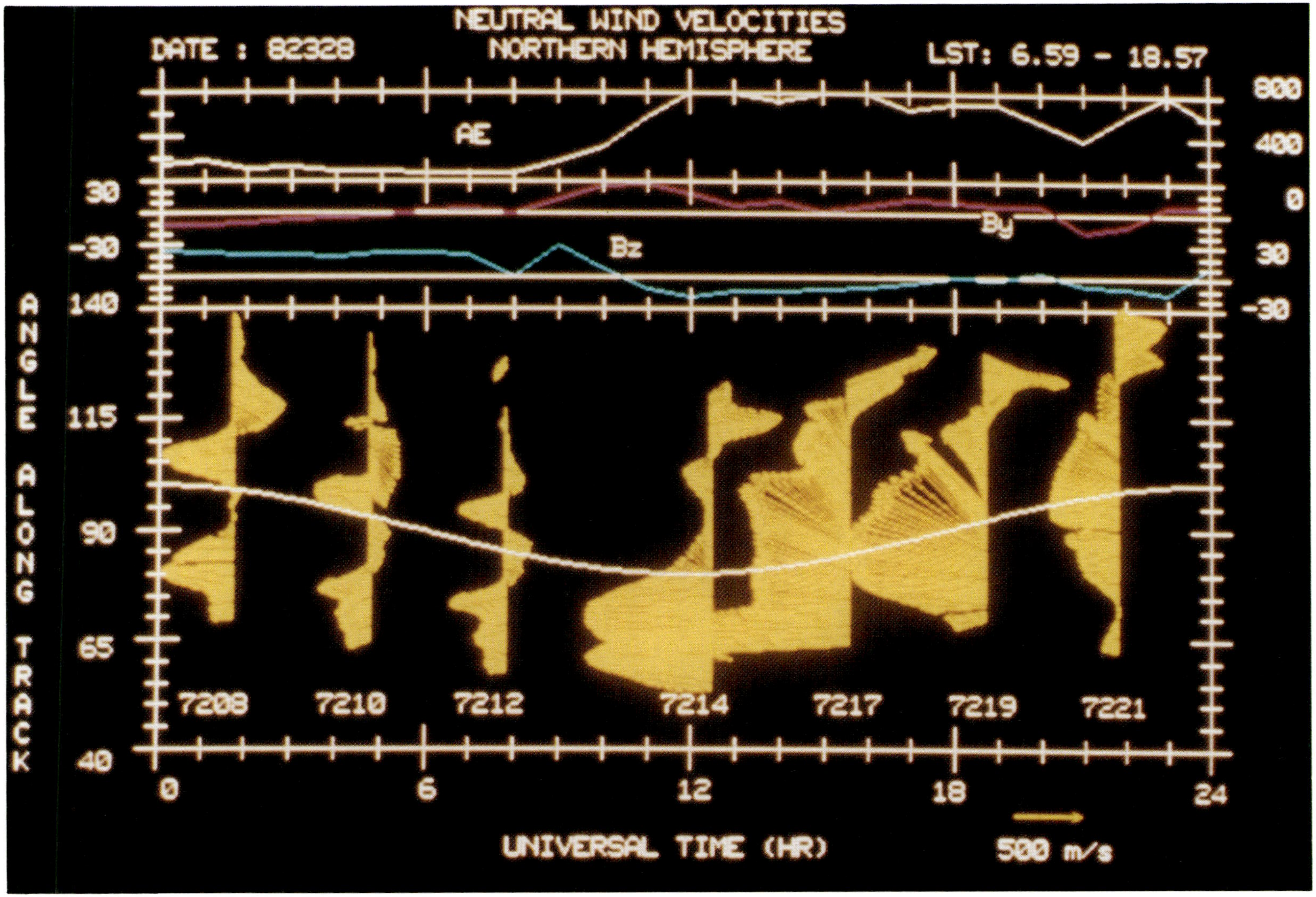

Fig. 17. Neutral wind observations made on the seven referenced DE 2 orbital passes over the northern hemisphere polar region just prior to and following the major geomagnetic disturbance of November 28, 1982. The winds are plotted such that the sunward direction is toward the right of the figure and dusk toward the top. Top panels show the variation of the auroral electrojet (*AE*) index and the interplanetary magnetic field (IMF) north-south (B_z) and east-west (B_y) components during the period.

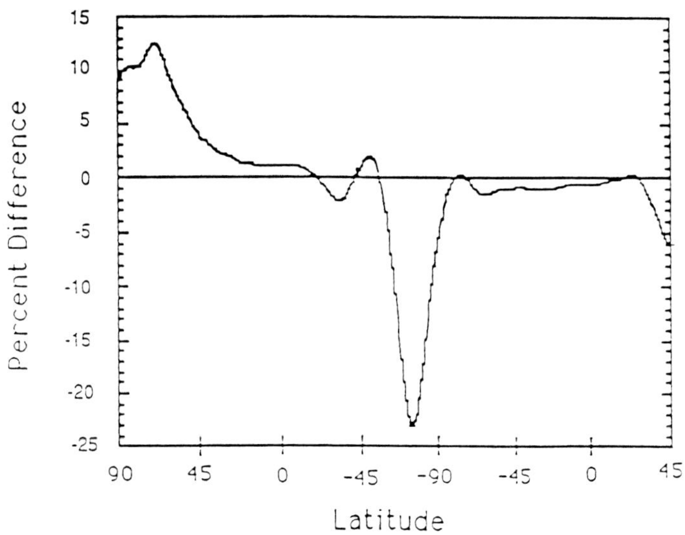

Fig. 18. Orbital drag perturbations for a DE-2 orbital pass on July 14, 1982 (F10.7=269; Ap=153) during a geomagnetic storm. The orbit was selected since its orbital plane (02:00 - 14:00 LT) was such as to have the greatest effect on drag. Excursions of up to 20-30% are evident that are localized to the high-latitude regions of enhanced neutral wind.

function of space, time, and geomagnetic and solar activity levels.

Unfortunately, the combination of progress in scientific understanding and progress in quantitative semi-empirical modeling, that was described briefly above, has not yet led to operational models of thermospheric density (providing mean values or standard deviations about the mean) with accuracies of much better than ~15% in the altitude range 100-300 km ([Hedin, 1987], see Figure 5). Furthermore, recent efforts to improve the accuracy of the operational models by incorporating additional experimental data (e.g., the DE neutral compositional data set) into the semi-empirical models have not markedly improved the situation [Marcos, 1985; Hedin, 1987; Hedin and Mayr, 1987]. The operational model accuracies are somewhat better at low latitudes and for geomagnetic quiet times (typically <10%), but density uncertainties at high latitudes during geomagnetic storm periods often exceed ~50%.

The reasons for the rather unsatisfactory performance of the operational models are complex. In general, the semi-empirical models themselves are limited by several factors including: 1) the inherent accuracy (or lack thereof) of the contributing sensors providing data for the model fitting procedures; 2) possible errors or inappropriate assumptions in the extrapolation procedures used to interpolate between data points and to extend the range of the empirical model to altitude regions where the data are sparse; 3) limited temporal and spatial resolution in the formulation of the empirical model (for example, MSIS-86 uses 3-hourly Kp values to determine the geomagnetic activity effect which observations show can be much more rapid in reality); 4) limited experimental coverage; and 5) the inherent stochastic nature of atmospheric variability, which might be expected to provide a fundamental limit on the accuracy with which any given geophysical situation can be quantitatively modeled. Any serious attempt to improve the accuracy of semi-empirical

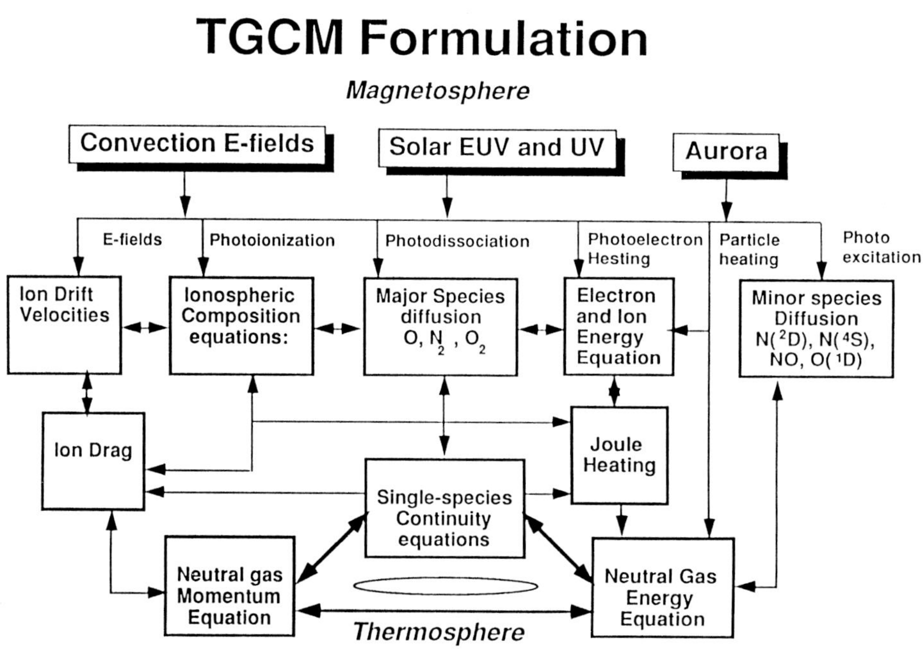

Fig. 19. Schematic illustrating the principal components of the NCAR TGCM (after Roble et al. [1988]). The primary magnetospheric inputs to the model are shown at the top of the diagram. Thermospheric variables are calculated as a function of space and time by solving the three conservation equations shown at the bottom of the diagram.

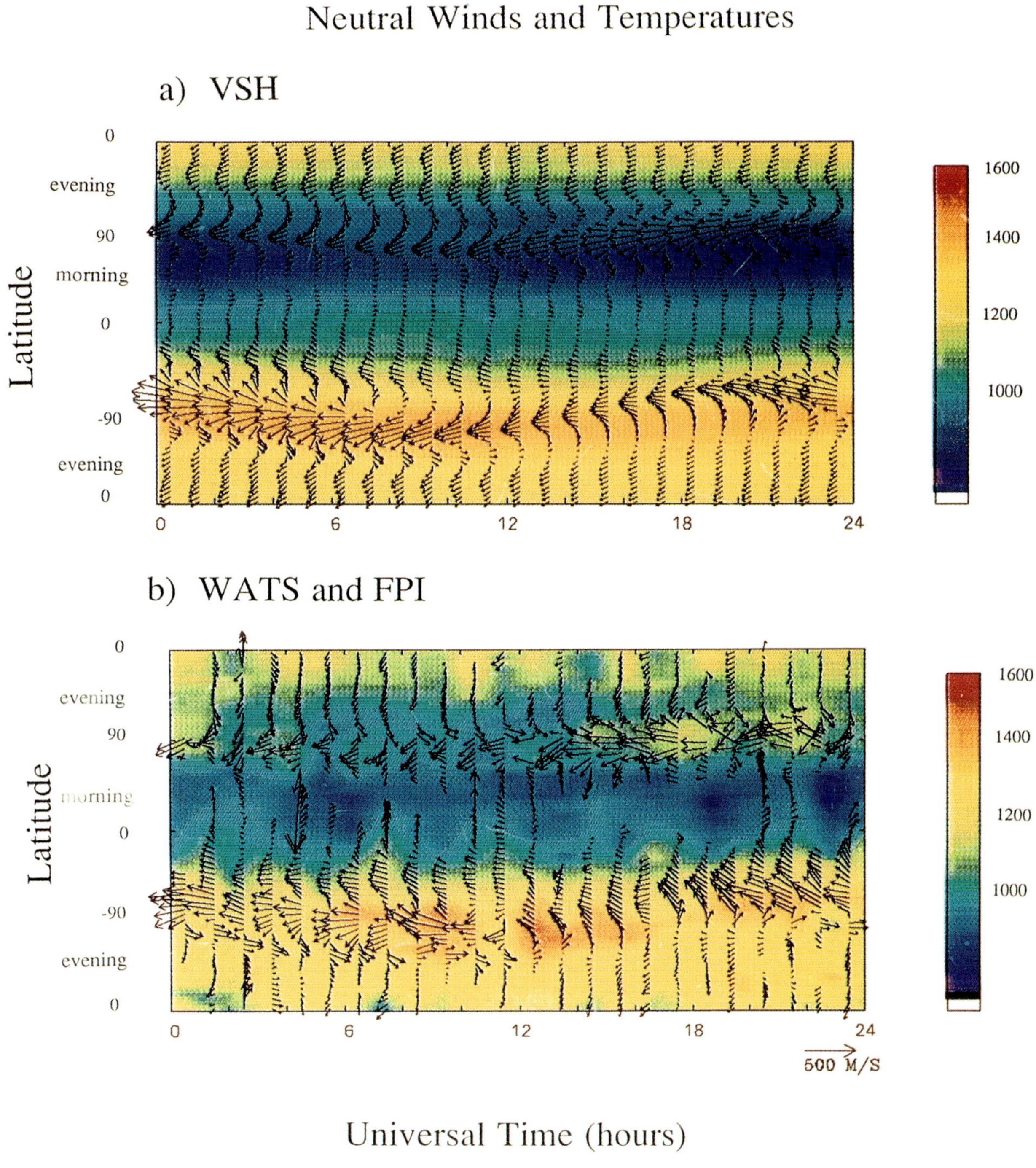

Fig. 20. Thermospheric wind and temperature plots for December solstice solar maximum conditions. (a) VSH model predictions; (b) averaged DE-2 winds and temperatures collected between November 1981 and January 1982. The plots are in geographic latitude and Universal Time.

models of thermospheric density will have to discuss the relative importance of these various factors.

In recent years a new approach to density specification modeling has attained some measure of success in reducing errors. This is the Vector Spherical Harmonic (VSH) model [Killeen et al., 1987]. The VSH model is a semi-empirical "hybrid" model that is based on numerical simulations of the NCAR-TGCM and available experimental data from DE-2 and SETA. The NCAR-TGCM solves the primitive equations of dynamic meteorology adapted to thermospheric altitudes, including equations for the conservation of energy and momentum and the individual species continuity equations. The physics inherent in the model's parameterizations and input prescriptions are those appropriate to thermospheric altitudes. Figure 19 is a block diagram that illustrates the major inputs to and internal calculations performed within the TGCM. Not shown are the forcings from below, due to the tidal and gravity wave processes described above. The basic TGCM model and subsequent major developments have been described previously [Dickinson et al., 1981; Roble et al., 1982; Dickinson, 1984; Fesen et al., 1986; Roble et al., 1988]. Most recently, the model was further extended [Roble et al., 1988] to include a self-consistent aeronomic scheme for the thermosphere and ionosphere.

The VSH model uses output from various "generic" runs of the TIGCM to develop a library of truncated spectral coefficients that represent the global thermospheric wind, temperature and neutral composition fields for a set of different geophysical conditions (solar maximum, solar minimum, equinox, solstice, etc.). The 3-dimensional wind field is represented by a vector spherical harmonic expansion in the horizontal, a Fourier expansion in Universal Time and a polynomial expansion in altitude. The global temperature field differs in that a scalar spherical harmonic expansion is used in the horizontal and cubic splines are used in the vertical. Neutral composition fields are described using logarithmic scalar harmonic expansions. The thermospheric state variables (wind, temperature and composition) may then be calculated simply using a computer subroutine which reads in the coefficients and regenerates the required field as a function of space and time and geophysical condition (F10.7 value and *Ap*). Recently, neutral-wind data from DE-2 and density data from SETA instruments have also been incorporated into the VSH model description using an objective analysis scheme to merge the experimental data and theoretically-calculated TIGCM gridded output fields. In this objective analysis scheme, measurements are used whenever available, and TIGCM model results are used in the absence of experimental data, with the whole process being carried out in the spectral domain to provide for smooth merger of the experimental and theoretical results.

VSH wind and density predictions

An example of a comparison between the global output of the VSH model and averaged DE-2 data is shown in Figure 20 [Killeen et al., 1990]. The lower part of this figure (20b) shows the averaged neutral wind field from DE-2 as a function of UT and latitude, for data collected between November 1981 and January 1982. The thermospheric temperatures from DE-2 are coded as colors according to the scale at the right. As can be seen, the largest wind speeds are observed in the two high-latitude regions (southern hemisphere polar regions towards the lower part of the plot and northern hemisphere polar regions towards the upper part). The measured thermal structure indicates a relatively hot summer polar region (temperatures up to ~1500K) and a relatively cold low latitude region (blue band in middle of plot - temperatures ~900K). Figure 20a shows the equivalent output from the VSH model, where the

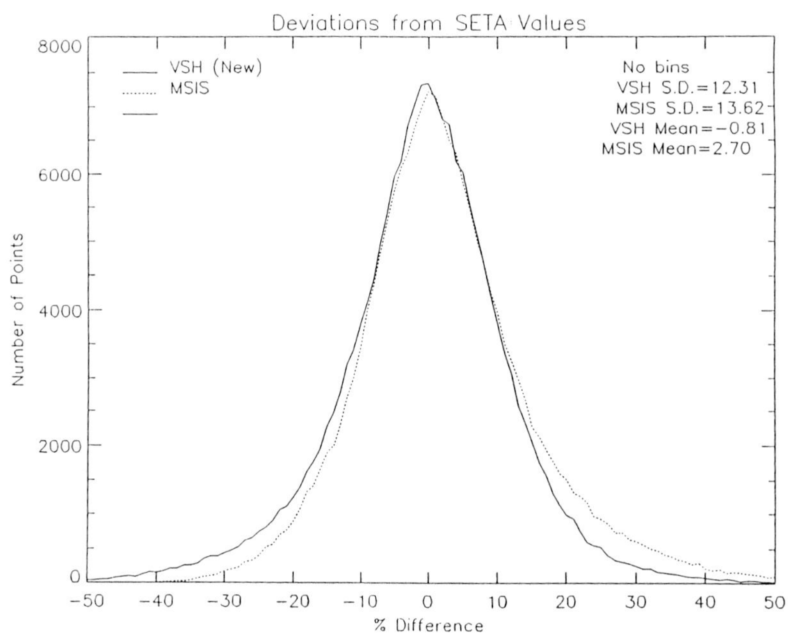

Fig. 21. Statistical comparison of density predictions by the VSH model (dotted line) and the MSIS-86 model (solid line) for the empirical data set of 186,000 individual measurements from the SETA-2 instrument. The data encompasses all latitudes and is concentrated at lower thermospheric altitudes. The distributions are plotted as a function of percent difference (model minus data, normalized to data). The calculated moments of the distributions are shown at upper right.

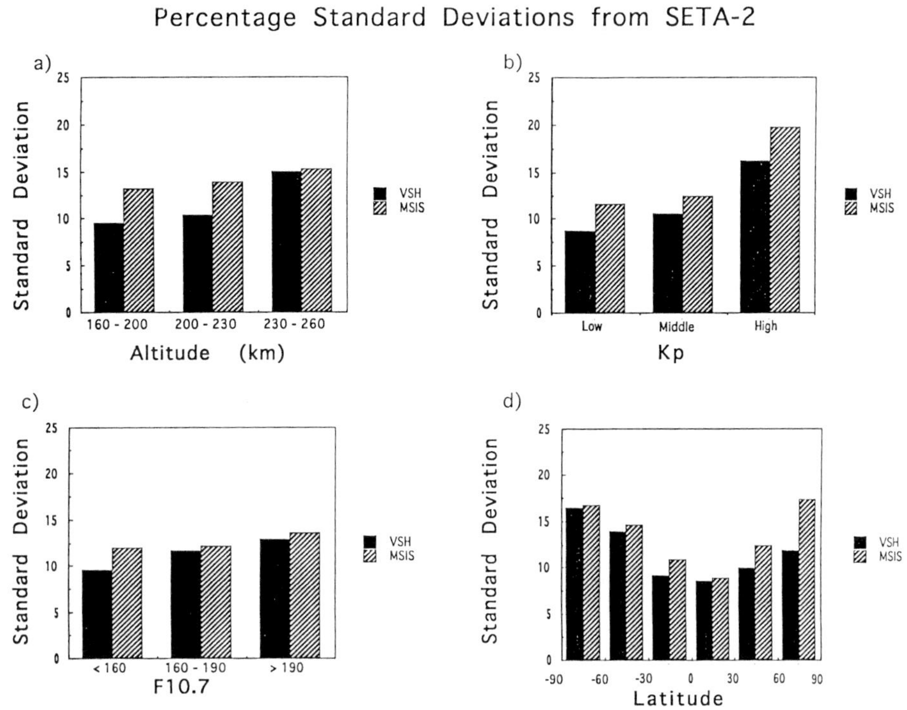

Fig. 22. Relative accuracies of VSH and MSIS-86 for density comparisons with the SETA-2 data base as a function of a) altitude; b) geomagnetic activity level (Kp); c) solar activity level (F10.7); and d) latitude. Plotted are the are percentage standard deviations from the mean for each bin.

model was interrogated to provide calculated December solstice wind and temperature fields in the same format as the experimental data. As can be seen, there is reasonable agreement between the model and the global-scale experimental data on the broad scale, with evident discrepancies at smaller scales. In particular, the experimental data shows a relatively hot northern hemisphere (winter) polar cap of ~1200K between 16:00 and 24:00 hrs UT, while the model predicts a significantly cooler region (by <200K). The model predicts the locations of the high-latitude wind reversals from sunward to antisunward flow with reasonable accuracy, but does not mimick the degree of localized variability seen in the measurements.

The VSH density predictions have also been compared extensively with experimental accelerometer results from the SETA family of instruments. A total of 186,000 individual density measurements were used to test the accuracy of the VSH and MSIS-86 models for density specification. In order to make the comparison with a totally independent set of experimental results, a version of the VSH model was used that contained no SETA densities. Thus, the VSH densities in this case are entirely reflective of the TIGCM calculations used to derive the VSH model. Figure 21 shows the results of the statistical comparison. The quantity plotted is the percent deviation for each model given by:

$$\frac{\rho_{MODEL} - \rho_{SETA}}{\rho_{SETA}} x\ 100$$

As can be seen, both VSH and MSIS models provide overall accuracies of ~12-13% for density, given by the calculated standard deviations of the two distributions. For the purposes of a more detailed statistical comparison, the data were binned according to level of geomagnetic activity and latitude interval. Figure 22 shows the calculated standard deviations for each bin as a function of a) altitude; b) geomagnetic activity level (*Kp*); c) solar activity (*F10.7*); and d) latitude. The accuracies of the models are best at mid- and low-latitudes and during geomagnetically quiet times, as would be expected. The VSH model performance is roughly equivalent to MSIS-86 for most bins, with substantial improvements at high latitudes and for both high and low geomagnetic activity. Studies such as these demonstrate that the hybrid VSH model may lead to significant improvements in density specification.

6. Conclusions

Perturbations to the orbital elements of satellites flying in low Earth orbit occur due to changes in thermospheric density and wind that are caused by a number of processes. These processes include gravity waves, tides, solar cycle, seasonal, and semi-annual oscillations, as well as storm-time processes. Satellite drag is linearly dependent on atmospheric density and quadratically dependent on the difference between the spacecraft velocity and the in-track wind. This paper has reviewed many of the principal causes of variations in satellite drag from the perspectives of both specification models and experimental measurements.

Gravity waves are non-predictable and can lead to local density fluctuations of 10's of percent. Semi-diurnal tides lead to changes of similar magnitude in the lower thermosphere, but are dissipated rapidly higher in the thermosphere, where the diurnal tide dominates. A detailed climatology of global tides is needed to determine these variations more accurately.

Geomagnetic storm time effects typically cause larger density perturbations than those associated with tides and gravity waves; the magnitudes of the perturbations depends on altitude and latitude and a number of other geophysical variables. Neutral winds can also significantly affect local drag (up to ~30%) in regions associated with the high-latitude wind jets.

The new VSH thermospheric density specification "hybrid" model, based on a combination of numerical model and empirical data offers some potential for attaining improved accuracies for density specification and prediction.

Acknowledgements. The authors are grateful to Dr. R. G. Roble and colleagues for providing access to the NCAR TGCM and for numerous discussions. This work was supported by Phillips Laboratory contract AFGL F196288-89K and by NSF grants ATM-8918476, ATM-9096134 and ATM-912115 and by NASA grants NAG5-465 and NAGW-1535 to the University of Michigan.

REFERENCES

Burns, A.G. and T.L.K.illeen, The Equatorial Neutral Thermospheric Response to Geomagnetic Forcing, *Geophys. Res. Lett., 19,* 977-980, 1992.

Burns, A.G., T.L. Killeen, and R.G. Roble, Processes responsible for the compositional structure of the thermosphere, *J. Geophys. Res., 94,* 3670-3686, 1989.

Burns, A.G., T.L. Killeen, and R.G. Roble, A Theoretical Study of Thermospheric Composition Perturbations during an Impulsive Geomagnetic Storm, *J. Geophys. Res., 96,* 14153-14167, 1991.

Campbell, I.M., and C. N. Gray, Rate constants for $O(^3P)$ recombination and association with $N(^4S)$, *Chem. Phys. Lett., 18,* 607-609, 1973.

Colegrove, F.D., Atmospheric composition in the lower thermosphere, *J. Atmos. Terr. Phys., 37,* 1563-1570, 1966.

Crowley, G., Dynamics of the Earth's thermosphere: A review, *U.S. National Report to International Union of Geodesy and Geophysics,* American Geophysical Union, 1991.

Dickinson, R.E., E. C. Ridley, and R. G. Roble, Thermospheric general circulation with coupled dynamics and composition, *J. Atmos. Sci., 41,* 205-219, 1984.

Dickinson, R.E., E.C. Ridley, and R.G. Roble, A three-dimensional, time-dependent general circulation model of the thermosphere, *J. Geophys. Res., 86,* 1499-1512, 1981.

Fesen, C.G., R.E. Dickinson, and R.G. Roble, Simulation of the thermospheric tides at equinox with the National Center for Atmospheric Research thermospheric general circulation model, *J. Geophys. Res., 91, A4,* 4471-4489, 1986.

Fitzpatrick, P.M., *Principles of Celestial Mechanics,* Academic press, New York, 1970.

Harris, I., and J. Jastrow, An interim atmosphere derived from rocket and satellite data, *Planet. Space Sci., 1,* 20-26, 1959.

Hedin, A., MSIS-86 thermospheric model, *J. Geophys. Res., 92,* 4649-4662, 1987.

Hedin, A.E., A revised thermospheric model based on mass spectrometer and incoherent scatter data: MSIS-83, *J. Geophys. Res., 88, A12,* 10170-10188, 1983.

Hedin, A.E., The atmospheric model in the region 90 to 2000 km, *Adv. Space Res., 8,* 9-26, 1988.

Hedin, A.E., M.A. Biondi, R.G. Burnside, R.M. Johnson, T.L. Killeen, C. Mazaudier, J.W. Meriwether, R.J. Sica, R.W. Smith, N.W. Spencer, V.B. Wickwar, and T.S. Virdi, Revised global model of thermosphere winds using satellite and ground based observations, *J. Geophys. Res., 96, A5,* 7657-7688, 1991.

Hedin, A.E., and H.G. Mayr, Characteristics of wavelike fluctuations in Dynamics Explorer neutral composition data, *J. Geophys. Res., 92,* 11159-11172, 1987.

Herrero, F.A., Satellite drag coefficients and upper atmosphere densities: present status and future directions, *AIAA/AAS Proc.,* 1607-1623, 1987.

Jacchia, L.G., Static diffusion models of the upper thermosphere with empirical temperature profiles, *Smithsonian Contrib. to Astrophysics, 8,* 215-257, 1965.

Jacchia, L.G., New static models of the thermosphere and exosphere with empirical temperature profiles, *SAO Special report, 170,* 1, 1970.

Jacchia, L.G., Revised static models of the thermosphere and exosphere with empirical temperature profiles, *SAO Special report, 332,* 1971.

Jacchia, L.G., Thermospheric temperature, density and composition, *SAO Special Report, 375,* 1, 1977.

Johnson, H.S., Gas phase kinetics of neutral oxygen species, *NBS-NSRSDS-20,* U.S. Government Printing Office, Washington, D.C., 1968.

Johnson, R.M., and E.P. Martin, Seasonal variations of lower-thermospheric neutral winds at Chatanika, Alaska (1976 - 1982), submitted to *J. Geophys. Res.,* 1990.

Killeen, T.L., and R.G. Roble, Thermosphere dynamics: Contributions from the first 5 years of the Dynamics Explorer Program, *Rev. Geophys., 26,* 329-367, 1988.

Killeen, T.L., R.G. Roble, and N.W. Spencer, A computer model of global thermospheric winds and temperatures, *Adv. Space Res., 7, 10,* 207-215, 1987.

Killeen, T.L., F.G. McCormac, A.G. Burns, J.P. Thayer, R.M. Johnson, and R.J. Niciejewski, On the dynamics and composition of the high-latitude thermosphere, submitted to *J. Atmos. Terr. Phys.,* 53, 797-815, 1991.

Killeen, T. L., B. Nardi, P. N. Purcell, R. G. Roble, T. J. Fuller-Rowell, and D. Rees, Neutral winds in the lower thermosphere from Dynamics Explorer 2, *Geophys. Res. Lett., 19,* 1093-1096, 1992.

King-Hele, D.G., Density of the atmosphere at heights between 200 km and 400 km from analysis of artificial satellite orbits, *Nature, 183,* 1224-1227, 1959.

Knechtel, E.D., and W. C. Pitts, Normal and tangential momentum accomodation for Earth satellite conditions, *Astonautica Acta, 18,* 171, 1973.

Liu, J.J.F., Advances in Orbit Theory for an Artificial Satellite with Drag, *J. Astronaut. Sci., XXXI, 2,* 165-188, 1983.

Manson, A.H., C.E. Meek, H. Teitelbaum, F. Vial, R. Schminder, D. Kurschner, M.J. Smith, G.J. Fraser, and R.R. Clark, Climatologies of semidiurnal and diurnal tides in the middle atmosphere (70-110 km) at middle latitudes (40-50° N), *J. Atmos. Terr. Phys., 51,* 579-594, 1989.

Marcos, F.A., Requirements for improved thermospheric neutral density models, *AIAA/AAS proc., 85, 312,* 1-21, 1985.

Marcos, F.A., T. L. Killeen, A. G. Burns, and R. G. Roble, Evaluation of new atmospheric drag modeling techniques, *Adv. Astronaut. Sci., 71,* 763-782, 1989.

Marcos, F.A., and J.M. Forbes, Thermospheric winds from the satellite electrostatic triaxial accelerometer (SETA) system, *J. Geophys. Res., 90,* 6543-6552, 1985.

Mayr, H.G., I. Harris, and F. A. Herrero, Global excitation of wave phenomena in a dissipative, multi-constituent medium, 1, Transfer function of the Earth's thermosphere, *J. Geophys. Res., 89,* 10929-10959, 1984.

Moe, K., and M.M. Moe, A dynamic model of the neutral thermosphere, *MDC G5891,* McDonnel Douglas, 1975.

Rees, D., T.J. Fuller-Rowell, and H. Rishbeth, The use of mass spectrometer measurements to derive thermospheric temperature and density, *Planet. Space Sci., 36,* 281-290, 1988.

Reese, K.W., R.M. Johnson, and T.L. Killeen, Long-term seasonal variations of lower thermospheric neutral densities determined from Chatanika incoherent scatter radar, unpublished manuscript, 1990.

Roble, R.G., R.E. Dickinson, and E.C. Ridley, Global circulation and temperature structure of thermosphere with high-latitude plasma convection, *J. Geophys. Res.*, *87*, 1599-1614, 1982.

Roble, R.G., E.C. Ridley, A.D. Richmond, and R.E. Dickinson, A coupled thermosphere/ionosphere general circulation model, *Geophys. R. Lett.*, *15*, 1325-1328, 1988.

Schunk, R.W., Mathematical structure of transport equations for multispecies flows, *Rev. Geophys. Space Phys.*, *15*, 429-445, 1977.

Seidl, M., and E. Steinheil, Measurement of momentum accomodation coefficients on surface characterized by Auger Spectroscopy, SIMS and LEED, in: *Proc. of Rarefied Gas Dynamics, 9th International Symposium*, edited by M. Becker and M. Fiebig, *Vol. II*, pp. E.9-1 - E.9-12., DFVLR Press, Germany, 1974.

Snow, D.E., and J.J.F. Liu, Atmospheric variations observed from orbit determination, *AIAA/AAS Proc.*, *492*, 1-21, 1991.

Vial, F., and J.M. Forbes, Recent progress in tidal modeling, *J. Atmos. Terr. Phys.*, *51*, 663-671, 1989.

Volland, H., C. Wulf-Mathies, and W. Priester, On the annual and semi-annual variations of the thermospheric density, *J. Atmos. Terr. Phys.*, *34*, 1053-63, 1972.

Walterscheid, R.L., Solar Cycle effects on the upper atmosphere: Implications for satellite drag, *J. Spacecraft and Rockets*, *26*, *6*, 442, 1989.

A. G. Burns, R. M. Johnson and T. L. Killeen, Space Physics research Laboratory, University of Michigan, 2455 Hayward Street, Ann Arbor, MI 48109-2143

F. A. Marcos, Phillips Laboratory/LIS, Hanscom Air Force Base, Massachusetts 01731-5000.

Surface Force Modeling for Precision Orbit Determination

J. C. Ries, C. K. Shum, and B. D. Tapley

Center for Space Research, The University of Texas at Austin

The accuracy requirements for the ephemerides of modern geodetic and oceanographic satellites impose significant demands on the modeling of the forces on the satellite. When tracking data is sparse, when long integration spans are being used, or when orbit predictions are required, the orbit accuracy is primarily limited by the fidelity of the modeling of the forces which influence the motion of the satellite. Considerable progress has been made in the modeling of both the static and temporally varying components of the gravitational forces, but attempts to improve the modeling of the surface forces have been less successful. Atmospheric drag and solar radiation pressure are the primary concerns, but highly accurate long arc solutions also require consideration of the forces due to the radiation from the Earth and the re-emission of heat by the satellite. The progress in modeling these effects is reviewed, and the influence on orbit improvement is examined.

Introduction

Historically, the errors in the modeling of the Earth's gravitational effects on satellite orbits have been the primary limitation in the quality of the orbit determination process. With recent improvements in the models for the geopotential, the computation of precisely determined satellite orbits is now often limited by the accuracy of modeling of the surface forces acting on the satellite. For satellites such as Lageos and Starlette, the effect of surface forces is minimized because the satellites are small, dense and spherical. Nevertheless, the nongravitational forces are still significant, and it is necessary to model these "nuisance" forces accurately if improved determination of the constant and temporally varying components of the Earth's gravity field is to be accomplished. For oceanographic satellites such as Seasat, Geosat, ERS-1, and TOPEX/POSEIDON, the usefulness of altimeter measurements collected by these satellites may be limited by the accuracy of the orbits, which often depends strongly on the accuracy of the surface force models. Surface forces are also a major limitation in the accuracy of the Global Positioning System (GPS) satellite orbits, which directly affects the positioning and baseline accuracies.

The modeling of atmospheric drag on artificial satellites has been the object of considerable research ever since satellites were first launched. Drag is a dissipative process which secularly reduces the satellite's orbital energy, and the accurate prediction of satellite orbits in the presence of drag is limited by the stochastic nature of the atmospheric density variations. Even for satellites as high as 1500 km, drag can be a significant modeling problem, particularly during periods of high solar activity. The pressure of radiation from the Sun, which, as seen in Table 1, is actually larger in magnitude than drag for most geodetic satellites of interest, has also been studied extensively. Solar radiation pressure effects, when there is no eclipsing by the Earth or Moon, tend to average out over an orbit and do not perturb the orbital elements as dramatically as drag. During eclipsing, however, solar radiation pressure can cause significant changes in the orbital elements. The effects on satellite orbits due to reflected sunlight and heat from the Earth are less well understood, and only recently has serious consideration been given to the problem of imbalanced radiation of satellite heat.

Atmospheric Drag

The force model associated with the effects of atmospheric drag is one of the most important for near-Earth satellites. Atmospheric drag will lead to significant effects on the satellite's orbit, producing secular changes in its semimajor axis and eccentricity which will cause the orbit to decay. The dominant feature of atmospheric resistance for most satellites is a drag force in the direction opposite to the relative wind. The drag acceleration, $\bar{a}$, of a composite satellite can be approximately modeled as:

$$\bar{a} = -\frac{1}{2}\frac{C_d A}{m}\rho v \bar{v} \qquad (1)$$

where $\bar{v}$ is the satellite velocity relative to the atmosphere, v is the magnitude of $\bar{v}$, ρ is the atmospheric density, C_d is the drag coefficient, A is the cross-sectional area perpendicular to $\bar{v}$, and m is the satellite mass. The parameter $(C_d A/m)$ is sometimes referred to as the ballistic coefficient. The drag coefficient is a

Environmental Effects on Spacecraft Positioning and Trajectories
Geophysical Monograph 73, IUGG Volume 13

TABLE 1. Average Accelerations Due to Surface Forces (based on current UT/CSR orbit determination models)

Satellite	Altitude, km	Atmospheric drag, (m/s^2)	Solar radiation pressure, (m/s^2)	Terrestrial radiation pressure, m/s^2)	Thermal radiation pressure, (m/s^2)
ERS-1	770	3×10^{-8}	7×10^{-8}	2×10^{-8}	–
Starlette	800–1100	9×10^{-10}	3×10^{-9}	8×10^{-10}	–
TOPEX/POSEIDON	1330	4×10^{-10}	6×10^{-8}	9×10^{-9}	5×10^{-9}
Lageos	5900	1×10^{-12}	4×10^{-9}	2×10^{-10}	3×10^{-12}

function of the geometry and surface properties of the spacecraft, the satellite speed, and the atmospheric constituents. For nonspherical satellites, the C_d and A will change if the satellite changes orientation, producing time variations in the ballistic coefficient which can be difficult to model. There also may be "lift" effects, from solar panels for example, generating accelerations in directions other than along $\bar{v}$ [Milani et al., 1987]. Klinkard et al. [1990] have investigated Monte-Carlo ray tracing techniques for computing free-molecular aerodynamical drag, lift and torque on a spacecraft of arbitrary complexity. Even for spherical satellites, there are fundamental problems in the determination of appropriate drag coefficients [Imbro et al., 1975; Herrero, 1988; Gaposchkin and Coster, 1988]. There is uncertainty about whether the atmosphere can be assumed to rotate at the same velocity as the Earth at high altitudes, and there is the problem of zonal, meridional and vertical winds [King-Hele and Walker, 1983; Marcos and Forbes, 1985; Killeen and Roble, 1988; Hedin et al., 1988]. The greatest uncertainty, however, is in modeling the upper atmospheric density, which is usually represented by empirical tables or by some analytical model for orbit determination purposes.

In the past two decades, numerous efforts to model the atmospheric density have followed two principal approaches: (1) Physical models of the thermosphere have been developed, based on the assumption of conservation of energy, momentum and mass and incorporating the chemical processes of atmospheric constituents, and (2) semiempirical models based on simplified physical concepts, inferred from in situ measurements and from the vast amount of data collected on the motions of satellites, have been developed in order to describe the behavior of the thermosphere.

Density Model Comparison

Since the development of a semiempirical thermospheric model for the upper atmosphere by Nicolet [1961], a number of different types of thermosphere models have been produced. A partial list of these models include the Harris-Priester model [Harris and Priester, 1962], the Jacchia models [Jacchia, 1970, 1971, 1977], the USSR COSMOS satellite-derived models [Elyasberg et al., 1972], the Mass Spectrometer-Incoherent Scatter (MSIS) models [Hedin et al., 1974, 1977*a*, 1977*b*, 1979; Hedin 1983, 1987], the Drag Temperature Model (DTM) [Barlier et al., 1978], and the ESRO4 models which are based on the ESRO4 gas analyzer data [Laux and Von Zahn, 1979]. The Jacchia models set fixed boundary conditions at 90 km and incorporate a diurnal variation, geomagnetic activity effects, a seasonal-latitudinal variation in the boundary conditions, and the seasonal migration of helium for altitudes from 90 to 2500 km. In the MSIS, DTM and ESRO4 models, the temperature and concentration variations at 120 km are represented by expansions in terms of spherical harmonics, in the formulation first proposed by Hedin et al. [1974], assuming an analytical temperature profile. The density for each major atmospheric constituent is then computed as a function of solar angle, latitude, solar flux and geomagnetic activity. Barlier et al. [1979], Kockarts [1981] and Barlier and Berger [1983] have reviewed different semiempirical thermospheric models, and the comparisons among some of the more recent models generally give good agreement; however, none of the models can represent correctly all the geophysical properties of the terrestrial thermosphere. Comparisons with accelerometer data taken on low satellites from 1974 to 1982 indicate that there has been little improvement in the accuracy during the last two decades [Marcos, 1988; Marcos et al., 1990]. At altitudes above 700 km, the density models can exhibit significant departures from the actual density [Gaposchkin and Coster, 1988].

Research conducted over the last two decades has been concerned with some basic characteristics of the structure of the upper atmosphere. These characteristics are:

· Diurnal Bulge. A nearly diurnal variation is caused by solar heating, which produces a nighttime minimum and a daytime maximum of exospheric temperatures in the global atmosphere. The atmospheric bulge is found to have a lag angle of 45 to 60 deg from the axis of the Earth-Sun line.

· Solar Activities. Solar activity has an important influence on disturbances in the upper atmosphere. The F10.7-cm solar flux index is generally used as an indication of the level of extreme ultraviolet solar radiation which heats up the Earth's atmosphere. The primary sources of these disturbances are solar flares and solar plasma events. These phenomena are associated with the 27-day solar rotation period as well as the 11-year sun spot cycle.

· Geomagnetic Activities. The interaction of geomagnetic activity with the charged particles in the upper atmosphere is most pronounced near the geomagnetic pole and reduces when approaching the equator. The planetary geomagnetic indices, K_p or A_p, are generally used as measures of geomagnetic activity. It should be noted that the practice of using the F10.7-cm flux and the geomagnetic indices to correlate with the thermosphere temperature does not imply that they necessarily represent the real physical mechanisms responsible for temperature and density changes [Barlier et al., 1978]. The precipitation index, which is used as an indicator of magnetospheric activity, may also reflect

TABLE 2. Satellite Orbital Characteristics

Satellite:	Starlette	Seasat	Spot-2	ERS-1
Launch date	1975	1978	1990	1991
Altitude (km)	800–1100	790	830	770
Mean eccentricity	0.020	.001	.001	.001
Mean inclination	50°	108°	98°	98°

the processes which tend to heat the atmosphere [Coster and Gaposchkin, 1988].

· Semiannual. A semiannual variation is well observed, but the causes of the effect are not understood.

· Other Variations. Longitudinal effects are observed to be associated with magnetic activity levels [Hedin et al., 1979]. Latitudinal variations are caused by the seasonal migration of helium.

In this study, the atmospheric models chosen for comparison are semiempirical models that account for most of the variations mentioned above. These models are the analytic version of the Jacchia 1971 (J71) model [Jacchia, 1971] based on Roberts's formulation [Roberts, 1971], the analytic version of the Jacchia 1977 (J77) model [Jacchia, 1977; Lafontaine and Hughes, 1981], the DTM [Barlier et al., 1978], and the MSIS-86 model [Hedin, 1987]. The J77 and DTM models were derived with a vast amount of satellite drag data and in situ measurements covering almost two solar cycles. MSIS-86 is based on in situ data from satellites and rocket probes, ground-based incoherent scatter stations, and recent in situ temperature and composition measurements made by the Dynamics Explorer B satellite. In most cases, however, the data become sparse above 600-km altitude.

Table 2 lists the orbital characteristics for the geodetic satellites which are used in the following density model evaluations. Starlette, Seasat and ERS-1 are tracked with laser ranging, although altimeter crossover data were also used for Seasat to provide additional orbit accuracy information. SPOT-2 was tracked with a new system called DORIS (Doppler Orbitography and Radiopositioning Integrated by Satellite) [Nouél et al., 1988]. A dense set of Spot-2 data from a global network of DORIS tracking stations could be very useful for investigating the effects of drag on a satellite of complex shape.

Figure 1 illustrates how much the various density models can differ over a satellite orbit. In Figure 1*a*, the atmospheric density is calculated using the Jacchia 71, Jacchia 77, DTM and MSIS-86 models for one revolution of ERS-1 during a period of high solar activity. The altitude of ERS-1 is fairly constant, so the once-per-revolution variation is the result of the change in density as the satellite passes through the atmospheric bulge caused by solar heating. In contrast, the large variations in the atmospheric density for one revolution of Starlette shown in Figure 1*b* are caused by the change in altitude, since Starlette has a relatively large orbital eccentricity. In both cases, it can be seen that the density predicted by the various models differs in amplitude as well as in the location of the peak density. Differences approaching 100% are possible between some of the models, and it is reasonable to assume that the models can differ from the actual density by similar amounts. This mismodeling of the

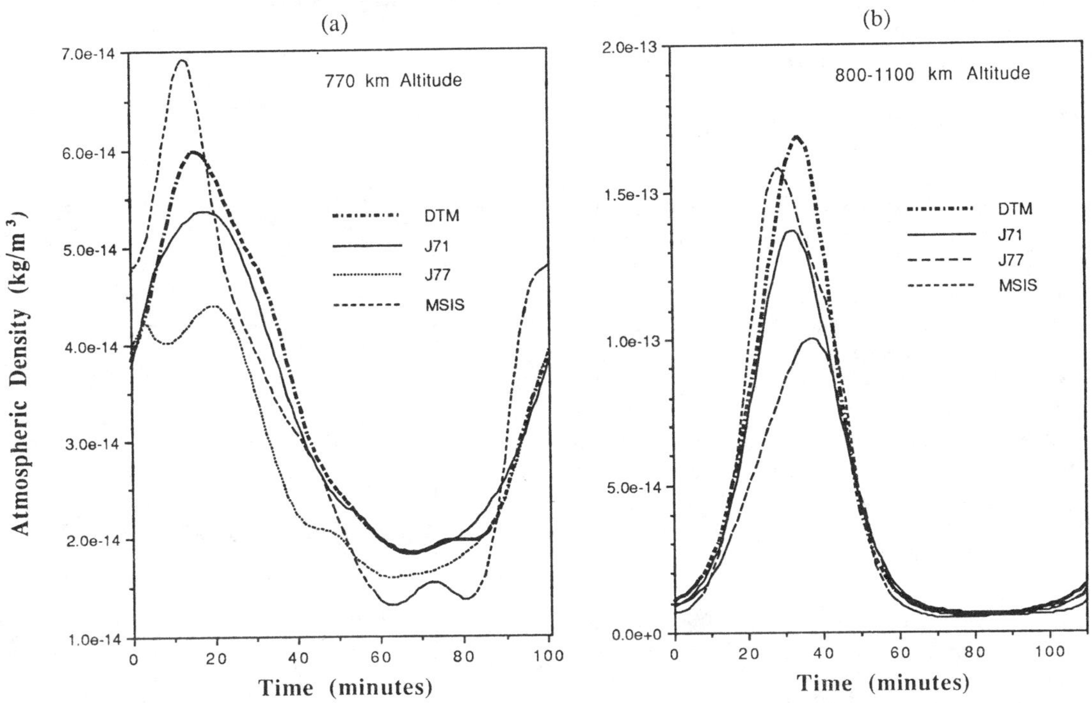

Fig. 1. Atmospheric density calculated using the Jacchia 71, Jacchia 77, DTM and MSIS-86 models for one revolution of (a) ERS-1 and (b) Starlette.

TABLE 3. Density Model Comparison Using Starlette Laser Range Orbit Fits

Models	Low solar activity		High solar activity	
	5-day C_d Average residual rms, cm	Daily C_d Average residual rms, cm	5-day C_d Average residual rms, cm	Daily C_d Average residual rms, cm
Jacchia 1971	18.0	14.4	74.7	14.7
Jacchia 1977	18.1	14.4	74.0	13.6
DTM	18.0	14.4	83.6	13.5
MSIS-86	18.1	14.4	72.4	14.7

Three 5-day orbit fits for Starlette in 1984 and 1989.

amplitude and phase of the density variation leads to errors in the determination of the orbital semimajor axis and eccentricity, and thus to errors in the along-track and radial components of the orbit. Similarly, errors in the magnitude of the drag and the rotation rate of the atmosphere can lead to errors in the inclination and ascending node of the satellite orbit, which appear as cross-track orbit errors [Cheng, 1988].

In spite of the differences between the density predictions, the performance of the various models in orbit determination is remarkably similar. A set of three 5-day Starlette arcs in 1984 were used to test the models for orbit determination purposes during a period of low solar activity. A similar set of arcs was chosen in 1989 to test the models during high solar activity. In each arc, only the initial conditions and the drag coefficient C_d were estimated. The drag coefficient was estimated once for the whole arc and also once each day. Estimation of C_d is an effective way to eliminate the average density model error over the estimation interval, resulting in considerably reduced along-track orbit errors.

The results of the orbit tests are summarized in Table 3. During low solar activity, the orbit fits with a 5-day C_d are nearly identical, with a root mean square (rms) fit to the laser data of 18 cm. The fits are improved when daily (or sub-arc) C_d are estimated, but the improvement is not dramatic, and the contribution from gravity modeling error is probably limiting the fits to the 14-cm level. During high solar activity, however, there is a great difference between estimating a single C_d and daily C_d, indicating that there is considerable density variation within 5 days that none of the models are accommodating. Greater differences are also observed between the models. This is to be expected since the model errors will become larger as the atmospheric density increases, and the density models generally include little, if any, data from periods of intense geomagnetic activity. With the estimation of a daily C_d, the four models show equivalent performance in terms of similar rms values for the laser range residuals.

The density of the DORIS tracking on SPOT-2 offers a unique view into the the level of orbit error that mismodeled atmospheric drag can cause within short periods of time. The data set used here contained up to 100 passes per day with an average precision of approximately 0.5 mm/sec. Tracking data taken from 3 May to 18 May 1990 were chosen for study because, during this interval, the F10.7-cm solar flux rose from a moderate value of 120 to a high value of 250. Five 3-day arcs were fit using the DTM density model and the University of Texas Center for Space Research (UT/CSR) TEG-2B gravity model (which incorporated DORIS data on SPOT-2) [Tapley et al., 1991]. A sub-arc drag coefficient as well as a single reflectivity coefficient was estimated for each arc. Two sub-arc lengths for the drag coefficient were studied: 1 day and 6 hours. The residuals were analyzed to estimate the level of along-track error remaining in the orbit for each pass of tracking data. As can be seen in Figure 2, the level of orbit error with daily C_d is much larger than when 6-hourly C_d are estimated. Clearly, a great deal of orbit error can accumulate within the period of a day due to drag during high solar activity. Reducing the sub-arc length for C_d to three hours provided only slight additional improvement, probably indicating the limiting accuracy of the TEG-2B gravity model.

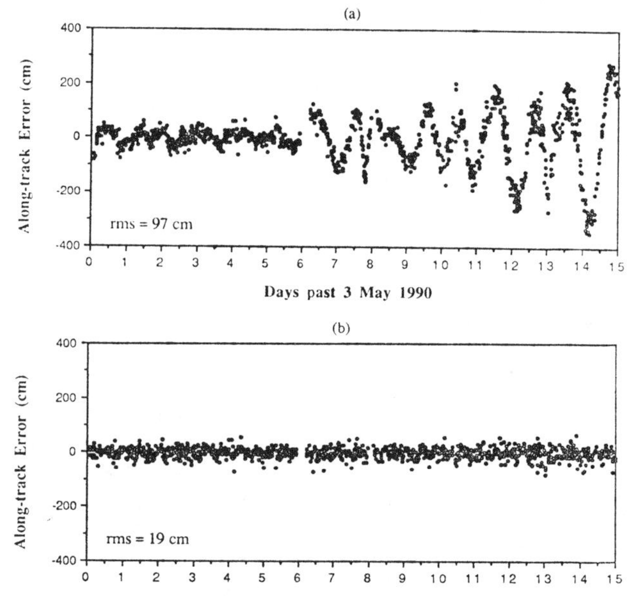

Fig. 2. Estimated along-track orbit errors in five 3-day Spot-2 arcs using DORIS tracking data and solving for (a) daily C_d and (b) 6-hourly C_d.

Density Correction

The present thermospheric density models cannot account for all the variations in the atmosphere described in the previous sections. The estimation of the drag coefficient, C_d, has been shown to be effective in reducing the along-track orbit errors. To achieve the radial orbit precision goals for altimetric satellites, it will be shown that some scheme for correcting the phase of the density model error may have to be applied as well.

In Shum et al. [1986], a number of density correction parameters were investigated, and it was found that the most effective yet minimal set of parameters for estimating drag model corrections was of the form

$$\vec{a} = -\frac{1}{2}\left[C_d + C_1 \cos u + C_2 \sin u\right] \frac{A}{m} \rho v \vec{v} \qquad (2)$$

where C_d is the drag coefficient as before, u is the argument of latitude (the sum of the argument of perigee and the mean anomaly), and C_1 and C_2 are correction parameters associated with the once-per-revolution density variation. The average along-track orbit errors can be largely removed by estimating sub-arc values for C_d, and estimating the density correction parameters C_1 and C_2 is very effective in reducing errors in the radial direction due to their effect on the orbital eccentricity.

To illustrate how the additional density parameters can benefit the orbit accuracy for an altimetric satellite, three 6-day Seasat orbits were chosen beginning 28 July 1978, a period of moderate solar activity. The initial conditions, the satellite reflectivity and the drag parameters were estimated using laser ranging and altimeter crossover data. Using differenced altimeter data at the points where the satellite ground tracks intersect eliminates the nontemporal portion of the geoid and ocean surface topography from the measurement and provides a measure of the accuracy of the radial component of the orbit. The laser ranging residuals tend to be dominated by the along-track component of the orbit error. As seen in Table 4, estimating a single C_d for each arc results in very poor fits for both the laser and crossover data. When a daily C_d is estimated, the fit to the laser data improves greatly, but the crossover fit has improved only slightly. This indicates that the error in the radial component of the orbit has not been improved. When a single pair of the density correction parameters C_1 and C_2 are also estimated, however, the crossover residual is decreased by more than a factor of 2, and there is also some improvement in the laser range residuals. Again, the four density models show equivalent performance in terms of similar rms values for the range and crossover residuals.

The parameters C_1 and C_2 used above are not necessarily the best choice for density model correction. There is significant correlation with C_d and also with the odd zonal harmonics of the geopotential model. Experiments have been performed to tune existing models by improving the modeling of the persistence of a geomagnetic event [Rothwell and Moore, 1991] or by estimating, in the orbit determination process, selected parameters in the density models [Nuth, 1991], but these tests met with limited success. Efforts in this area must be continued since it appears that the estimation of drag model correction parameters will always be required in the determination of precise satellite orbits.

Radiation Pressure

Solar Radiation Pressure

Interest in the effect of solar radiation pressure on satellite orbits began with the early Vanguard, Echo, and Pageos satellites [e.g., Musen, 1960; Parkinson et al., 1960; Muhleman et al., 1960; Bryant, 1961; Kozai, 1961]. Considering only spherical or axis-stabilized, rapidly spinning, nonspherical satellites, the long-term effect of solar radiation pressure on the orbit in the absence of shadowing tends to be minor because the force is approximately constant in inertial coordinates [Anselmo et al., 1983*a*]. When shadowing occurs, however, there can be significant effects on the orbital eccentricity, semimajor axis, and other orbit elements [Lála and Sehnal, 1969; Sehnal, 1970; Lála, 1971; Aksnes, 1976]. Rubincam and Weiss [1985] examine the effect of lunar shadowing on the Lageos satellite and find that it is possible for the semimajor axis to change at the centimeter level. For satellites with more complex shapes, the effects of solar radiation pressure can be very difficult to model accurately. For example, in spite of the extensive effort to accurately model the pressure of solar radiation, and to a lesser degree the thermal emission, on GPS satellites, the effects of radiation pressure during eclipse seasons are still not well understood, and it is recommended that GPS satellites experiencing eclipsing should not be used for precise geodetic work [Fliegel et al., 1992]. In general, the difficulty in precise modeling of radiation pressure on nonspherical shapes has led to the practice of treating most satellites as spheres.

For spherical satellites such as Lageos and Starlette, modeling the effect of solar radiation pressure is fairly straightforward. The

TABLE 4. Density Model Comparison Using Seasat Laser Range and Altimeter Crossover Orbit Fits

Density models	6-day C_d Average residual rms, cm	Daily C_d Average residual rms, cm	C_1, C_2, and daily C_d Average residual rms, cm
Jacchia 1971	270 / 70	68 / 69	50 / 24
Jacchia 1977	303 / 64	61 / 61	46 / 27
DTM	273 / 67	67 / 66	49 / 24
MSIS-86	343 / 72	83 / 70	50 / 25

Three 6-day orbit fits for Seasat in 1978.
Laser range / altimeter crossover.

usual model for the acceleration due to radiation pressure is of the form

$$\bar{a} = -\frac{\Phi}{c}\frac{C_r A}{m}\bar{s} \tag{3}$$

where Φ is the radiation flux at the satellite (taking into account shadowing by the Earth or Moon in the case of solar radiation), c is the speed of light, A is the cross-sectional area, m is the satellite mass, C_r is the reflectivity coefficient, and $\bar{s}$ is the unit vector along the direction from the satellite to the radiation source. The solar irradiance at 1 Au is nearly constant, varying with the solar cycle between 1367 and 1368 W m^{-2} [Wigley and Raper, 1990]. This value must be corrected to the actual distance from the Sun to obtain Φ, which introduces an annual variation of about 46 W m^{-2}. The cross-sectional area and mass of the satellite are usually known, but the reflectivity coefficient, which represents the combination of the specular and diffuse reflectivities, is generally not known and must be estimated in the orbit determination procedure. The reflectivities of satellites like Lageos and Starlette are expected to change very little with time, and the variations observed in short-arc estimates for the reflectivity usually indicate that other modeling errors are being absorbed. For nonspherical satellites which are treated as if spherical, the estimation of the reflectivity parameter can compensate to some degree for mismodeling the cross-sectional area.

The effect of solar radiation pressure on the orbit is greater when the satellite encounters eclipsing of the Sun by the Earth or Moon, so the model for this effect is important. A conical shadow model (umbra and penumbra) is generally used, although the numerical integration step-size is usually much larger than the duration of partial shadowing, and the effect is essentially a discontinuity in the solar radiation pressure force. Procedures have been developed to more accurately account for the effect of shadowing in the numerical integration process [Feulner, 1990]. Another concern is the appropriate radius of the Earth to be used in the shadow calculations. The atmosphere of the Earth attenuates the solar radiation as the satellite approaches shadowing, but this effect can be approximated by increasing the effective radius of the Earth's shadow by about 20 km [Haley, 1974]. Some orbit programs also take into account the oblateness of the Earth's shape when calculating the shadow radius.

A precise calculation of the acceleration of a satellite of arbitrary shape due to radiation pressure can be obtained by subdividing the surface of the spacecraft into elemental flat plates. The contribution of each surface element to the total acceleration is then given by

$$\bar{a}_i = -\frac{\Phi}{c}\frac{A_i\cos\theta_i}{m}\left[\left[\frac{2}{3}\delta_i + 2\beta_i\cos\theta_i\right]\bar{n}_i + (1-\beta_i)\,\bar{s}\right] \tag{4}$$

where Φ, c, m and $\bar{s}$ are as before, A_i is the area of the ith surface element, $\bar{n}_i$ is the surface normal unit vector, θ_i is the angle between the $\bar{n}_i$ and $\bar{s}$, β_i is the specular reflectivity (percent of total incident radiation), δ_i is the diffuse reflectivity (percent of total incident radiation), and Lambert's cosine law of diffuse reflection has been assumed [Milani et al., 1987]. In practice, only a finite number of surface elements can be used, and there can be considerable uncertainty in the reflective properties of the various surface materials. Even if known at the beginning of the satellite mission, the surface properties can change considerably due to exposure to the space environment. There are also the problems of modeling the orientation of each surface with respect to the radiation source and accounting for shadowing or reflections by other parts of the spacecraft. The widespread practice of treating most satellites as spheres is predominately due to the uncertainties, complications and computational burden of this level of modeling.

Terrestrial Radiation Pressure

The Earth is the second largest source of radiant energy, and the contribution of reflected sunlight and infrared radiation is estimated at 10% to 25% of the direct solar radiation effect for most satellites [Wyatt, 1963; Sehnal, 1970]. The infrared component of terrestrial radiation pressure is dominantly radial and fairly uniform since it is nearly independent of the day/night terminator [Barlier et al., 1977]. The Earth albedo component is mostly diffuse reflection, with only about 1% to 10% of the radiation being reflected specularly by the oceans [Wyatt, 1963; Baker, 1966; Sehnal, 1970]. The effect of reflected sunlight is also largely radial, but significant along-track effects can be generated when the terminator is visible. High eccentricity orbits and satellites with large area-to-mass ratios are particularly sensitive to terrestrial radiation pressure, and the along-track accelerations which do not average out over the orbit can interfere with the recovery of drag coefficients [Prior, 1972; Lautman, 1977; Knocke et al., 1988]. The along-track effect of terrestrial radiation pressure on the Lageos satellite has been investigated by several groups [Anselmo et al., 1983*b*; Rubincam and Weiss, 1986; Barlier et al., 1986; Rubincam et al., 1987; Lucchesi and Farinella, 1992].

Smith [1970] examines the development and drawbacks of purely analytical approaches to determining the orbital effects of terrestrial radiation pressure and concludes that complete treatments can probably only be carried out numerically. The model used at UT/CSR, illustrated in Figure 3, is a typical numerical model in that it integrates the heat and diffusely reflected sunlight from the visible disk of the Earth [Knocke et al., 1988] based on an approach devised by McCarthy and Martin [1977]. The resulting radiation fluxes are used in Eq. (3), although the reflectivity for optical wavelengths may differ from the infrared reflectivity. Borderies and Longaretti [1990] have developed a new treatment in which the surface integrals over the illuminated portion of the Earth are reduced to one-dimensional integrals. The two procedures have been shown to converge to equivalent results when a sufficient number of surface elements are used [J. Guinn, personal communication, 1990]. However, the primary uncertainty in any model for terrestrial radiation pressure is the magnitude and nature of the reflection assigned to each element of the Earth's surface.

The average and seasonal variations in the Earth's albedo and infrared emission are given by second-degree zonal

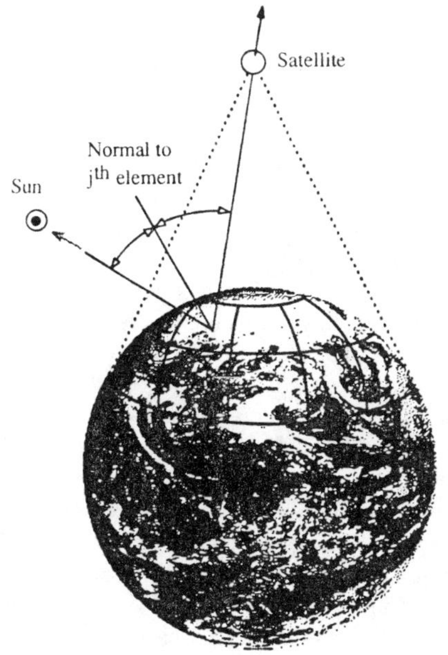

Fig. 3. UT/CSR terrestrial radiation pressure model.

representations in the UT/CSR model, which models the latitudinal variation and a small seasonally dependent hemispherical asymmetry. More complex representations of the Earth's albedo are available which take into account the land and water distribution of the Earth [e.g. Sehnal, 1979], but because of cloud cover and its variability from hour to hour, it is not clear whether there is any benefit from the higher resolution models. Ocean areas may change from diffuse to specular, and the reflectance for most terrestrial surfaces appears to have some dependence on the solar zenith angle [Pinker and Razgaitis, 1983]. As a consequence, it seems that any practical model for Earth radiation pressure will contain a great deal of uncertainty, even for satellites without complex geometries.

Thermal Radiation Pressure

Radiation from the Sun, Earth and internal sources will heat the exposed portions of a spacecraft, and any imbalance in the reradiated heat pattern generates a "photon thrust." The anomalous along-track acceleration of Lageos, which is discussed in detail later, may be due in large part to the non-isotropic distribution of temperatures over the satellite. For most satellites, however, the uncertainties of atmospheric drag and solar radiation pressure tend to mask the effects of thermal radiation forces, and the thermal forces tend to be accommodated in part by the estimation of the drag and solar radiation pressure coefficients.

For a satellite of arbitrary shape, the contribution of a surface element to the total acceleration due to the emission of externally and internally generated heat can be expressed as

$$\bar{a}_i = -\frac{2}{3}\,\frac{\sigma}{c}\,\frac{A_i \varepsilon_i}{m}\,T_i^4\,\bar{n}_i \qquad (5)$$

where c, m, A_i and $\bar{n}_i$ are as before, ε_i is the surface emissivity coefficient, T_i is the surface temperature in degrees Kelvin, σ is the Stefan-Boltzman constant, and Lambertian diffusion is again assumed [Marshall et al., 1991]. While a considerable force may be generated by some surfaces, it is the imbalance in the thermal thrusts over the spacecraft that is the concern. Considering a solar panel, for example, the very large thermal force generated by the sunlit side of the panel could be largely offset by the radiation of heat by the opposite side if the heat flow through the solar panel material is rapid. The main satellite body, although usually smaller in area than the solar panels, may generate a larger thermal thrust because of the much greater temperature differences involved. The cooling and heating as the satellite goes in and out of shadowing are also important components of the effect of thermal radiation pressure on the orbit.

The GPS satellites have demonstrated that thermal radiation forces can be a significant part of the radiation force modeling problem. The original ROCK-4 solar pressure model for GPS represents the spacecraft by 13 distinct flat and cylindrical surfaces [Fliegel et al., 1985]. The model does not account for the reradiation of heat absorbed by the satellite. Since most of the energy is reradiated by the same surfaces that absorbed it, the net effect is to increase the total radiation force from what the ROCK-4 model predicts. This increased force has been accommodated in new approximating functions for the GPS radiation force [Fliegel et al., 1992]. During eclipsing, however, the thermal radiation component would linger while the vehicle cooled, an effect not accounted for in the models. A similar problem occurs when the vehicle heats up after leaving the Earth's shadow. These modeling deficiencies are not easily absorbed by estimating a scaling parameter for the nominal model, and GPS orbits determined during eclipse seasons are generally regarded as being less reliable. Another problem for GPS has been the suspicion that heat is preferentially radiated from louvers on one side of the spacecraft. This has led to the now-standard practice of solving for a Y-bias acceleration when computing precise orbits. It may be that not all of the Y-bias is due to thermal radiation, however. Asymmetry caused by misalignment of the solar panels could also be contributing to the Y-bias acceleration [Fliegel et al., 1992], demonstrating that the problems of modeling solar, terrestrial and thermal radiation pressure are not necessarily completely separable.

Surface Force Modeling for Topex/Poseidon

The 13-cm radial orbit accuracy requirement for the joint NASA/CNES Ocean Topography Experiment (TOPEX/POSEIDON), planned for launch in mid-1992, imposes unique demands on the orbit determination process. With the recent improvements of the geopotential models, accurate representations of the surface forces has become a major concern. The Colorado Center for Astrodynamics Research at the University of Colorado (CCAR), the NASA Goddard Space Flight Center (GSFC), the Jet Propulsion Laboratory (JPL) and UT/CSR have invested considerable effort into the development of models which can meet the orbit accuracy requirements while still remaining computationally reasonable. Some of the results of this research are discussed here as an example of strategies for modeling surface forces on satellites with complex shapes. More detailed discussions can be found in Marshall et al. [1991] and Antreasian and Rosborough [1992].

As seen in Table 1, solar radiation pressure is by far the largest nongravitational effect at the TOPEX/POSEIDON altitude of 1330 km. Terrestrial and thermal radiation pressure are each about an order of magnitude smaller, and atmospheric drag is another order of magnitude below that. The area-to-mass ratio for the TOPEX/POSEIDON spacecraft, illustrated in Figure 4*a*, is expected to vary between 0.004 $m^2 kg^{-1}$ and 0.015 $m^2 kg^{-1}$. To demonstrate the inadequacy of using a "cannonball" to represent the TOPEX/POSEIDON spacecraft, a simulation was performed in which the atmospheric drag and solar, terrestrial and thermal radiation forces were generated with simple models which approximated the actual shape and orientation of the satellite. When only initial conditions, C_r and daily C_d were estimated in a 10-day fit to the simulated data using the cannonball model, with no other modeling errors introduced, the residual radial orbit error had an rms of 17 cm. The 10-day arc length has been adopted to match the ground track repeat cycle. Analysis of the trajectory differences indicated that the radial error was primarily the result of an unmodeled secular trend in the eccentricity of the orbit. Since the real surface forces will be more complex than the models used here, 17 cm is probably an optimistic accuracy estimate for orbit determination if a sphere is used to represent the TOPEX/POSEIDON spacecraft. Clearly, a more accurate model for the surface forces is required, or additional parameters must be chosen for estimation which can accommodate the modeling deficiencies.

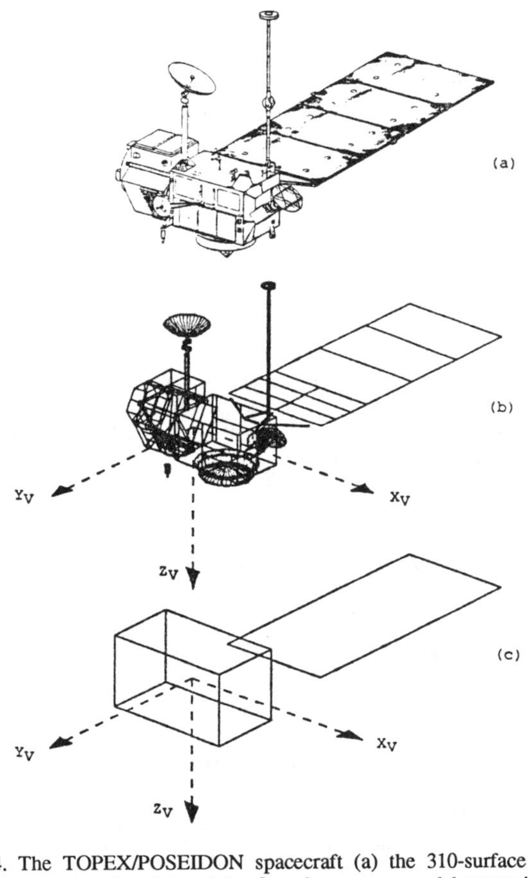

Fig. 4. The TOPEX/POSEIDON spacecraft (a) the 310-surface micro-model approximation, (b) and the 8-surface macro-model approximation, and (c) [figure from Marshall et al., 1992].

A more detailed representation of the TOPEX/POSEIDON spacecraft has been developed in which the main body of the spacecraft is approximated by a 6-sided box and the solar panel is represented by a flat plate. This 8-sided "box-wing" model, shown in Figure 4*c*, is expected to be able to reproduce the surface forces on the satellite to the few percent level if appropriate values for the surface properties are obtained. For each surface of the box-wing model, the area, the specular and diffuse reflectivity, the emissivity and the temperature function are required in order to evaluate Eqs. (4) and (5), but these quantities must represent the integrated or "lumped" effect over the corresponding spacecraft surfaces. The lumped effect of hundreds of satellite surfaces is not easily obtained, and a method of verifying that the resulting model represents the spacecraft accurately is essential.

The problem of obtaining the necessary parameters for the 8-surface "macro-model" has been addressed by employing a precise finite-element model of the spacecraft. This detailed thermal radiative "micro-model" of the spacecraft, illustrated in Figure 4*b*, consists of 310 elements and serves as the best estimate of the true surface forces, but it is computationally intensive and impractical for the precision orbit determination process. The accelerations which are generated by the "micro-model", however, can be used in a least-squares adjustment of the optical and thermal parameters for the macro-model. Thus, the macro-model is "tuned" to represent the micro-model, and the resulting macro-model should give a fairly accurate representation of the radiation pressure without excessive computational effort. The micro-model also provides the reference model which can be used to determine the level of agreement obtained with the macro-model or with any other approximation, but because of approximations and uncertainties in the satellite surface properties, temperatures, and orientations, the micro-model cannot be considered to be absolutely accurate. The solar panel, for example, may not be pointed precisely in the direction assumed in the models, which, depending on the nature of the misalignment, could lead to significant long-period orbit errors.

The results to date indicate that the macro-model should be effective for TOPEX/POSEIDON. Comparisons to the micro-model for several different orientations of the orbit with respect to the Sun indicate that the effects of solar radiation pressure are duplicated by the macro-model to better than 1%. The thermal component of the macro-model, which is considerably more complicated due to heating and cooling which occurs during eclipsing, approximates the micro-model to about 10%. To account for the possibility of an overall bias in the macro-model due to deficiencies in the micro-model, scaling parameters will be available for estimation in the orbit determination process. As a result, the total contribution of the surface forces to the rms radial orbit error is expected to remain below the 6-cm level.

An alternative to the modeling described above is available in case the macro-model does not perform to the required specifications. As seen in the Seasat orbit fits discussed earlier, the estimating the once-per-revolution density correction parameters C_1 and C_2 is effective in reducing the radial orbit errors, which is the primary concern for an altimetric mission. When the TOPEX/POSEIDON simulation described above included the estimation of a pair of the C_1 and C_2 parameters every 2.5 days of the 10-day arc, the rms radial orbit error was only 1 cm. Since drag is not the major cause of orbit error for TOPEX/POSEIDON, the density correction parameters are apparently effective in absorbing most of the unmodeled radiation pressure effects. In fact, it can be shown that the estimation of either a radial or an along-track acceleration with a once-per-revolution variation is able to accommodate a secular trend in the orbit eccentricity over the fit interval and remove the corresponding effect on the radial component of the orbit. This demonstrates that estimating appropriately chosen empirical orbit correction parameters can significantly reduce the level of orbit error without employing complex surface force models. There may be difficulties, however, caused by high correlations between the various parameters being estimated, particularly when the tracking data are sparse.

Surface Forces on Lageos

Since the beginning of the Lageos mission, an anomalous along-track acceleration of the satellite has been observed which is reducing the semimajor axis by approximately 1 mm per day. The mean values of the drag-like acceleration determined every 15 days by UT/CSR from a 14.1-year arc are plotted in Figure 5. The mean of the acceleration over the entire arc is about −3.5 picometers sec^{-2} with fluctuations that are sometimes as large as the mean. The largest variations (or spikes) always occur when the Lageos satellite is experiencing eclipsing of the Sun by the Earth, although every eclipsing interval does not necessarily generate a spike. Thus both the mean and the variations require explanation. Although the spikes were difficult to explain, it was considered likely that the neutral and charged particle drag on the Lageos satellite could explain the average semimajor decrease [Rubincam, 1980; Afonso et al., 1985]. The effect of sunlight reflected by the Earth has also been considered [Anselmo et al., 1983b; Rubincam and Weiss, 1986; Barlier et al., 1986; Rubincam et al., 1987; Lucchesi and Farinella, 1992]. It will be shown that there may be an explanation for the mean acceleration as well as for the spikes which occur during eclipse seasons. It is possible, however, that a significant part of the residual variations in the Lageos along-track acceleration is due to sunlight diffusely or specularly reflected by the Earth.

Thermal Drag

A thermal drag model, a variant on the Yarkovsky effect, has been proposed by Rubincam [1987, 1988] which is able to account for much of the observed average along-track acceleration. The infrared radiation from the Earth is absorbed by the laser retroreflectors, and, because Lageos is spinning rapidly, the heat distribution is uniform longitudinally but not latitudinally.

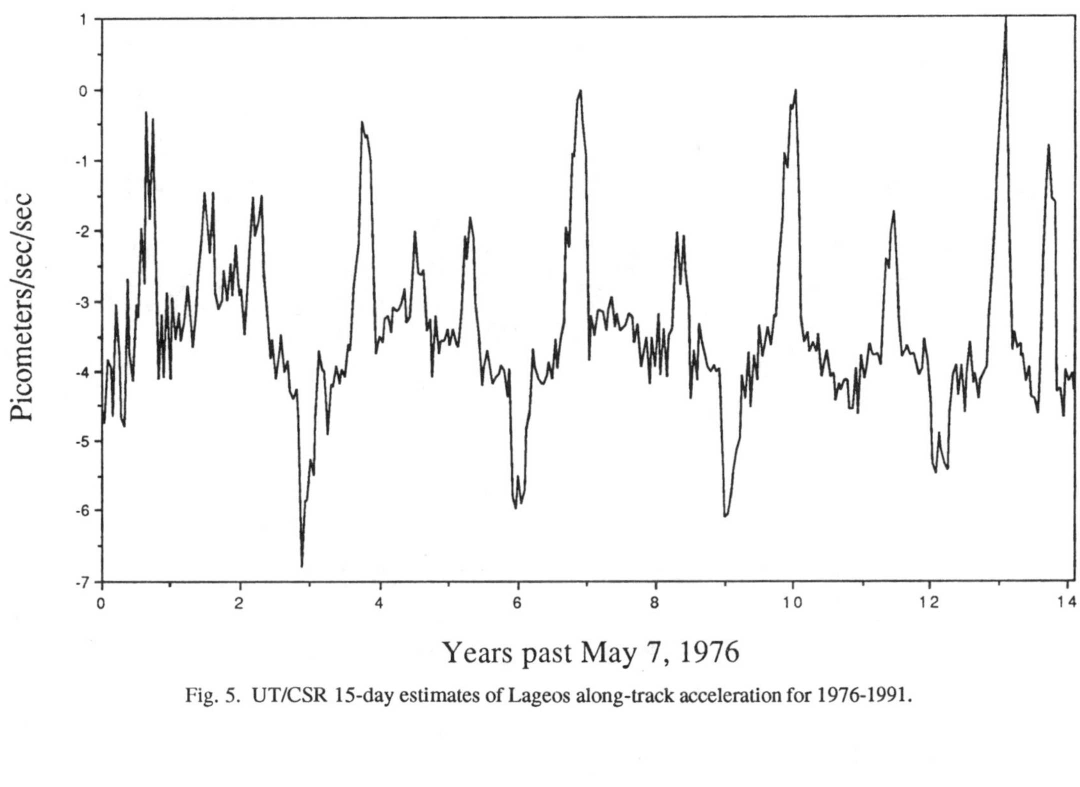

Fig. 5. UT/CSR 15-day estimates of Lageos along-track acceleration for 1976-1991.

This creates a temperature imbalance between the Lageos northern and southern hemispheres, generating a thrust along the spin axis as the heat is re-radiated. If the heat was emitted immediately, the thrust would have no net effect averaged over one orbital revolution of the Lageos satellite, but due to the combination of the geometry of the reflector mounting and the thermal properties of the materials, there is enough thermal inertia that some of the heat is not emitted immediately. Instead, the emission is delayed until the satellite has moved a significant distance along its orbit. This delay causes the thermal thrust to have a non-zero average in the along-track direction if the spin axis of the satellite is oriented in or close to the orbital plane. The proposed model accounts for about 70% of the observed drag, and the remainder of the average drag most likely consists of a combination of neutral particle drag and charged particle drag [Rubincam, 1990]. This model also predicts long-period variations about the mean with frequencies of once and twice per node revolution of the Lageos orbit (1050- and 525-day periods).

The effect of thermal drag due to heating by the Earth was simulated for the first 12.4 years of the Lageos orbit. The 15-day averages of the accelerations generated by the nominal Rubincam model, augmented by 1 picometer sec^{-2} to account for atmospheric drag [Rubincam, 1990], is compared in Figure 6 to the observed accelerations. Ignoring the spikes that occur in the observed drag during eclipse seasons, the periodic variation in the estimated drag appears to have the approximately same magnitude and phase as the observed drag.

Close examination of Figure 6 reveals that the nominal model (Rubincam + atmospheric drag) appears to diverge from the observed accelerations in the last part of the 12 year arc in both amplitude and phase (continuing to ignore the spikes during periods of shadowing). The once-per-node variation in the thermal drag is proportional to the cosine of the spin axis declination, and the phase will depend on the right ascension [Rubincam, 1987]. Thus, Figure 6 appears to suggest that the spin axis of Lageos is evolving from its original orientation (where the agreement between the nominal model and observed drag is good) and becoming aligned with the Earth's poles. A similar conclusion is reached by Rubincam [1990]. Rubincam [1987] suggested that the spin axis of the Lageos satellite should eventually align itself with the Earth's spin axis. This has been confirmed by Bertotti and Iess [1991], who have analyzed the effect of gravitational and magnetic torques on the satellite as the eddy current dissipation reduces the spin rate. In the UT/CSR model, a model for the spin axis evolution was determined empirically so that the thermal drag fits the observed drag better. In Figure 6, it can be seen that the modified model is able to maintain good agreement in amplitude and phase of the periodic variation throughout the 12 year arc. A few measurements of the actual Lageos spin vector are critically needed, since this effect, and the effects described below, depend strongly on the orientation of the spin axis.

Examination of the residuals in the orbit inclination provides additional evidence supporting the thermal drag model. In UT/CSR Lageos long-arcs where the thermal drag was not modeled, the residual errors in the orbit inclination have exhibited a slope of approximately 1.3 to 1.5 mas/year. The slope in the Lageos inclination residuals has been a mystery for some time, since there seemed to be no reasonable force which could generate this particular signal. The effect of a co-rotating atmosphere at

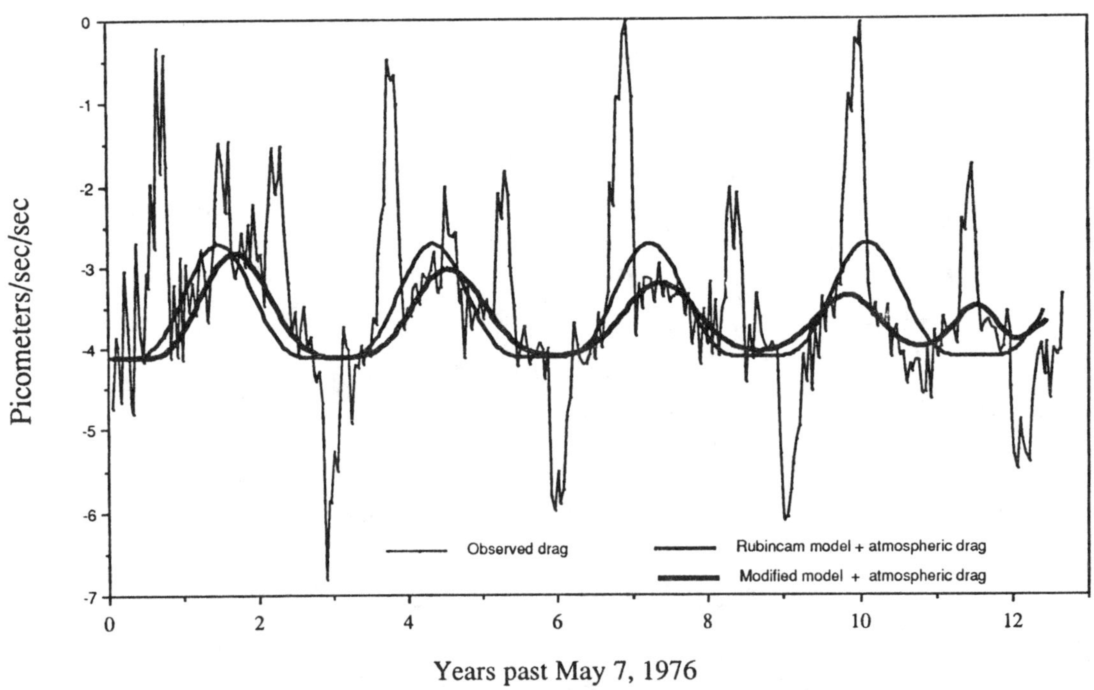

Fig. 6. Thermal drag models compared to observed along-track acceleration.

Lageos altitude was considered, but it was found that even if one assumed that 100% of the drag was atmospheric drag, a co-rotating atmosphere could account for less than 10% of the inclination slope. By incorporating the thermal drag model in the latest long-arc, the inclination slope is only 0.3 mas/year. It is very convincing evidence that the thermal drag model, which was based on engineering data and not on the observed accelerations, is able to explain much of the average drag, the variation at the 1050- and 525-day periods, and most of the peculiar inclination slope. If the magnitude of the thermal drag force was increased to about 90% of the observed drag, then the remaining inclination slope would be explained also. Any other explanation for the inclination slope is still unknown.

Yarkovsky-Schach Effect

A similar thermal thrust mechanism, due to heating by the Sun, has been proposed to account for at least part of the large variations that occur only during eclipse seasons [Rubincam, 1982; Slabinski, 1988; Afonso et al., 1989; Scharroo et al., 1991; Ries et al., 1991]. Since the spin axis orientation is essentially fixed with respect to the Sun over an orbit, the hemisphere of the Lageos satellite which is experiencing summer (tilted towards the Sun) will be warmer than the opposite hemisphere at every point in the orbit. This temperature difference will be essentially constant except for the additional influence of the Earth heating. Ignoring the Earth's heat and the effect of shadowing, it can be seen that, since the thermal thrust generated by solar heating will be uniform over the orbit, there is no significant net effect on the orbit. Consequently, between eclipse seasons, the thermal drag effect and atmospheric drag are able to account for the systematic part of the observed drag. The high-frequency fluctuations between eclipse seasons may be evidence of errors in the modeling of Earth reflected sunlight.

During shadowing, however, the solar thermal force does not average out, and there is a net along-track acceleration. If the temperature gradient relaxation time is long compared to the eclipse duration, the perturbation will be primarily a positive along-track effect [Rubincam, 1982; Slabinski, 1988], but both positive and negative accelerations can be generated if the relaxation time is short [Afonso et al., 1989; Scharroo et al., 1991]. A model for this effect was included in the force model for Lageos, and the estimated drag was compared to the observed drag. The maximum magnitude of the Yarkovsky-Schach acceleration (80 picometer sec^{-2}) was empirically chosen to give peaks with amplitudes comparable to the observed drag. Afonso et al. [1989] and Scharroo et al. [1991] independently analyzed the expected temperatures and arrived at accelerations with similar magnitudes. The spin axis orientation in the UT/CSR model was based on the modified spin axis model when calculating the "season" for each hemisphere. Unfortunately, it has not been possible to find a set of parameters for this model alone which can predict the correct series of peaks.

Asymmetric Reflectivity

An additional mechanism is proposed that, when combined with the Yarkovsky-Schach effect, appears to account for the spikes observed during the eclipse seasons. If it is assumed that the two halves of the satellite do not have the same effective reflectivity, then there will be an asymmetry in the solar radiation pressure on the satellite that, averaged over the spin period, will be along the direction of the spin vector [Rubincam et al., 1987]. The cause of this asymmetry is open to speculation. Scharroo et al. [1991] speculates that the two halves of the satellite may have been finished differently and finds that the northern hemisphere of Lageos need only be about 1% more reflective than the southern one. Ries et al. [1991] suggests that the non-symmetrical distribution of the infrared corner-cubes may be the cause. There are three infrared cubes in the southern hemisphere and only one in the northern hemisphere, and since they appear opaque at optical wavelengths, the northern hemisphere is likely to be more reflective than the southern hemisphere. Whatever the source of the asymmetry, augmenting the thermal forces with an imbalance in the hemispheric reflectivity leads to a model that is capable of generating accelerations similar to those observed on the Lageos satellite over the first twelve years, as shown in Figure 7. The agreement, however, is far from perfect, which illustrates that the proposed models are not definitive and still require additional analysis. Such analysis may need to include consideration of the effects of Earth-reflected sunlight since, clearly, the observed along-track acceleration of Lageos represents a complex combination of several different phenomena.

Lageos Spin Vector

As the Lageos spin vector evolves with time, the models described above will become less reliable. Analysis of the along-track accelerations estimated in the latest Lageos long-arc, which incorporated the UT/CSR surface force models described above, indicates that the agreement after the first twelve years is degrading. It is critical for the modeling of these surface forces that measurements of the spin axis are obtained. Eventually, however, the spin rate will slow enough to result in chaotic behavior. When this occurs, it is not clear whether the thermal forces will average out and thus be attenuated, or simply become more difficult to model.

Conclusions

The increasing orbit accuracy requirements for geodetic and oceanographic satellites places stringent demands on the mathematical models which are used to represent the dynamical effects in the orbit determination process. Great progress has been achieved in modeling the gravitational effects, and the accuracy of the models for the nongravitational effects has become very important. Considerable effort has been expended over the last two decades in developing atmospheric density models, but there appears to be little progress in the accuracy of the density models for precision orbit determination applications. It may well be that the atmospheric density at satellite altitudes can never be computed to sufficient accuracy due to the complex and random nature of the density variations. In addition, data are limited on the gas-surface interactions at satellite altitudes and

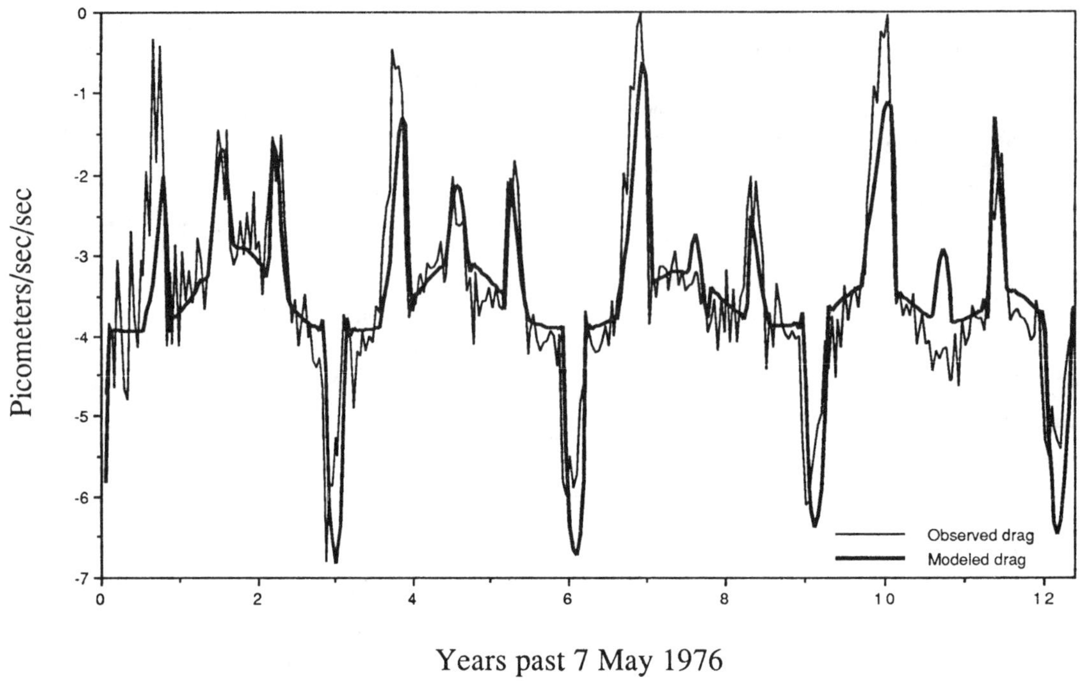

Fig. 7. Combined models for thermal drag, Yarkovsky-Schach, asymmetric reflectivity and atmospheric drag compared to observed acceleration.

velocities. The effects of radiation pressure and thermal thrusts on a satellite of complex shape can also be very difficult to model. Even the most detailed and computationally intensive model for these surface forces will be limited by uncertainties in the orientation, thermo-optical properties and temperature of the satellite surfaces (which are all generally changing with time). Consequently, while more complex models for the surface forces are required, the estimation of appropriate parameters in the force models must also be considered if the accuracy of the tracking data and gravitational models is to be fully exploited. Satellite missions such as ERS-1 and TOPEX/POSEIDON will provide excellent opportunities to test models and strategies for determining precise orbits for satellites with large surface forces.

Acknowledgments. This research was supported partially by the Jet Propulsion Laboratory under contract JPL-956689, and NASA Headquarters, under contract NAGW-1330. Additional computing resources for this work were provided by the University of Texas System Center for High Performance Computing.

References

Afonso, G., F. Barlier, C. Berger, F. Mignard, and J. J. Walch, Reassessment of the charge and neutral drag of Lageos and its geophysical implications, *J. Geophys. Res., 90,* 9381, 1985.

Afonso, G., F. Barlier, M. Carpino, P. Farinella, F. Mignard, A. Milani, and A. M. Nobili, Orbital effects of Lageos seasons and eclipses, *Ann. Geophys., 7,* 501-504, 1989.

Aksnes, K., Short-period and long-period perturbations of a spherical satellite due to direct solar radiation pressure, *Celest. Mech., 13,* 89-104, 1976.

Anselmo, L., B. Bertotti, P. Farinella, A. Milani, and A. M. Nobili, Orbital perturbations due to radiation pressure for a spacecraft of complex shape, *Celest. Mech., 29,* 27-43, 1983*a*.

Anselmo, L., P. Farinella, A. Milani, and A. M. Nobili, Effects of the Earth-reflected sunlight on the orbit of the Lageos satellite, *Astron. Astrophys., 117,* 3-8, 1983*b*.

Antreasian, P. G., and G. W. Rosborough, A prediction of radiant energy forces on the TOPEX/POSEIDON spacecraft, *J. Spacecraft and Rockets, 29,* 81-90, 1992.

Baker, R.M.L., Jr., Radiation on a satellite in the presence of partly diffuse and partly specular reflecting body, *Trajectories of Artificial Celestial Bodies (1965),* J. Kovalevsky (Ed.), Springer-Verlag, 85-150, 1966.

Barlier, F., Y. Boudon, J. L. Falin, R. Futaully, J. P. Villain, J. J. Walch, A. M. Mainguy, and J. P. Bordet, Preliminary results obtained from the low-g accelerometer CACTUS, *COSPAR Space Research, XVII,* M. Rycroft and A. Stickland (Eds.), Pergamon Press, 341-347, 1977.

Barlier, F., C. Berger, J. Falin, G. Kockarts, and G. Thuillier, Atmospheric model based on satellite drag data, *Ann. Geophys., 34,* 9-24, 1978.

Barlier, F., C. Berger, J. Falin, G. Kockarts, and G. Thuillier, Comparisons between various semiempirical thermospheric models of the terrestrial atmosphere, *J. Atmosph. Terrest. Phys., 41,* 527-541, 1979.

Barlier, F., and C. Berger, A point of view on semiempirical thermospheric models, *Planet. Space Sci., 31,* 945-966, 1983.

Barlier, F., M. Carpino, P. Farinella, F. Mignard, A. Milani, and A. M. Nobili, Non-gravitational perturbations on the semimajor axis of Lageos, *Ann. Geophys., 4A,* 193-210, 1986.

Bertotti, B. and L. Iess, The rotation of Lageos, *J. Geophys. Res., 96,* 2431, 1991.

Borderies, N., and P. Longaretti, A new treatment of the albedo radiation pressure in the case of a uniform albedo and of a spherical satellite, *Celest. Mech., 49,* 69-98, 1990.

Bryant, R. W., The effect of solar radiation pressure on the motion of an artificial satellite, *Astron. J., 66,* 430, 1961.

Burns, J. A., P. L. Lamy, and S. Sater, Radiation forces on small particles in the solar system, *Icarus, 40,* 1-48-48, 1979.

Cheng, M. K., Time variations in the Earth's gravity field from Starlette orbit analysis, *CSR-88-6*, 193 pp., Center for Space Res., Univ. of Tex. at Austin, 1988.

Coster, A. J., and E. M. Gaposchkin, Evaluation of new parameters for use in atmospheric models, *Astrodynamics 1987, Vol. 65, Adv. Astron. Sci.*, J. K. Solder et al. (Eds.), 1669-1682, 1988.

Elyasberg, P., B. Kugaenko, V. Synitsyn, and M. Voiskovsky, Upper atmosphere density determination from the COSMOS satellite deceleration results, *Space Res., XII*, 1972.

Farinella, P., A. M. Nobili, F. Barlier, and F. Mignard, Effects of thermal thrust on the node and inclination of Lageos, *Astron. Astrophys., 234*, 546-555, 1990.

Feulner, M. R., The numerical integration of near-Earth satellite orbits across SRP boundaries using the method of modified back differences, *CSR-TM-90-03*, 106 pp., Center for Space Res., Univ. of Tex. at Austin, August 1990.

Fliegel, H. F., W. A. Feess, W. C. Layton, and N. W. Rhodus, The GPS radiation force model, in *Proc. of the First International Symposium on Precise Positioning with the Global Positioning System*, National Geodetic Survey, NOAA, Rockville, Md., 113-119, 1985.

Fliegel, H. F., T. E. Gallini, and E. R. Swift, Global Positioning System radiation pressure force model for geodetic applications, *J. Geophys. Res., 97*, 559-568, 1992.

Gaposchkin, E. M., and A. J. Coster, Evaluation of recent atmospheric density models, *Astrodynamics 1987, Vol. 65, Adv. Astron. Sci.*, J. K. Solder et al. (Eds.), 1701-1720, 1988.

Haley, D., An expanded effective Earth radius for use in Earth shadow and solar radiation pressure calculations, NASA Contract NAS5-11933, 12 pp., Wolf Res. and Devel. Corp., Riverdale, Md., September 1974.

Harris, I., and W. Priester, Theoretical models for the solar cycle variations of the upper atmosphere, *NASA TN D-144*, Goddard Space Flight Center, Greenbelt, Md., 1962.

Hedin, A. E., H. G. Mayr, C. A. Reber, N. W. Spencer, and G. R. Carignan, Empirical model of global thermospheric temperature and composition based on data from the OGO 6 quadrupole mass spectrometer, *J. Geophys. Res., 79*, 1974.

Hedin, A. E., J. E. Salah, J. V. Evans, C. A. Reber, G. P. Newton, N. W. Spencer, D. C. Kayser, D. Alcaydé, P. Bauer, L. Cogger, and J. P. McClure, A global thermospheric model based on mass spectrometer and incoherent scatter data, MSIS 1, N2 density and temperature, *J. Geophys. Res., 82*, 2139, 1977*a*.

Hedin, A. E., C. A. Reber, G. P. Newton, N. W. Spencer, H. C. Brinton, H. G. Mayr, and W. E. Potter, A global thermospheric model based on mass spectrometer and incoherent scatter data, MSIS 2, composition, *J. Geophys. Res., 82*, 2148, 1977*b*.

Hedin, A. E., C. A. Reber, N. W. Spencer, H. C. Brinton, and D. C. Kayser, Global model of longitude/UT variations in thermospheric composition and temperature based on mass spectrometer data, *J. Geophys. Res., 84*, 1, 1979.

Hedin, A. E., A revised thermospheric model based on mass spectrometer and incoherent scatter data: MSIS-83, *J. Geophys. Res., 88*, 10,170-10,188, 1983.

Hedin, A. E., MSIS-86 thermospheric model, *J. Geophys. Res., 92*, 4649-4662, 1987.

Hedin, A. E., N. W. Spencer, and T. L. Killeen, Empirical global model of upper thermospheric winds based on atmosphere and Dynamics Explorer satellite data, *J. Geophys. Res., 93*, 9959-9978, 1988.

Herrero, F. A., Satellite drag coefficients and upper atmosphere densities: Present status and future directions, *Astrodynamics 1987, Vol. 65, Adv. Astron. Sci.*, J. K. Solder et al. (Eds.), 1607-1623, 1988.

Imbro, D. R., M. M. Moe, and K. Moe, On fundamental problems in the deduction of atmospheric densities from satellite drag, *J. Geophys. Res., 80*, 3077-3086, 1975.

Jacchia, L., New static diffusion models of the thermosphere and exosphere with empirical temperature profiles, *Spec. Rep. 313*, 87 pp., Smithsonian Astrophys. Obs., Cambridge, Mass., 1970.

Jacchia, L., Revised static models of the thermosphere and exosphere with empirical temperature Profiles, *Spec. Rep. 332*, 113 pp., Smithsonian Astrophys. Obs., Cambridge, Mass., 1971.

Jacchia, L., Thermospheric temperature, density, and composition: New models, *Spec. Rep. 375*, 107 pp., Smithsonian Astrophys. Obs., Cambridge, Mass., 1977.

Killeen, T. L., and R. G. Roble, Thermospheric dynamics: Contributions from the first five years of the Dynamics Explorer program, *Rev. Geophys., 26*, 329-367, 1988.

King-Hele, D. G., and D. M. C. Walker, Upper-atmosphere zonal winds from satellite orbit analysis, *Planet. Space Sci., 31*, 509-535, 1983.

Klinkard, H., Ch. Koeck, and P. Renard, Precise satellite skin-force modeling by means of Monte-Carlo ray tracing, *ESA Journal, 14*, 1990.

Knocke, P., J. C. Ries, and B. D. Tapley, Earth radiation pressure effects on satellites, in *Proc. AIAA/AAS Astrodynamics Conference*, Minneapolis, Minn., August 15–17, 1988.

Kockarts, G., Some recent advances in thermospheric models, *Adv. Space Res., 1*(197), 1981.

Kozai, Y., The effect of solar radiation pressure on the motion of an artificial satellite, *Spec. Rep. 56*, Smithsonian Astrophys. Obs., Cambridge, Mass., 1961.

Lafontaine, J., and P. Hughes, An analytic version of Jacchia's 1977 model atmosphere, *Celest. Mech., 7*, 1981.

Lála, P., and L. Sehnal, The Earth's shadowing effects in the short-periodic perturbations of satellite orbits, *Bull. Astron. Inst. Chechosl., 20*, 327-330, 1969.

Lála, P., Semianalytical theory of solar pressure perturbations of satellite orbits during short time intervals, *Bull. Astron. Inst. Chechosl., 22*, 63-72, 1971.

Lautman, D. A., Perturbations of a close-Earth satellite due to sunlight diffusely reflected from the Earth. I: Uniform albedo, *Celest. Mech., 15*, 387-420, 1977.

Laux, U., and U. von Zahn, Longitudinal variations in thermospheric composition under geomagnetically quiet conditions, *J. Geophys. Res., 84*, 1942, 1979.

Lucchesi, D., and P. Farinella, Optical properties of the Earth's surface and long-term perturbations of Lageos' semimajor axis, *J. Geophys. Res., 97*, 7121-7128, 1992.

Marcos, F. A., and J. M. Forbes, Thermospheric winds from the satellite electrostatic triaxial accelerometer system, *J. Geophys. Res., 90*, 6543-6552, 1985.

Marcos, F. A., Accuracy of satellite drag models, abstract in *Astrodynamics 1987, Vol. 65, Adv. Astron. Sci.*, J. K. Solder et al. (Eds.), 1721, 1988.

Marcos, F. A., T. L. Killeen, A. G. Burns, and R. G. Roble, Evaluation of new atmospheric drag modeling techniques, *Astrodynamics 1989, Vol. 71, Adv. Astron. Sci.*, C. L. Thornton et al. (Eds.), 763-770, 1990.

Marshall, J. A., S. B. Luthcke, P. G. Antreasian, and G. W. Rosborough, Modeling radiation forces acting on TOPEX/POSEIDON for precision orbit determination, Tech. Memo. 104564, NASA Goddard Space Flight Center, Greenbelt, Md., 1992.

McCarthy, J. J., and T. V. Martin, A computer efficient model of Earth albedo satellite effects, *Planet. Sci. Rep. 012-77*, NASA Goddard Space Flight Center, Greenbelt, Md., 28 pp., 1977.

Milani, A., A. M. Nobili, and P. Farinella, *Non-gravitational Perturbations and Satellite Geodesy*, Adam Hilger, 125 pp., 1987.

Musen, P., The influence of solar radiation pressure on the motion of an artificial satellite, *J. Geophys. Res., 65*, 1391-1396, 1960.

Muhleman, D. O., D. B. Holdridge, R. L. Carpenter, and K. C. Oslund, Observed solar radiation pressure perturbations of Echo I, *Science, 131*, 1487, 1960.

Nicolet, M., Density of the heterosphere related to temperature, *Spec. Rep. 75*, Smithsonian Astrophys. Obs., Cambridge, Mass., 1961.

Nouél, F., J. Bardina, C. Jayles, Y. Labrune, and B. Truong, DORIS: A precise satellite-positioning doppler system, *Astrodynamics 1987, Vol. 65, Adv. Astron. Sci.*, J. K. Solder et al. (Eds.), 311-320, 1988.

Nuth, V., On the estimation of atmospheric drag parameters for precision orbit determination, *CSR-TM-91-02*, 85 pp., Center for Space Res., Univ. of Tex. at Austin, 1991.

Parkinson, R. W., H. M. Jones, and I. I. Shapiro, Effects of solar radiation pressure on Earth satellite orbits, *Science, 131*, 920-921, 1960.

Pinker, R. T., and J. A. Razgaitis, Estimating the solar zenith dependence of the clear-sky planetary albedo for land surfaces from the GOES satellite, *J. Geophys. Res., 88*, 6007-6011, 1983.

Prior, E. J., Observed effects of Earth-reflected radiation and hydrogen drag on the orbital accelerations of balloon satellites, *The Use of Artificial Satellites for Geodesy*, S. Henriksen et al. (Eds.), 197-207, American Geophysical Union, Washington, D.C., 1972.

Ries, J. C., R. J. Eanes, and M. M. Watkins, The along-track acceleration of Lageos, *Eos Trans. AGU, 72*(17), Supplement p. 95, 1991.

Roberts, C. E., An analytical model for upper atmospheric densities based upon Jacchia's 1970 model, *Celest. Mech., 4*, 368-377, 1971.

Rothwell, D. A, and P. Moore, Atmospheric density parameters for Seasat laser ranging, *Adv. Space Res., 11*, 131-136, 1991.

Rubincam, D. P., Atmospheric drag as the cause of the secular decrease in the semimajor axis of Lageos's orbit, *Geophys. Res. Lett., 7*, 468-470, 1980.

Rubincam, D. P., On the secular decrease in the semimajor axis of Lageos's orbit, *Celest. Mech, 26*, 361-382, 1982.

Rubincam, D. P., Lageos orbit decay due to infrared radiation from Earth, *J. Geophys. Res., 92*, 1287-1294, 1987.

Rubincam, D. P., Yarkovsky thermal drag on Lageos, *J. Geophys. Res., 93*, 13,805-13,810, 1988.

Rubincam, D. P., Drag on the Lageos satellite, *J. Geophys. Res., 95*, 4881-4886, 1990.

Rubincam, D. P., and N. R. Weiss, The orbit of Lageos and solar eclipses, *J. Geophys. Res., 90*, 9399-9402, 1985.

Rubincam, D. P., and N. R. Weiss, Earth albedo and the orbit of Lageos, *Celest. Mech., 38*, 233-296, 1986.

Rubincam, D. P., P. Knocke, V. R. Taylor, and S. Blackwell, Earth anisotropic reflection and the orbit of Lageos, *J. Geophys. Res., 92*, 11,662-11,668, 1987.

Scharroo, R., K. F. Wakker, B. A. C. Ambrosius, and R. Noomen, On the along-track acceleration of the Lageos satellite, *J. Geophys. Res., 96*, 729-740, 1991.

Sehnal, L., Radiation pressure effects in the motion of artificial satellites, *Dynamics of Satellites (1969)*, B. Morando (Ed.), Springer-Verlag, 262-272, 1970.

Sehnal, L., The Earth albedo model in spherical harmonics, *Bull. Astron. Inst. Chechosl., 30*, 199-204, 1979.

Shum, C. K., J. C. Ries, B. D. Tapley, P. Escudier, and E. Delaye, Atmospheric drag model for precision orbit determination, *CSR-86-2*, 37 pp., Center for Space Res., Univ. of Tex. at Austin, 1986.

J. C. Ries, C. K. Shum, and B. D. Tapley, Center for Space Research, The University of Texas at Austin, Austin, Texas.

Drag Effects on the SPOT 2 Satellite as Measured by the DORIS Tracking System

F. NOUEL - JP. BERTHIAS - A. GUITART - A. PIUZZI - C. VALORGE

Centre National d'Etudes Spatiales - Toulouse - France

ABSTRACT

The second French Earth observation satellite SPOT 2 has on board an accurate tracking system called DORIS (Doppler Orbitography and Radiopositioning Integrated by Satellite). The DORIS ground network presently consists of 38 worldwide deployed beacons, while SPOT is in a Sun synchronous circular orbit at 832 km altitude.

Since the launch (Feb. 1990), the SPOT 2 orbit has been computed from DORIS tracking data with a precision of a few meters in the along track component. 1990 was a very active year in terms of solar activity reacting on the atmosphere and creating on the SPOT 2 orbit large atmospheric drag variations which were emphasized by the size of the platform. Results as well as computation strategies are presented to deal with the atmospheric effect on the SPOT trajectory. These computations took advantage of the quasi continuous tracking coverage, which was part of the DORIS concept.

Two atmospheric density models are compared, namely DTM and MSIS 86. The satellite shape is modeled taking into account the following components: solar panel and bus and their attitude. In the orbit computation software named ZOOM, several possibilities are examined to apply correcting parameters for minimizing the orbit errors. These options range from a scale factor for several days to one every three hours, or to insure continuity in the acceleration.

INTRODUCTION

The second French Earth observation satellite, SPOT 2, is equiped with a very accurate tracking system, DORIS (Doppler Orbitography and Radiopositioning Integrated by Satellite) [Dorrer et al., 1990]. DORIS consists of a receiver on board the spacecraft and 38 dual-frequency beacons deployed worldwide, all of them built around high quality Ultra Stable Oscillators. The worldwide distribution of the beacons ensures an almost continuous coverage of the SPOT 2 orbit.

Since the launch of the spacecraft in February 1990, its orbit has been computed on a regular basis using the DORIS data and a sophisticated orbit determination software package, ZOOM, developed by CNES for the future TOPEX/POSEIDON mission. Accuracies of a few meters along track are routinely obtained [Nouël et al., 1991]. The year 1990 and the first six months of 1991 have been particularly rich in terms of high solar activity. The interaction of the high solar fluxes and of the strong geomagnetic storms with the atmosphere triggered periods of highly fluctuating density. Due to its large size and its relatively low orbit (830 km altitude) SPOT 2 is very sensitive to density variations, and methods had to be devised to maintain the accuracy of the orbit determination. In return, some of the coefficients determined in the process provide a good insight into the actual density fluctuations.

DATA SELECTION

For the data analysis we have selected the period 13 to 28 March 1991 which exhibits an interesting range of solar activity. Plots of solar flux and of geomagnetic activity (3 hour planetary index Kp) for this period are given in Figure 1 and 2, respectively . We have divided the data into 3 five days arcs (13 to 17 March, 18 to 22 March, 23 to 28 March 1991) (17 March 91 = day 15050).

Each arc is characteristic of a different type of phenomena. The first arc corresponds to a quiet period. The flux is high overall, with a peak near 250.10^{-22} W m^{-2} Hz^{-1}. The geomagnetic index remains close to 2. During the second arc, the solar flux decreases to about 230.10^{-22} W m^{-2} Hz^{-1}. Meanwhile a very short geomagnetic storm brings the index to a high level (Kp = 6) for about 3 hours. The third arc contains an exceptionally powerful magnetic storm. The geomagnetic index reaches 9, the top of the scale, 3 days in a row. Meanwhile the solar flux drops sharply to about 200.10^{-22} W m^{-2} Hz^{-1}.

During this 2 week period the DORIS system functioned flawlessly. Over 75 000 measurements were obtained in more

Environmental Effects on Spacecraft Positioning and Trajectories
Geophysical Monograph 73, IUGG Volume 13

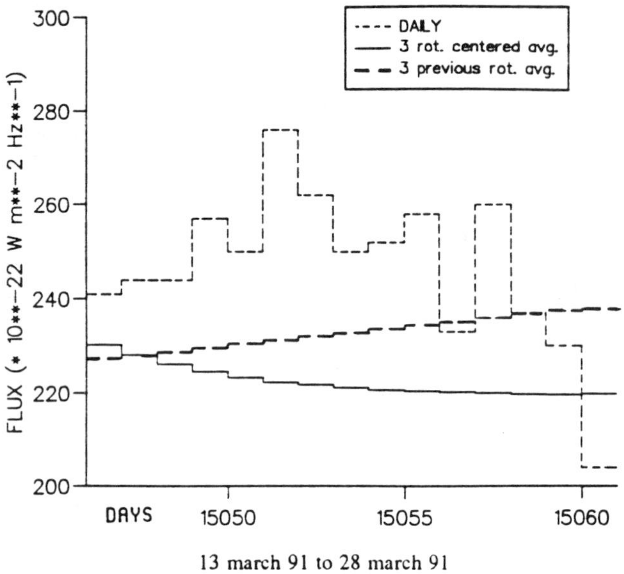

Fig. 1. Solar flux

than 2 500 passes. After removal of blunder points and application of an elevation cut-off criterion (10 degrees from 13 to 23 March and 15 degrees after removal of passes below 25 degrees from 24 to 28 March 91), we are left with nearly 55 000 high-quality data points, or, on average, 450 measurements per 3 hour period.

MODEL SELECTION

The CNES high-accuracy orbit determination software, ZOOM, is a versatile tool which allows the user to select the input models and to specify the spacecraft geometry. For this study we have selected the following inputs:

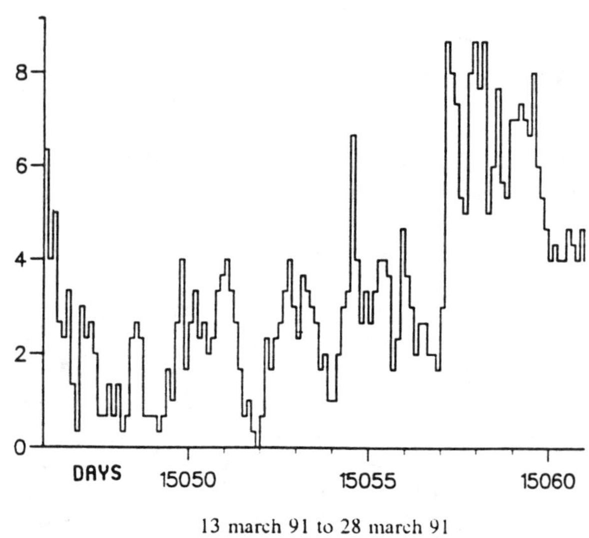

Fig. 2. Geomagnetic activity (Kp)

* Dynamical models:

- gravity field PGS-4405 obtained from the Goddard Space Flight Center (GSFC) (full 50th degree and order field);
- lunar and solar perturbations, both direct and indirect, with complete Wahr model for solid Earth tides;
- Schwiderski ocean tide model, truncated to degree 13 and order 3;
- planetary perturbations (planetary, solar and lunar ephemerides are provided by the Bureau des Longitudes);
- direct solar radiation pressure taking into account the bus and solar panel geometry of the spacecraft and its attitude. A force scale factor is solved for;
- atmospheric drag models for which the details are given below.

* Measurement models:

- ionospheric delay computed from the dual-frequency measurements;
- tropospheric delay computed with the CNET1 model provided by the Centre National d'Etudes en Télécommunications (CNET). This model uses the meteorological data received from the beacons;
- station coordinates set JCOD1 certified by the Institut Géographique National (IGN);
- variable data weight giving less weight to lower elevation measurements. As an example:

cut-off	10°	15°	20°	30°	90°
weight	0,53	0,79	0,89	0,95	0,99

- correction to the beacon frequencies assumed constant per pass and solved for.

ATMOSPHERIC DRAG MODELING

Two cases have been investigated. The first case uses the DTM model [Barlier et al., 1977] extended by the introduction of hydrogen, as provided by the Centre d'Etudes et de Recherches Géodynamiques et Astronomiques (CERGA). The second case uses the MSIS model [Hedin, 1989] received from GSFC. Both models are empirical 3 dimensional descriptions of the atmospheric density developed in terms of spherical harmonics. Their predictions are similar in standard conditions, but they exhibit very different behaviors in the presence of high geomagnetic activity.

The data sets processed by the authors of DTM and MSIS do not include periods of intense geomagnetic activity. Therefore, if we apply these models during a magnetic storm, we use them outside their domain of validity. To circumvent this problem we have modified the inputs to our models in such a way that we use the models in a range where we can trust their predictions. For DTM we fixed the Kp index at a value of 2, which corresponds to the largest fraction of the data that entered in the construction of the model. For MSIS we simply switched off the contribution from the geomagnetic activity. Concurrently, we introduced a set of scale factors which multiply the density predicted by the model. These scale factors are assumed constant for each 3 hour interval and are solved for from the tracking data. By doing so we generate empirical correction factors, which primarily represent the effect of geomagnetic activity on the atmosphere and which are very easy to compare with the indices measured over the same 3 hour period.

RESULTS

We found that the standard deviation of the post fit range-rate residuals is small, which is a proof of the good quality of the resulting ephemeris. For the first two 5 day arcs the r.m.s. is close to 1.8 mm s^{-1}, and nearly the same for cases 1 and 2. Expessed in along-track position error per pass, the residuals are less than 2 meters. For the third arc the r.m.s. is higher, 4 mm s^{-1} for case 1 and 5 mm s^{-1} for case 2. The corresponding along-track residual error per pass is at most 5 and 6 meters, respectively . For comparison, when the orbit for the third arc is evaluated using only 1 drag coefficient per day, there are 20 passes (for a total duration of 0.3 day) with residuals greater than 22 meters on 24 March, the day of the storm. The 3 hour drag scale factors are therefore necessary for accurate orbit determination in periods of rapidly varying geomagnetic activity.

The consistency between the solutions of case 1 and case 2 has been checked by comparing the resulting ephemerides (Figure 3a-3b-3c). As expected from a least square fit we observe a quadratic effect in the along-track direction and a bow-tie effect in the radial direction. The intense magnetic storm of 24 March is still visible with a differential effect less than 2 m along-track, 0.5 m radially and 0.05 in cross-track.

The drag scale factors are well determined, with a standard deviation of about 0.07 for values ranging from 0.5 to around 3. Figures 4 and 5 display the coefficients obtained for each of the 3 arcs for cases 1 and 2 respectively. The visual correlation between these plots and that of the geomagnetic index (Figure 2) is striking.

Variations in magnetic activity on the scale of a day correspond to similar variations of the coefficients. Shorter term fluctuations are not always followed by an effect on the scale factors. However, when sharp and strong increases in activity take place they give rise to an immediate response. The behaviour of the coefficients during the last arc is more surprising. After about 1 day at a high level, the density disturbance produced by the magnetic storm on 24 March seems to level off, despite the fact that the geomagnetic activity remains very high for 3 more days.

To be able to compare our experimental results with the predictions of the models, we have computed the ratio of the density evaluated using the true value of the geomagnetic index to the density evaluated with either the constant index (for DTM) or no index (for MSIS). Plots of this ratio computed along the satellite orbit are given in Figure 6 for DTM and 7 for MSIS. To help in comparing these ratios with the empirical drag factors we plot on Figures 8 and 9 both the factors and the average values of the ratios over the 3 hour intervals.

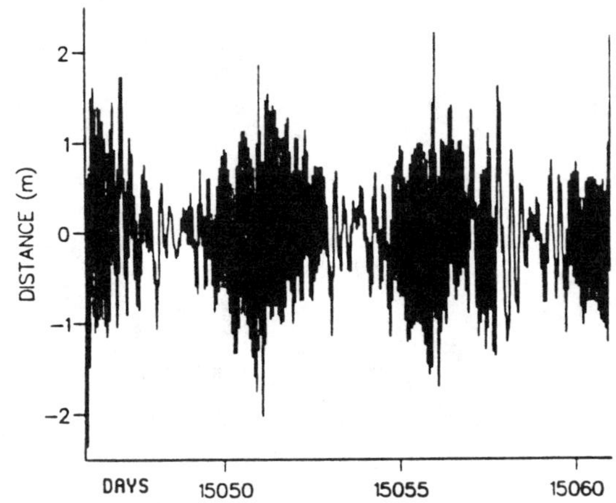

Fig. 3a. Along-Track distance

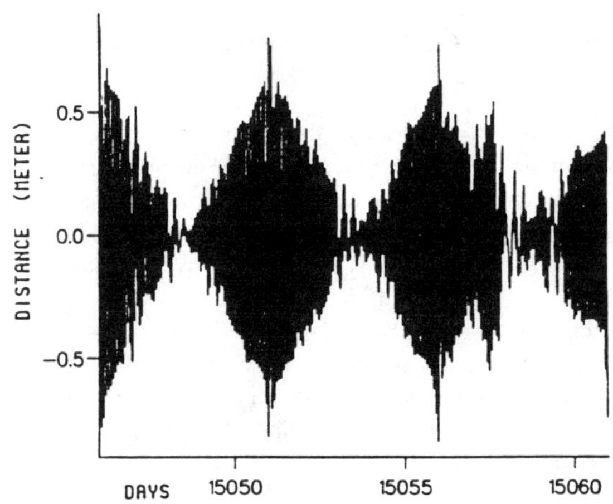

Fig. 3b. Radial distance

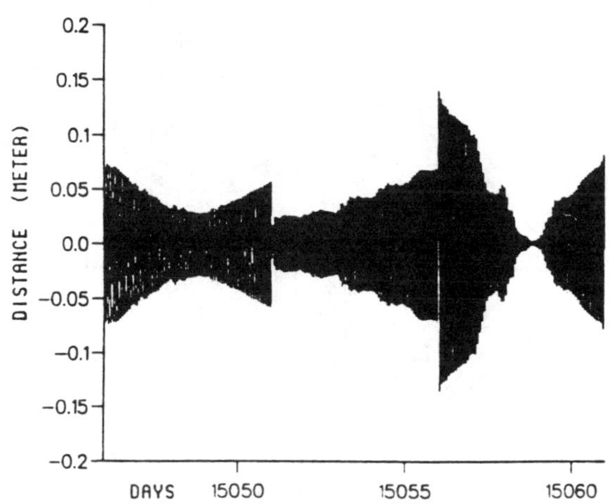

Fig. 3c. Cross-Track distance

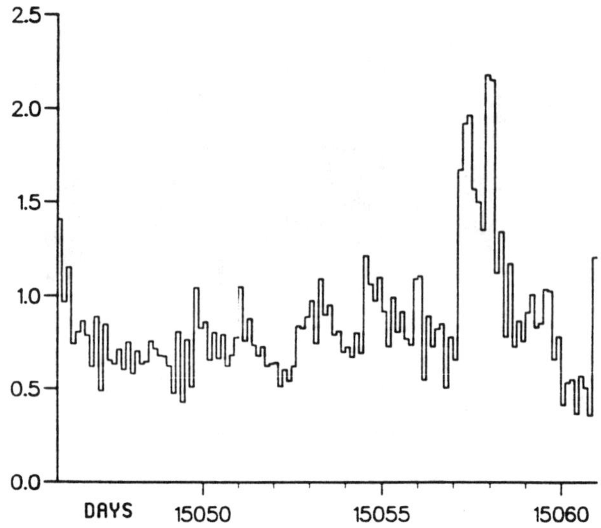

Fig. 4. Empirical drag coefficients with DTM Model

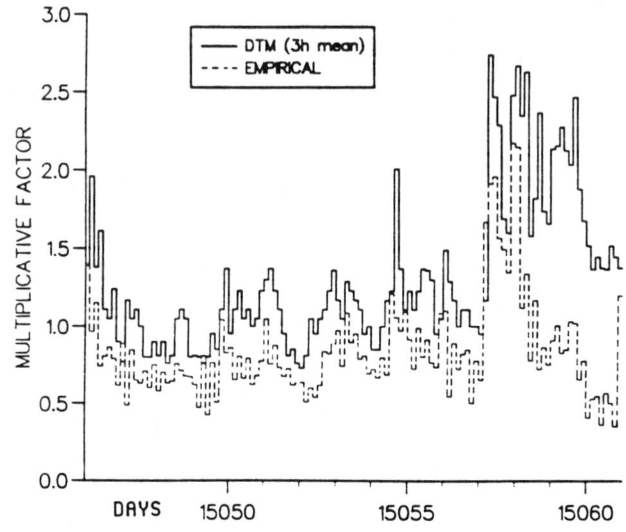

Fig. 6. Atmospheric response to geomagnetic activity with DTM model

These curves clearly show the good qualitative agreement between model and experimental results during periods of low geomagnetic activity. They also show that the empirical factors react faster to sharp increases than the models. In addition, there is an overall trend for the DTM ratio to overestimate the drag factors, while the MSIS ratio underestimates the experimental coefficient.

The rapid fall of the scale factors after the storm concurrent with the drop in solar flux hints at the possibility that the source of the difference is the poor modeling of the density response to flux variations.

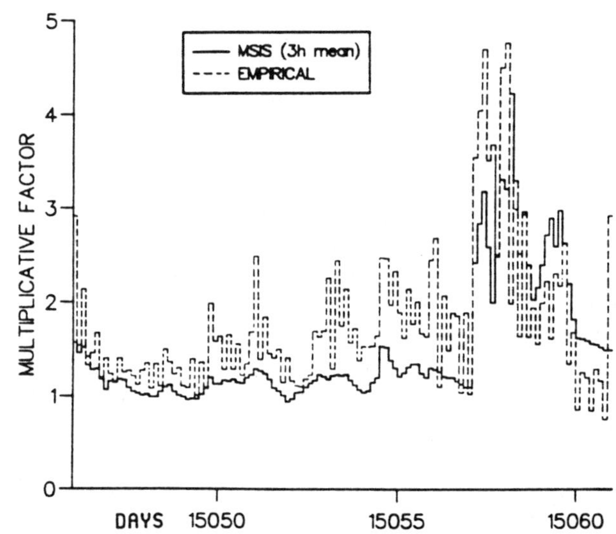

Fig. 7. Atmospheric response to geomagnetic activity with MSIS model

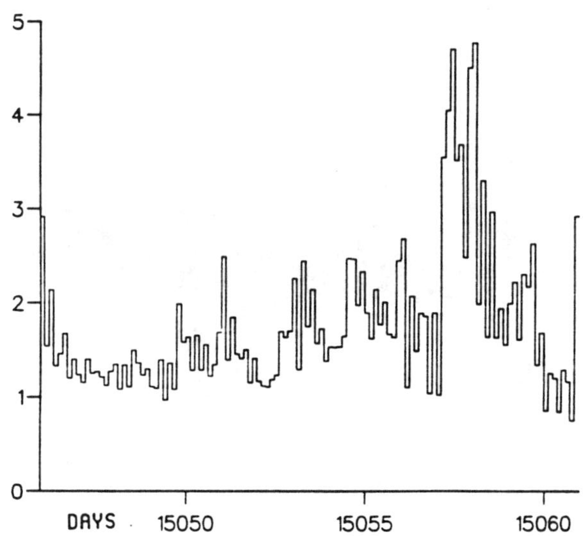

Fig. 5. Empirical drag coefficients with MSIS modem

It should be noted that the mean ratios and scale factors represent different types of averaging. Therefore, one has to be careful when comparing them. The mean ratios are arithmetic averages of densities over 3 hour periods. The scale factors represent the averaged effect of the density on the spacecraft during the same time. However, as can be seen in Figure 6 and 8, during periods of low geomagnetic activity the model ratios are nearly constant, and therefore the two averaging procedures cannot lead to significantly different results.

CONCLUSION

Because of the continuous orbital coverage provided by the DORIS tracking system it is possible to solve for empirical

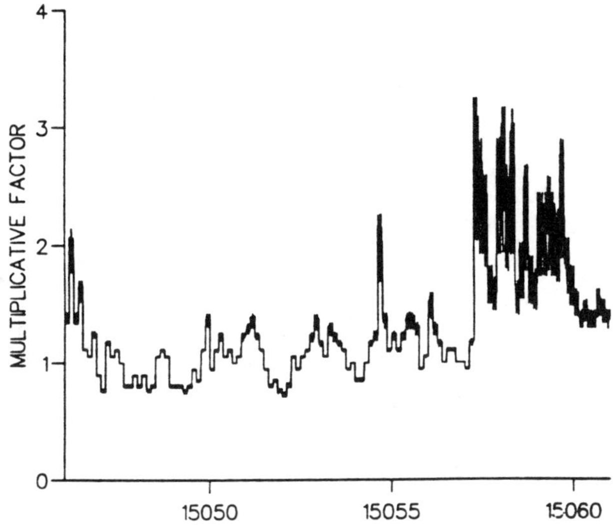

13 march 91 to 28 march 91

Fig. 8. DMT density ratio

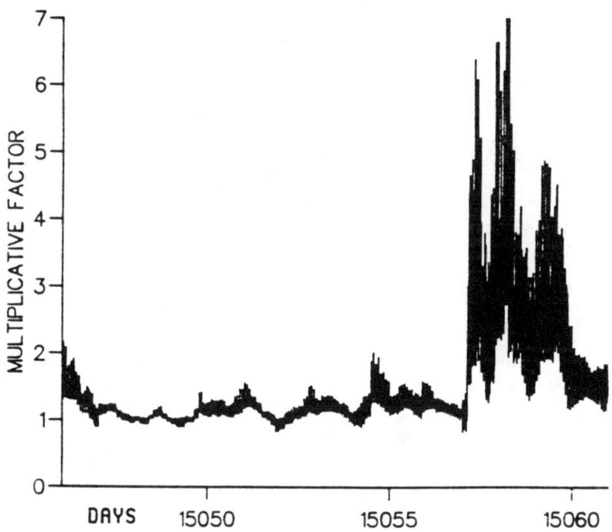

Fig. 9. MSIS density ratio

corrections to the modeled atmospheric density. With a coefficient per 3 hour period, the quality of the resulting ephemeris is good, even during periods of high solar activity. In practice, as we have done, the user can switch off the part of the atmosphere models which requires the input of the geomagnetic indices. Hence, there is no need to wait for the publication of definitive indices in order to obtain a good ephemeris.

However, this technique has its limitations. Using two different underlying density models, DTM and MSIS, we obtain two significantly different ephemeris, even after solving for 3 hour density correction factors. As we have no way to decide which ephemeris is best, we are left with an uncertainty in our resulting product which is higher than anticipated on the basis of the measurements.

As a byproduct, the 3 hour scale factors enable us to probe the short-term response of the atmosphere. Of course, these coefficients absorb all the deficiencies of the dynamical models, including gravity field imperfections. However, the correlations that we have found between the variations of the scale factors and that of the solar activity are strong indicators that they actually absorb density modeling errors.

In order to improve the accuracy of orbit determination for low altitude orbiters, atmosphere density models need to be improved. The DORIS data provides us with new information that could greatly benefit the development of improved density models. So further collaboration between the two disciplines seems to be the way to go.

Acknowledgements. The work described in this paper was carried out by the Centre National d'Etudes Spatiales with support from the DORIS project. Special thanks are due to B.H. Putney, S. Nerem, F. Lerch of the Goddard Space Flight Center for providing us with the PGS-4405 gravity field before publication, and for continuing the level of scientific openness and cooperation between our institutions that was established by J.G. Marsh.

REFERENCES

Barlier F., Berger C., Falin JL., Kockarts G. and Thuillier G., A thermospheric model based on satellite drag data, Aeronomica Acta, A185, 1977.

Dorrer M., Laborde B., and Deschamps P., DORIS : system assessment results with DORIS on SPOT 2, IAF Dresden, October 1990.

Hedin AE., MSIS-86 thermospheric model, J. Geophys R., 92, 1987.
Hedin AE., The atmosphere model in the region 90 to 2000 km, COSPAR 1989.

Nouël F., Berthias JP., Broca P., Deleuze M., Guitart A., Laudet P., Pierret C., Piuzzi A. and Valorge C., Precise orbit determination of SPOT platform with DORIS, AAS paper 91-409, AAS/AIAA Astrodynamics Specialist Conference, Durango, CO, August 1991.

F. Nouël, JP. Berthias, A. Guitart, A. Piuzzi, C. Valorge, CNES, 18, Avenue Edouard Belin, 31055 Toulouse Cedex France.

Particle and Thermal Drag Forces on Geodetic Satellites at High Altitudes

FRANÇOIS BARLIER AND FRANÇOIS MIGNARD

Observatoire de la Côte d'Azur and Centre d' Etudes de Géodynamiques et d' Astrométrie , Avenue Copernic 06130 Grasse (France)

At the LAGEOS altitude, particle and thermal drag acting on a satellite are significant perturbations of the trajectory.The recent launch of the geodetic satellites ETALON calls for an evaluation of the drag force of non-gravitationnal origin. In this presentation we review the existing modelling developed for LAGEOS and extend it to the case of ETALON. We show that observations of both the electric and so-called thermal drag are within the capabilities of the laser tracking of ETALON as has been the case for LAGEOS for many years.

INTRODUCTION

The advent of third generation laser tracking stations allows us to measure routinely the distance of a geodetic satellite to cm-accuracy. Almost fifteen years of continuous observations of LAGEOS have permitted a close investigation of the long term change in the orbital parameters. In particular numerous studies have been devoted to the origin of the unmodelled average along-track acceleration of -3.3 x 10^{-12} ms^{-2} (Fig. 1) [Smith and Dunn, 1980, Afonso et al., 1985, Rubincam, 1987, Mignard et al.,1990, Rubincam, 1990]. Particle and thermal drag have been proved to be of importance at high altitudes. Several periodic effects have since been recognized and discussed, for example, the Earth's albedo effect [Anselmo et al., 1983, Barlier et al., 1986, Rubincam et al., 1987, Luchesi and Farinella, 1991] or the thermal thrust - the Yarkovsky - Schach effect - [Slabinski, 1988, Afonso et al., 1989, Rubincam, 1988, Ries et al., 1991] or again an albedo asymmetry on the satellite [Rubincam et al., 1987, Scharroo et al.,1991, Rubincam, 1990]. Although the debate is still open as to the relative importance of these effects and despite the noise that arises from the superimposition of many small non-gravitational effects, a model can be proposed which fits the available observations reasonably well. All the results depend on the spin position of the satellite and in the future, this orientation will change as the satellite will be despun by eddy currents [Bertotti and Iess, 1991]. In this paper we review and extend the modeling developed for LAGEOS to evaluate the implications on the orbit of the Soviet-built satellites ETALON due to the drag and to the thermal thrust. Already along-track accelerations of the order of few 10^{-13} ms^{-2} have been reported by Reigber, (1991) and Bass and Tatevian , (1991). LAGEOS and ETALON orbital parameters are given in Table 1.

Environmental Effects on Spacecraft Positioning and Trajectories
Geophysical Monograph 73, IUGG Volume 13

NEUTRAL DRAG

Neutral drag has been included as a perturbing force acting on a satellite since the launch of the first artificial satellite. It results from encounters between the atoms and molecules of the upper atmosphere and the surface of the satellite. From a long tradition in this field the drag is usually expressed as a drag acceleration by :

$$\gamma = \frac{1}{2} C_d \rho \frac{S}{M} V_S^2$$

where C_d is the drag coefficient, usually on the order of 2 to 5 and V_S the satellite velocity in Earth-fixed rotating frame. The density is ρ and S/M is the satellite area-to-mass ratio. At the altitude of a geodetic satellite like LAGEOS the atmosphere is mainly composed of hydrogen atoms, which outnumber other species such as helium and oxygen by at least two orders of magnitude. Because of the low mass of the hydrogen atoms (m_H), their thermal velocity, $V_T = (8kT/\pi m_H)^{1/2}$ is comparable or larger than the satellite orbital velocity. T stands for the temperature of the gas, k for the Boltzman constant. The apparent motion of the gas seen from the satellite is no longer that of a nearly monodirectional stream, but rather of a slightly asymmetric random motion with more particles colliding from the forward direction. The detailed computation of this effect is given in Afonso et. al. , 1985. They show that the drag coefficient is :

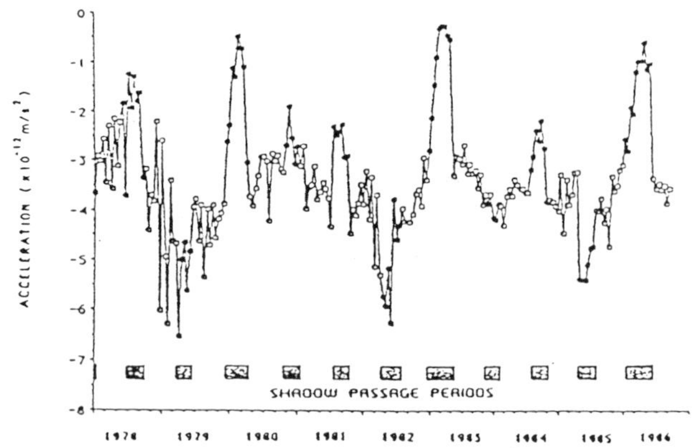

Fig. 1. Along-track acceleration residuals in LAGEOS' orbit determination carried out at GSFC (presented at Bologna in May 1986 at the Wegener/Medlas meeting by P. Dunn et al. and reproduced by courtedy of the authors). One constant acceleration value per every 15 days orbital arc was solved for. Time intervals when the orbit was subjected to eclipses are shown, and the corresponding points are represented as full squares. The along-track acceleration averages about to -3.4 x 10^{-12} ms^{-2}, but is modulated by a pattern of long-term oscillations, with eclipses corresponding in most cases to extrema of the curve.

$$C_d = \delta \left[\frac{8}{3} \left(\frac{V_T}{V_S} \right) + \frac{8}{15} \left(\frac{V_S}{V_T} \right) \right] \quad (V_T > V_S)$$

and applies both to the neutral and charged particles. In this expression δ is the accomodation factor equal to 1.4 for hydrogen and 1.3 for helium. This coefficient would be 1 for a perfect absorber and 2 for elastic collisions.

Temperature and particle density are not precisely known and depend on the position of the satellite, on the date, and also on the local time [Hedin et al., 1977a,b]. However one can use average conditions to evaluate the magnitude of the neutral drag for LAGEOS and extrapolate to ETALON by assuming a diffusive equilibrium in the atmosphere. There are more uncertainties in the case of ETALON, first, because the physical conditions of the exosphere at the altitude of ETALON are poorly known, and also because the model used to compute the drag coefficient assumes that the velocity distribution of the particles is isotropic. This is no longer the case at an altitude of 20000 km, since the particles are in the ballistic regime and have preferentially an upward velocity.

With the figures listed in Table 2, the neutral drag on LAGEOS comes to $0.3 \times 10^{-12} ms^{-2}$, that is to say about 10 percent of the observed drag. The neutral drag on ETALON, turns out to be a hundred times smaller and is not detectable.

ELECTRIC DRAG.

In the case of LAGEOS it was proposed as early as 1980, that the main source of drag would be produced by the collisions with charged particles, rather than neutral particles.There were three reasons to support this statement:

- The charged particles (H^+,O^+, He^+) are more numerous than the neutral ,
- The temperature is higher,
- The collisional cross section is larger than the geometric cross section.

Through collisions with the electrons of the plasmasphere a conducting satellite acquires a net negative charge and accordingly in the present case a electric potential on the order of -1 to -3V [Al'Pert, 1974]. The trajectories of the protons are bent toward the satellite as a result of the electrostatic force directed toward the center of the satellite. Thus collisions may occur with particles having an impact parameter larger than the satellite radius. In addition some particles are simply scattered without impacting the satellite; they do however exchange linear momentum and contribute to some extent to the drag (Fig.2).

Afonso et al., (1985) have modelled the electric potential around LAGEOS by computing a spherically symmetric solution of the Poisson equation characterized by the Debye length and the potential at the surface of the satellite. In this case the potential at a distance, r, from the center of the satellite is given by :

$$U(r) = U_0 \frac{R}{r} \exp [(R-r)/D]$$

where D is the Debye's length

TABLE 1. LAGEOS and ETALON orbital parameters

		LAGEOS	ETALON-1
Date		4 May 1976	14 Jan 1989
Semi Major axis	a (km)	12270	25503
Mean motion	n (rev/day)	6.3866	2.1315
Eccentricity	e	0.0044	0.0006
Inclination	i (deg)	109.75	64.87
Long. node	Ω (deg)	28.8	175.65
Arg. Perigee	ω (deg)	261.3	254.35
Node rate	dΩ/dt (deg/day)	+0.343	-0.033
Perigee rate	dω/dt (deg/day)	-0.214	-0.004

TABLE 2. Physical parameters for LAGEOS and ETALON related to the neutral drag.

	LAGEOS	ETALON-1	RATIO E/L
Radius (m)	0.300	0.647	2.16
Mass (kg)	470	1344.5	4.65
S/M ($m^2 kg^{-1}$)	6.9 x 10^{-4}	9.9 x 10^{-4}	1.30
Velocity (m/s)	5705	3950	0.692
V_T (m/s)	4800 (1)	4800(1)	1
C_d (H)	4.0	4.5	1.1
ρ (H) (m^{-3})	4x10^9	8x10^7	0.02
γ (m s^{-2})	3x10^{-13}	5x10^{-15}	0.016

[1] Hydrogen , T = 1100K

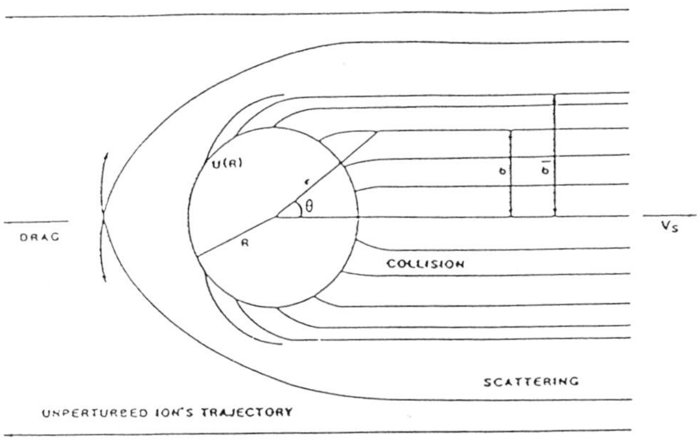

Fig. 2. Charge drag effect. Trajectories of ions in the neigbourhood of a negatively charged satellite. (After Afonso et al., 1985).

$$D=\sqrt{\frac{\varepsilon_0\ kT}{e^2 n_e}}$$

with $\varepsilon_0 = 8.84 \times 10^{-12}$ F m^{-1}, the permittivity of the vacuum and n_e the electron density, e is the elementary charge,R the satellite radius. $U_0 = U(R)$ denotes the electric potential of the satellite surface, and is of the order of -1 to -3V. With this potential it is possible to compute the trajectories of a proton as a function of the initial velocity and impact parameter and to determine the collisional cross section. Let $\bar{b}$ be the largest impact parameter yielding a collision for a particular initial velocity. The relative collision cross section $\bar{b}^2/R^2$ shown in Fig. 3 depends on the ratio of the electrostatic energy, e.U(R), to the kinetic energy of the particles E and also on the ratio of the satellite radius to the Debye length (R/D). For LAGEOS and ETALON the quantity (e.U(R)/E) is of the order of -3 and the relative collision cross section which is always less than 1 - e.U(R)/E i.e less than 4.

Finally the charge drag can be expressed as:

$$\gamma = \frac{1}{2} N_0\, m\, \frac{\bar{b}^2}{R^2} C_D \frac{S}{M} V_S^2$$

where N_0 is the number density of ions and m their mass.

Data pertaining to LAGEOS and ETALON are summarized in Table 3. The drag coefficient C_d is computed in the same way as for the neutral particles, taking into account the higher thermal velocity of the ions and the smaller orbital speed.The plasmasphere parameters (temperature and ion concentration) need now be estimated.

Estimate of the Plasmasphere Parameters

Around the earth, the magnetosphere extends to a very large distance and displays a complex structure (see for example Rishbeth, 1969). Magnetospheric phenomena are often described with the help of a specific type of coordinate system. This is the (B,L) system of Mc Ilwain (1961) in which B is the field intensity and L the geocentric distance of a given

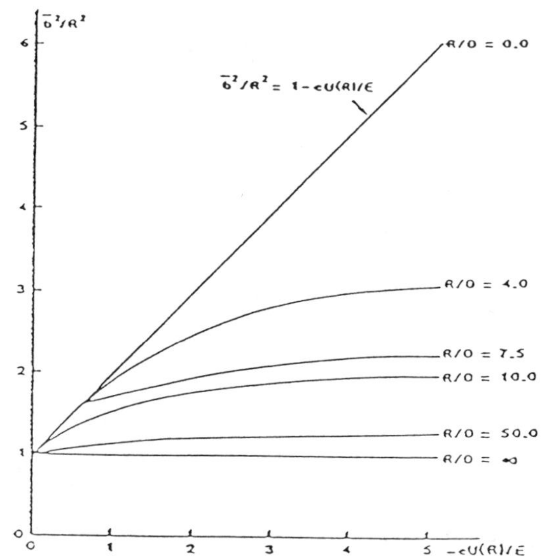

Fig. 3. For ions, collision cross section with respect to the neutral collision cross section in the case of a sphere ($\bar{b}^2/R^2$). The abscissa is the ratio of the electrostatic energy to the kinetic energy of the particle (eU(R)/E) for different values of the ratio of satellite radius to Debye length (R/D). It is assumed that the electronic temperature equals the ionic temperature. (After Afonso et al., 1985).

magnetic shell in the equatorial plane (surface of revolution of a field line around the dipole axis). L is measured in units of the Earth's radius. The plasmasphere is situated within the magnetosphere and is characterized by significant fluxes of protons [Van Allen and Frank, 1959]. It is bounded by the plasmapause where a rapid outward decrease of the concentration can be observed down to a factor of 0.1 at about L=4, (Fig. 4) while the temperature of the plasma jumps very significantly (Fig. 5). This is the "knee" of the plasmasphere. As a result LAGEOS and ETALON are within the plasmasphere in the equatorial region (Fig.6).

TABLE 3. Physical parameters for LAGEOS and ETALON related to the electric drag

	LAGEOS	ETALON-1	RATIO E/L
Temperature (K)	3000	10 000	3.3
n_e (m^{-3})	3.3×10^9	3.3×10^8	0.1
n (H^+)	3.0×10^9	3.0×10^8	0.1
n (He^+)	0.3×10^9	0.3×10^8	0.1
Debye length (m)	0.062	0.38	6.1
$\bar{b}^2/R^2$	2.4	3.1	1.3
C_d (H^+)	5.6	13.9	2.5
C_d (He^+)	3.4	7.2	2.1
γ (H^+ equator) ms^{-2}	7.6×10^{-13}	1.6×10^{-13}	0.21
γ (He^+ equator)	1.8×10^{-13}	0.3×10^{-13}	0.17
γ (equator)	9.4×10^{-13}	1.9×10^{-13}	0.20

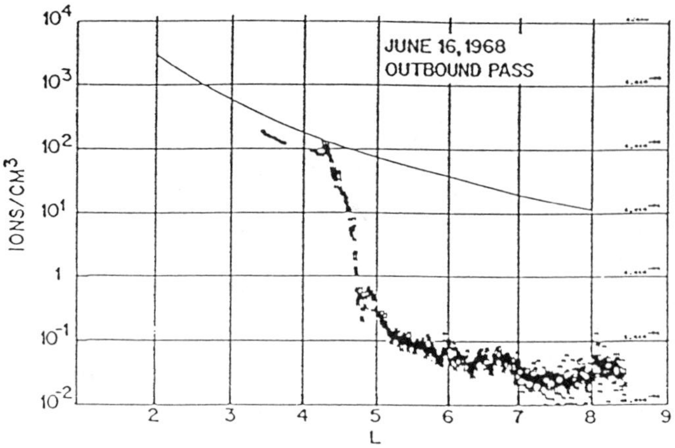

Fig. 4. An outbound pass made on 16 June, 1968. The measurements of H^+ ions as a function of L were made by the ion mass spectrometer on OGO 5 satellite in the 2100 LT region. The lower sensitivity of the instrument is approximately 0.1 ion/cm^3. (After Chappell et al., 1971).

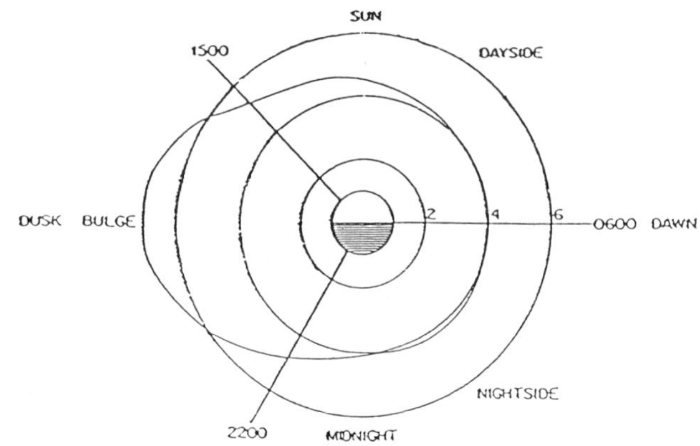

Fig. 6. An L local-time plot showing the different local-time sectors of the plasmasphere, dayside, nightside and bulge. The dark line shows the average plasmapause position determined from more than 150 OGO 5 profiles. (After Chappell et al., 1971).

It is of interest to note that according to Chappel et al., (1970) the plasmapause is correlated with the average geomagnetic activity and there is a general decrease in radius with increasing activity. There is also a diurnal effect, the plasmasphere expansion being a more noticeable on the dayside than on the nightside with a significant bulge at dusk [Chappel et al., 1971].

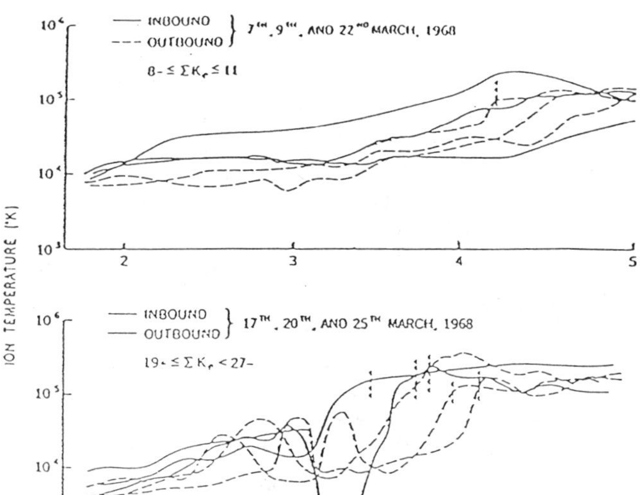

Fig. 5. A series of ion temperature profiles from OGO 5 satellite, as a function of L, grouped on the basis of the 3 hourly magnetic activity index K_p. The top panel (a) shows the measured profiles for relatively quiet times and the lower panel (b) for disturbed period (After Serbu et al., 1970).

Experimental data obtained in the plasmasphere (Temperature and ion concentrations) exhibit significant fluctuations depending on the solar time and on geomagnetic activity but average values and limits of variations can be determined from previous satellite experiments (from the OGO 5 satellite, [Serbu and Maier, 1970, Harris et al., 1970, Chapell et al., 1970], from ISIS 2 satellites, [Lanzerotti and Fukunishi, 1975] from PROGNOZ, [Gringauz, 1976], and from the work by Chiu et al, 1979. LAGEOS moves in the equatorial and mid-latitude regions at L = 2, where the temperatures ranges from 2000K to 15000K and H^+ concentration from 10^9 to 10^{10} m^{-3} and He^+ and O^+ around 2 x 10^8 m^{-3}. Mean values for the physical conditions at L=2 were chosen in the past as follows : T = 5000K for the temperature, n(H^+) = 3 x 10^9m^{-3}, n(He^+) = n(O^+) = 2 x 10^8m^{-3}.

More recently, new data have been obtained with the DYNAMICS EXPLORER data (DE-1). Mean values can be found in a statistical survey of plasmaspheric ion properties performed by Comfort et al. (1988). These averages represent more than 100 transits in two local time regions (7-11 hours and 18-22 hours), obtained in October and November 1981. It is of interest to note that the L parameter is not sufficient to determine the parameters which are also dependent on the altitude above the earth surface especially at L = 4. This means that when the satellite crosses the magnetic lines corresponding to L = 3 or 4 in the auroral regions, the H^+ concentration decreases so that the charge drag is significantly smaller in these high latitudes regions than in the equatorial or mid-latitude regions, and all the more so as the satellite crosses magnetic lines corresponding to L greater than 4. Finally in order to get an order of magnitude for the effects, we have performed the numerical calculation with :

LAGEOS :

L = 2 ; T= 3000K ; n(H^+) = 3 x 10^9 m^{-3}; n(He^+) = 3 x 10^8 m^{-3}

ETALON :

$L = 4$; $T=10000K$; $n(H^+) = 3 \times 10^8\ m^{-3}$; $n(He^+) = 3 \times 10^7 m^{-3}$

The consistency of the two sets of plasmaspheric parameters for LAGEOS, based on two independant space experiments ten years apart, gives an indication of the reliability of these data.

ESTIMATE OF CHARGE DRAG

The charge drag can now be estimated for the two previous sets of plasmaspheric parameters.With the old data set referred to above, we obtained for LAGEOS a drag acceleration of 13 x 10^{-13} ms^{-2}. Similar results based on the more recent estimates of the plasmasphere parameters are listed in Table 3, for LAGEOS and extropolated to ETALON.

An additional effect of 10% which depends on electric charge has to be included to take into account the scattering effect [Afonso et al., 1985]. For LAGEOS ,the total effect in the equatorial region is of the order of 1.0 x 10^{-12} ms^{-2} to 1.3 x 10^{-12} ms^{-2} while it amounts to 2.0 x 10^{-13} ms^{-2} for the ETALON satellites, according to the data set used.

It should be noted that the present estimate of the drag does not differ strongly from values derived earlier by Afonso et al. (1985), as long as the satellite lies in the equatorial plane. In reality both LAGEOS and ETALON spend half their time in the plasmasphere, while the ion density falls off rapidly outside it, when the satellites orbit in the upper latitudes.

As a result the average along track charge drag is probably half its value in the equator, namely 5 to 6.5 x $10^{-13}ms^{-2}$ for LAGEOS (~ 20% of the unmodelled acceleration), and only 1.0 x 10^{-13} ms^{-2} for ETALON.

Concerning LAGEOS, our conclusion is that altogether charge and neutral drag account for 30% of the total drag, in agreement with Rubincam (1990); the other major single source contributing to the secular decrease of the semimajor axis is mainly due to the Yarkosky thermal drag. A possible positive acceleration due to anisotropic solar heating could play a role but certainly not very important [Afonso et al., 1989, Mignard et al., 1990].

THERMAL DRAG

A surface at temperature T radiates energy in the form of photons which carry away momentum. If various parts of the surface of a satellite are not at the same temperature, a net force will act upon the satellite. In 1987, Rubincam has shown that such a process originated by the infrared terrestrial emission leads to an anistropic temperature distribution of the satellite's surface, the amplitude of which depends on the orientation of the spin axis with respect to the heating source.

More generally, this process has to be extended to any other radiating sources. On a spinning satellite like LAGEOS or ETALON there are seasons with a summer and a winter hemisphere at any time of the year. The difference of illumination between this hemisphere pointing toward the Sun and that pointing in the opposite direction, yields a net temperature difference between the two hemispheres. Once averaged over the spin period, this results in a cold and warm hemisphere, as for the Earth when north polar region are plunged in the dark during winter, while the south enjoys a mild temperature. More photons are radiated by the warmer region than by the cooler, which entails a net force directed along the spin axis toward the winter side. The effect is investigated in detail in Afonso et al. 1989 and only the essential results are referred to here.

The seasonal effect on LAGEOS and ETALON

The recoil acceleration is given by :

$$\gamma = - \alpha\ (\omega\ .\ s)\ \omega$$

where ω is the unit spin vector and s the sun pointing unit vector and α a numerical factor depending on the geometry and thermal properties of the satellite. In the absence of eclipse the above acceleration averages out to zero over one orbit. The exact balance between positive and negative along track acceleration breaks down when the satellite crosses the Earth's shadow. This occurs roughly during the eclipse season when the longitude of the sun and that of the satellites ascending node are close to zero and π. In this case the average along track force is preferentially positive, although there exists specific geometrical configurations which result in a negative force. For LAGEOS, the maximum possible acceleration is directed along the spin axis and was found by Afonso et al. (1989) to be:

$$\gamma_L = -\ 5.89\ 10^{-11}\ ms^{-2}\quad (\omega\ .\ s)$$

If only an order of magnitude is sought after, this value can be extrapolated to ETALON taking into account only the change in the area to mass ratio between the two satellites and the relative percentage of the surface of the retroreflectors, as well. This gives,

$$\gamma_E = -\ 6.80\ 10^{-11} ms^{-2}\quad (\omega\ .\ s)$$

This recoil acceleration must be averaged out over a complete orbit; only the drag component along the velocity vector is of interest here. Therefore we actually compute

$$\gamma_E = -\ 6.80\ 10^{-11}\ ms^{-2}\quad (\omega\ .\ s)\quad (\omega\ .\ v)$$

where **v** is the unit velocity vector. The above expression is integrated and averaged out over one orbital period, taking into account the passage through the shadow where the solar heating is turned off. The result is rather sensitive to the orientation of the spin axis. To carry out our numerical estimates we have used two possible solutions to determine the post-launch ω of ETALON-1. The first one takes ω aligned along the velocity vector at the apogee kick and gives in equatorial coordinates α = 110 deg, δ = 20 deg. The second solution is based on information provided to us by S. Tatevian (private communication), which indicates that at launch time, the spin axis of ETALON made an angle of 153 deg with the sun direction. This gives roughly α = 170 deg and δ = -15 deg. Results are shown on Fig. 7 for ETALON. The maximum effect

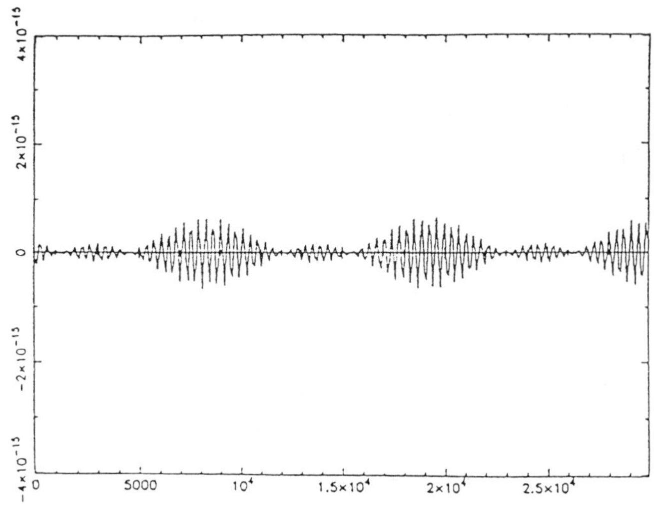

Fig. 7. Model of along-track acceleration of ETALON-1 due to the Sun seasonal anisotropic thermal effect as a function of the number of days since the launch date (Jan. 14, 1989). The equatorial coordinates of the spin axis are α = 110 deg, δ = 20 deg

remains very small, typically of 10^{-15} ms^{-2} while it is as large as 10^{-12} ms^{-2} for LAGEOS [Afonso et al., 1989] as the asymmetry introduced by the shadow is less pronounced for ETALON than for LAGEOS. The geocentric anglular size of the trajectory in the shadow decreases with increasing altitudes. At most the shadow crossing lasts 16 percent of the orbital period for LAGEOS, but only 8 percent for ETALON.

Drag due to the infrared terrestrial heating.

This situation is different for the infrared terrrestrial emission. According to Rubincam (1987), the component γ directed along the spin axis can be written as follows :

$$\gamma = -\,0.41 \frac{4\pi\varepsilon R_L \Delta F_{IR}^{0} f_0}{9 M_L c (1+\zeta^2)^{1/2}} \cos(n_L t - \delta)$$

with $\zeta = \tan\delta$

where δ is the thermal lag angle, c the speed of light, ε the retroreflector emissivity (0.9), $\Delta F_{IR}{}^{o}$ is the Earth infrared irradiance (62.5 Wm^{-2}), fo the fraction of the front face to the total surface area of the retroreflector (0.3). The factor 0.41 comes from the fact that the retroreflectors cover only 41% of the LAGEOS' surface. R_L stands for the radius of the satellite, M_L for its mass, n_L for its mean motion.

Then $$\gamma = S_{MAX} \frac{2}{\sin\delta} \cos(n_L t - \delta)$$

With $$S_{MAX} = -\,0.41 \frac{2\pi\,\varepsilon\, R_L^2\, \Delta F_{IR}^{0}\, f_0}{9 M_L c (1+\zeta^2)}\,\zeta$$

In a new model Rubincam (1988) proposed :

S_{MAX} = 2.781 x 10^{-12} ms^{-2} and δ = 55 deg

The recoil acceleration projected on the velocity vector must be averaged over an orbit. A mean value of -2.3 x 10^{-12} ms^{-2} is found.

The extrapolation to ETALON can be done by computing the Earth infrared irradiance at the ETALON altitude (about 15 Wm^{-2}) as well as the value of δ by assuming that the thermal response of the reflectors is the same as for LAGEOS. By taking into account the value of the radius of ETALON and the number of retroreflectors one find for S_{max} a value of 7.0 x 10^{-13} ms^{-2}. The average value will depend strongly on the position of the spin axis but the order of magnitude could be a few $10^{-13}$$ms^{-2}$ (Fig. 8a and Fig. 8b). For ω (110 deg, 20 deg), the along-track acceleration is now of the order of 10^{-13} ms^{-2}. By combining this value with the charge drag a total along-track deceleration

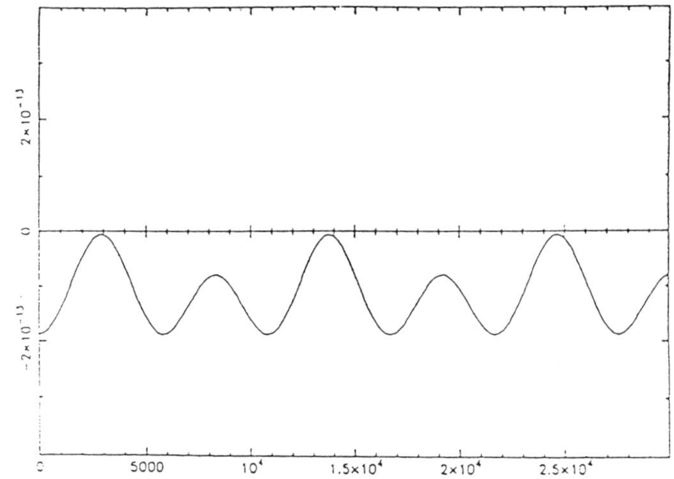

Fig. 8a Model of along-track acceleration of ETALON-1 (ms^{-2}) due to the infrared terrestrial anisotropic thermal effect (Yarkowsky - Schach effect) as a function of the number of days since the launch date (Jan. 14, 1989). The equatorial coordinates of the spin axis are : α = 110 deg, δ = 20 deg.

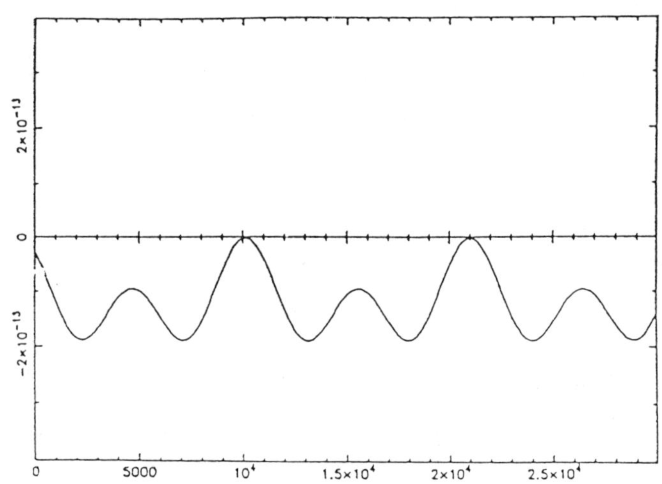

Fig. 8b Same as Fig. 8a but the coordinates of the spin axis are : α = 170 deg, δ = -15 deg.

of about 2 x 10^{-13} ms^{-2} is obtained in a reasonable agreement with the present observations B.Schutz (private communication). The model predicts that this average along-track acceleration should increase as a function of time for the years to come as a result of the change of the geometry of the orbital plane with respect to the Sun.

CONCLUSION

We have updated our previous estimates of the neutral, electric and thermal drag on the geodynamic satellite LAGEOS, using the most recent determinations of the plasmasphere ionic composition, density and temperature. These new estimates do not lead to significant changes compared to our previous computations . We end up with a neutral drag of 3 x 10^{-13} ms^{-2} and an average charge drag of 5 x 10^{-13} ms^{-2}. With the thermal effect due to the Earth's infrared radiation [Rubincam 1990] of 23 x 10^{-13} ms^{-2}, the total adds up to 31 x 10^{-13} ms^{-2}, very close to the observed unmodelled drag.

A similar modelling is applied to the soviet built ETALON, by extrapolating physical parameters from an altitude of 5800 km to 19000 km. Neutral drag as expected, is negligible being less than $10^{-14}ms^{-2}$, the charge drag amounts to 10^{-13} ms^{-2}, while the thermal effect averages out over duration of 30 years, corresponding to the nodal period, to 10^{-13} ms^{-2}. The current behaviour is highly dependent on the spin axis direction . We have considered two likely cases which give at the moment respectively a drag of 2 x 10^{-13} ms^{-2} and 1.3 x 10^{-13} ms^{-2} .

It is always difficult to assess precisely the level of uncertainty for these various drag effects. Regarding LAGEOS where the agreement with the observations for the secular trend is now rather good, the estimates can be expected to be accurate to 10 or 20% but for ETALON the level of uncertainty is certainly much higher on the order of 50%. Models which are Physically more realistic are required.

In any case, the total drag on ETALON is larger than $10^{-13}ms^{-2}$ and well within the capabilities of the laser tracking stations, provided long arc solution are built.

REFERENCES

Afonso, G., F. Barlier, C. Berger, and F. Mignard, Effet du freinage atmosphérique et de la traînée électrique sur la trajectoire du satellite LAGEOS, *C.R. Ac. Sci. Paris, 290* B 445-448, 1980.

Afonso, G., F. Barlier, C. Berger, F. Mignard, and J.J. Walch, Reassessment of the charge and neutral drag of Lageos and its geophysical implications, *J. Geophys. Res.*, 90, 9381-9398, 1985.

Afonso, G., F. Barlier, M. Carpino, P. Farinella, F. Mignard, A. Milani, and A.M. Nobili, Orbital effects of LAGEOS seasons and eclipses, *Ann. Geophysicae.*, 7, (5), 501-514, 1989.

Al'pert, Y.L., Waves and satellites in the near-earth plasma, in *Studies in Soviet Science*, pp. 45-47, Consultants Bureau, New York, 1974.

Anselmo, L., P. Farinella, A. Milani, and A. M. Nobili, Effects of the earth reflected sunlight on the orbit of the LAGEOS satellite, *Astron. Astrophys.*, 117, 3-8, 1983.

Barlier, F., M. Carpino, P. Farinella, F. Mignard, A. Milani, and A. M. Nobili, Non-Gravitational perturbations on the semimajor axis of LAGEOS, *Ann. Geophysicae*, 4A(3), 193-210, 1986.

Bass, V.P., S.K. Tatevian, Non-gravitational perturbations of the ETALON orbit, International Symposium ETALON satellites, laser data analysis (under press), Moscow, 1991.

Bertotti, B., and L. Iess, The rotation of LAGEOS, *J. Geophys. Res.*, 96, 2431, 1991.

Chappell, C.R., K.K. Harris, and G.W. Sharp, A study pf the influence of magnetic activity on the location of the plasmapause as measured by OGO 5, *J. Geophys. Res.*, 75, 50-56, 1970.

Chappell, C.R., K.K. Harris, G.W. Sharp, The dayside of the plasmasphere, *J.Geophys.Res.*, 76, 7632-7647, 1971.

Chiu, Y.T., J.G. Luhmann, B.K. Ching, and D.J. Boucher, An equilibrium model of plasmaspheric composition and density, *J. Geophys. Res.*, 84, 909-916, 1979.

Comfort, R.H., I.T. Newberry, and C.R. Chappell, Preliminary statistical survey of plasmaspheric ion properties from observations by DE1/RIMS, *Geophysical Monograph 44*, Modeling Magnetospheric plasma ed. by T.E. Moore, J.M. Waite, J.r. Editors 107-114, 1988.

Gringauz, K. I., The cold plasma throughout the magnetosphere, in *Physics of Solar Planetary Environments*, vol. 2, edited by D.J. Williams, pp. 672-682, AGU, Washington, D.C., 1976.

Harris, K.K., G.W. Sharp, and C.R. Chappell, Observations of the plasmapause from OGO 5, *J. Geophys. Res.*, 75, 219-224, 1970.

Hedin, A.E., J.E. Salah, J.V. Evans, C.A. Reber, G.P. Newton, N.W. Spencer, D.C. Kayser, D. Alcayde, P. Bauer, L. Cogger, and J.P. McClure, A global thermospheric model based on mass spectrometer and incoherent scatter data, MSIS 1, N_2 density and temperature, *J. Geophys. Res.*, 82, 2139-2147, 1977a.

Hedin, A.E., C.A. Reber, G.P. Newton, N.W. Spencer, H.C. Brinton, H.G. Mayr, and W.E. Potter, A global thermospheric model based on mass spectrometer and incoherent scatter data, MSIS 2, Composition, *J. Geophys. Res.*, 82, 2148-2156, 1977b.

Lanzerotti, L.J., and H.Fukunishi, Relationships of the characteristics of magnetohydro-dynamic waves to plasma density gradients in the vicinity of the plasmapause, *J. Geophys. Res.*, 80, 4627-4634, 1975.

Luchesi, D., and P. Farinella, Optical Properties of the earth's surface and long-term perturbations of LAGEOS' orbit, Sub. *J. Geophys. Res.*, 1991.

McIlwain, C.E., Coordinates for mapping the distribution of magnetically trapped particles, *J. Geophys. Res.*, 66, 3681-3691, 1961.

Mignard, F., G. Afonso, F. Barlier, M. Carpino, P. Farinella, A. Milani, and A.M. Nobili, LAGEOS: Ten years of quest for the non-gravitational forces, *Adv. Space Res.*, 10, (3) 221-(3) 227, 1990.

Ries, J.C., R.J. Eanes, and M.M. Watkins, The Along-Track Acceleration of Lageos, *EOS Transactions*, 72, 17, Supplement p.95, 1991.

Risbeth, H., and O.K. Garriott, Introduction to ionospheric physics, *International Geophysics series*, Vol. 14, Acad. Press, 1969.

Rubincam, D.P., Drag on the LAGEOS satellite, *J. Geophys. Res.*, 95 (B4), 4881-4886, 1990.

Rubincam, D.P., LAGEOS orbit decay due to infrared radiation from Earth, *J. Geophysical. Res.*, 92 (B2), 1287-1294, 1987.

Rubincam, D.P., P. Knocke, V.R. Taylor, and S. Blackwell, Earth anisotropic reflection and the orbit of LAGEOS, *J. Geophys. Res.*, 92 (B11), 662-668, 1987.

Rubincam, D.P., Yarkovsky thermal drag on LAGEOS, *J. Geophys. Res.*, 93 (B11), 13,805-13,810, 1988.

Reigber, C., A gravity model including ETALON data - GRIM-4. International Symposium ETALON Satellites, laser data analysis (under press), Moscow, 1991.

Scharroo, R., K.F. Wakker, B.A.C. Ambrosius, and R. Noomen, On the along-track acceleration of the LAGEOS satellite, *J. Geophys. Res.*, 96., 729-740, 1991.

Serbu, G.P., and E.J.R. Maier, Observations from OGO 5 of the thermal ion density and temperature within the magnetosphere, *J. Geophys. Res.*, 75, 6102-6113, 1970.

Slabinski, V., LAGEOS acceleration due to intermittent solar heating during eclipse periods, presented at the 10th meeting of the Division on Dynamical Astronomy, American Astronomical Society, Gaitherburg, Maryland, July 1988 (Abstract in Bull. Am. Astron. Soci. 20, 902, 1988).

Smith, D., and P. Dunn, Long term evolution of the LAGEOS orbit, *Geophys. Res. Lett.*, 7, 437-440, 1980 .

Van Allen, J.A., and L.A. Frank, Radiation around the earth to a radial distance of 107400km, Nature 183, 430-434, 1959.

F. Barlier and F. Mignard, OCA/CERGA, av. Copernic, 06130 Grasse, FRANCE.

Influence of Thermospheric Drag, Winds and Gravity Waves on a Gradiometer Mission at 200 km

CHRISTIANE BERGER AND FRANCOIS BARLIER

Observatoire de la Côte d'Azur
Centre d'Etudes et de Recherches Géodynamiques et Astronomiques
Avenue Copernic, 06130 Grasse, FRANCE

A dedicated mission at 200 km has been proposed by ESA and NASA to improve the short wavelengths of the gravity field expansion. This mission is based on gravity gradient measurement. Such a measurement is perturbed by air drag and its fluctuations induced by winds and gravity waves. The impact of air drag on the design of the mission and on the measurement is described as well as ways to overcome these problems.

INTRODUCTION

The purpose of a gradiometer mission is to measure the gravity tensor components in order to improve our knowledge of the gravity field of the Earth at short wavelengths, typically a few hundred kilometers when considering its expansion in spherical harmonics. To accomplish this, a dedicated mission lasting six months has been proposed by ESA and NASA for launch about 1998 at 200 km on a near polar orbit. It is the ARISTOTELES mission [ESA Report, 1990], including also geomagnetic field determination. The scientific importance of such a mission is described in Balmino et al. [1984] and Lambeck [1990].

The gradiometer consists of four ultra-sensitive accelerometers in which a proof mass is maintained inside a cavity without any mechanical contact by means of electrostatic forces [Bernard and Touboul, 1989]. The measurement of these electrostatic forces allows one to determine the components of the gravity tensor with a precision of 10^{-2} Eötvös (10^{-11} s^{-2}). If the distance between two accelerometers is one meter, it corresponds to a precision for the acceleration of 10^{-11} ms^{-2}. Unfortunately the gradiometer is not only sensitive to the gravity gradient existing in space but also to the non-gravitational forces acting on the satellite, mainly air drag. If the accelerometers were of ideal construction, it would be possible to separate in the gradiometer signal the part induced by the gravity gradient from the part induced by non-gravitational effects, but this can never be the case. Therefore, it is of primary interest to analyse the impact of the perturbations induced by air drag and its fluctuations due to winds and gravity waves on the gradiometer measurements as well as on the design of the general characteristics of the mission.

AIR DRAG EFFECTS

The problem has been recently studied [Touboul et al., 1991]. The purpose here is to recall the main results and to extend the analysis to new atmospheric features.

Role of Air Drag on the Characteristics of the Mission

Several aspects are successively considered.

a) The four three-axes accelerometers (i,j, k-axes in the accelerometer reference frame) are located in the corners of a plane normal to the orbit. The orbital variations of the acceleration with respect to the local three axes (x along the trajectory, y normal to the orbital plane and z along the descending vertical) are shown in Figure 1.

On the x-axis the acceleration induced by the air drag density varies slowly along the circular orbit due to altitude change above the Earth's surface. There are two minima, in absolute value, at the poles and two maxima at the equator, the maximum to minimum ratio being of a factor of two. The orbital variations of the acceleration on the y-axis are mainly due to the relative motion of the atmosphere with respect to the orbital plane. As the atmosphere is driven by the Earth's rotation in a first approximation, the acceleration is largest at the nodes (positive at the ascending node) and undergoes variations generated by horizontal winds. For a spherical satellite, the acceleration along the y-axis is less than 1% of the along orbit acceleration. But here, the largest acceleration along the y-axis is about ten times smaller than along the x-axis because of the area to mass ratio which is

Environmental Effects on Spacecraft Positioning and Trajectories
Geophysical Monograph 73, IUGG Volume 13

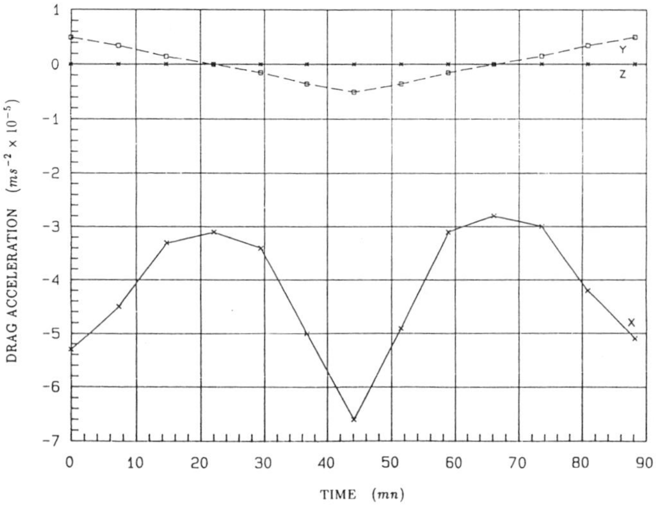

Fig. 1. Orbital variation of the satellite accelerations due to the aerodynamic drag with respect to the three local axes [Adapted from AERITALIA, 1990].

larger for the axis perpendicular to the orbit plane. The acceleration on the z–axis, generated by vertical winds, is smaller than along the y–axis.

b) Due to the shape of the satellite, the drag also entails external torques with maximal values of the order of 0.4 10^{-3} Nm. They must be counterbalanced by the attitude control system. Over the six month mission, the satellite is Earth–pointed with an accuracy better than 0.5° but the pointing stability is very stringent. It can be shown that the variations of the angular velocity and acceleration must be kept lower than 10^{-6} rad s^{-1} and 10^{-7} rad s^{-2}, respectively. Therefore, the external torques must be corrected very gently.

c) Air drag acting on a satellite at 200 km induces a sharp decrease of the altitude so that, for a six month mission, small thrusts must be periodically applied. This system entails critical problems such as to maintain the centre of mass of the satellite and the geometric centre within a few millimeters in the gradiometer plane or to limit movements of liquid (sloshing) in the hydrazine tanks which perturb the accelerometer measurements after each thrust.

d) The altitude at which the mission has to be performed must be chosen as low as possible in order to obtain maximum of sensitivity. However, a compromise must be found between the mean level of the air density, the volume of the hydrazine tanks and the duration of the mission.

e) As is shown in Figure 2, it is difficult to predict 7–8 years in advance the precise solar activity level. Uncertainties of 50 units in the solar flux are quite possible.They entail variations in temperature ranging from 125 K to 225 K depending on the mean level of the solar activity (the lower the solar activity, the higher the temperature variations). As a result, uncertainties in density can vary from 10% to 40% according to Table 1 which gives the total density versus

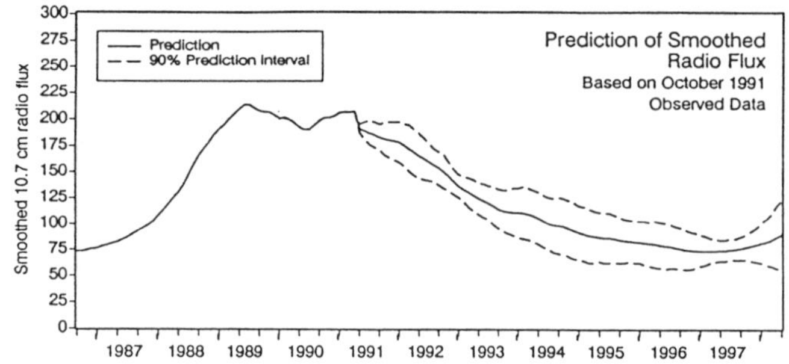

Fig. 2. SESC regression model prediction of smoothed 10.7 cm radio flux [from SESC, NOAA, Boulder, November 1991].

the mean global thermospheric temperatures corresponding to extreme conditions of the solar activity level. Moreover the prediction of a mean density level from existing thermospheric models is precise only within 10 or 20 %. Taking also into account the difficulty of giving in advance a date for the launch, it can be concluded that such uncertainties have to be considered as an input in the mission design. If the mean level of drag is too high with respect to the mission design, it would be possible either to shorten the mission duration or to perform the experiment at higher altitudes (210 km or 220 km, for example).

Influence of Air Drag on the Gradiometer Measurements

On an axis, for instance the i–axis, the measurement (γ_i) provided by any of the accelerometers can be written as follows [Touboul et al., 1991]:

$$\begin{aligned}\gamma_i = \;& (1+\varepsilon_i)\Gamma_i + (\varepsilon_i^j + \varepsilon_i^{j'})\Gamma_j + (\varepsilon_i^k + \varepsilon_i^{k'})\Gamma_k \\ & + \varepsilon_i^{i2}\Gamma_i^2 + \varepsilon_i^{ij}\Gamma_i\Gamma_j + \varepsilon_i^{ik}\Gamma_i\Gamma_k \qquad (1) \\ & + \text{bias} + \text{noise}\end{aligned}$$

where:
– $\Gamma_i, \Gamma_j, \Gamma_k$ is the sum of external accelerations, mainly the air drag ($\Gamma_{iD}, \Gamma_{jD}, \Gamma_{kD}$) and of the acceleration inside the satellite due to the non–uniformity of the gravity field ($\Gamma_{iG}, \Gamma_{jG}, \Gamma_{kG}$)
– ε_i represents the scale factor for the i–axis.

Solar Flux (10.7 cm)	70	110	160	225	290
T(K) / Altitude (km)	700 K	900 K	1100 K	1300 K	1500 K
200	1.61 E-10	2.38	3.02	3.53	3.94
210	1.10	1.72	2.26	2.70	3.07
220	0.77	1.27	1.72	2.10	2.43
230	0.55	0.94	1.32	1.66	1.95

TABLE 1. Densities (kg m^{-3}) from the J77 model as a function of altitude (km) and mean global temperature of the thermopause (K), as well as of the mean solar flux at 10.7 cm (expressed in 10^{-22} Wm^{-1} Hz^{-1}) [from Jacchia, 1977)].

- $\varepsilon_i^j, \varepsilon_i^k$ represent the effects of the non-perpendicularity of the j,k-axes with respect to the i-axis.
- $\varepsilon_i^{j'}, \varepsilon_i^{k'}$ represent the coupling between the i-axis and the j,k-axes.
- $\varepsilon_i^{i2}, \varepsilon_i^{ij}, \varepsilon_i^{ik}$ represent the non-linearity of the i-axis response under the action of $\Gamma_i, \Gamma_j, \Gamma_k$.

It is assumed that the satellite can be considered as a rigid body during the measurements and without rotation around its centre of mass; these hypotheses do not change the nature of the problem analyzed.

Let us assume Γ_k to be parallel to the x-axis undergoing the main air drag perturbation (Γ_{kD}), Γ_i normal to the orbital plane and Γ_j along the radius vector. The gravity tensor ($\delta\Gamma_{ijG}$) can be deduced from the difference between the measurements (γ_i) and (γ_j) performed by accelerometers considered two by two. In fact (γ_i) and (γ_j) are affected by biases and noise. The gradiometer can only yield, without biases, the variations $\Delta(\delta\Gamma_{ijG})$ along the trajectory between two times t_1 and t_2 provided the coefficients "ε" as well as the air drag acceleration ($\Gamma_{iD}, \Gamma_{jD}, \Gamma_{kD}$) are temporarily stable in the considered interval of time ($\simeq$200 s) ($\Gamma_{jD} < \Gamma_{iD} < \Gamma_{kD}$).

It is outside the scope of this paper to enter into the details of the gradiometer data processing but it can be shown that the accelerometer measurement (γ_i) will be perturbed by the main terms, as seen in equation 1:
- $\varepsilon_i^k\ \Gamma_{kD},\ \varepsilon_j^k\ \Gamma_{kD}$ (non-perpendicularity of the axes)
- $\varepsilon_i^{ik}\ \Gamma_{kD},\ \varepsilon_j^{jk}\ \Gamma_{kD}$ (cross non-linearity coefficients)
- $\varepsilon_i\ \Gamma_{iD},\ (\varepsilon_i^k + \varepsilon_i^{k'})\ \Gamma_{kD}$ (cross or temporal changes of scale factors and coupling between axes)
- $\varepsilon_i^{i^2}\ \Gamma_{iD}$ (non-linearity coefficients).

Touboul et al. [1991], showed that biases are introduced in the budget error by terms proportional to:
- $\Gamma_{iD}, \Gamma_{jD}, \Gamma_{kD}, \Delta\Gamma_{iD}, \Delta\Gamma_{jD}, \Delta\Gamma_{kD}$ through the $\varepsilon_i, \varepsilon_i^k$, $\varepsilon_i^{k'}$ coefficients effects
- $\Delta\Gamma_{kD}\ \Gamma_{iD},\ \Delta\Gamma_{kD}\ \Gamma_{jD}$ through the ε_i^{ik} coefficients effects
- $\Delta\Gamma_{iD}^2,\ \Delta\Gamma_{jD}^2$ through the ε_i^{i2} coefficients effects.

The symbol Δ stands for a temporal variation between two times t_1 and t_2.

Thus, the errors introduced are proportional to the mean level of the air drag, to its fluctuations and to the product of these quantities.

Let us note that the air drag accelerations will be deduced from the measurements of the four accelerometers. The main source of errors is due to the difficulty of estimating accurately enough the value of the "ε" coefficients or of keeping them "temporarily stable" or small enough.

Some numerical orders of magnitude have now to be given.

First, the in-orbit calibration of the accelerometers yields an estimate of the scale-factor (ε_i) with a 10^{-5} relative precision and an estimate of defects in orientation of the accelerometer axes with a 10^{-5} rad precision (ε_i^k) [Touboul et al., 1991]

As an example, along the i-axis normal to the orbital plane, an error $\delta\varepsilon_i$ can introduce an error of the form:

$$\delta\varepsilon_i(\Delta\Gamma_{iG} + \Delta\Gamma_{iD})$$

Generally speaking with $\delta\varepsilon_i$ of the order of 10^{-5}, $\delta\varepsilon_i\Delta\Gamma_{iG}$ will be negligible. But if Γ_{iD} is not constant, an error of $\delta\varepsilon_i\Delta\Gamma_{iD}$ will be introduced and can be non negligible. Identically, for another coefficient ε_i^k or $\varepsilon_i^{k'}$, the error would be $\delta\varepsilon_i^k\Delta\Gamma_{kD}$ or $\delta\varepsilon_i^{k'}\Delta\Gamma_{kD}$. Thus, if the sensitivity of 10^{-2} E.U. (10^{-11} s^{-2}) is required all these errors must be significantly less than 10^{-11} ms^{-2} due to their cumulative effect.

For $\Delta\Gamma_{iD}$=5. 10^{-7} ms^{-2} and $\delta\varepsilon$= 10^{-5} we get: $\delta\varepsilon\Delta\Gamma_{iD}$= 5. 10^{-12} ms^{-2}. This means that fluctuations of Γ_{iD} (or Γ_{jD}, Γ_{kD}) must be less than 5. 10^{-7} ms^{-2}. That can be generally the case in the plane perpendicular to the velocity vector but generally not along the velocity vector.

The other coefficients also induce significant errors. From studies performed at ONERA values of 100 m^{-1}s^2 for the ε_i^{i2} coefficient and of 1 m^{-1}s^2 for the ε_i^{ij}, ε_i^{ik} coefficients can be expected.

With the k-axis along the orbit we can have:
Γ_{kD}=5. 10^{-5} ms^{-2}, $\Delta\Gamma_{kD}$= 5. 10^{-6} ms^{-2} (10% of Γ_{kD}),
Γ_{iD}=5. 10^{-6} ms^{-2}, $\Delta\Gamma_{iD}$=5. 10^{-7} ms^{-2} (10% of Γ_{iD})

Then, the perturbing term on the accelerometer measurement is equal to:
$\varepsilon_i^{ik}\Delta\Gamma_{kD}\Gamma_{iD} + \varepsilon_i^{i2}\Delta\Gamma_{iD}^2$
= 2.5 10^{-11} ms^{-2} + 2.5 10^{-11} ms^{-2}
= 5. 10^{-11} ms^{-2}

If this term can be estimated with only 10 % accuracy, it represents a maximum of acceptable errors. Thus, in practice, the drag fluctuation effect has necessarily to be limited to less than 5. 10^{-7} ms^{-2} in the plane (i, j) and at least to less than 5. 10^{-6} ms^{-2} along the velocity vector. It would also be necessary to have the non-linearity coefficients as small as possible. As a result, the errors being proportional to the air drag fluctuations, in the ARISTOTELES mission devices reducing the fluctuations to less than about ± 8% have to be installed on board. In the useful interval of time it is shown later on that, in this case, the number of occurences where the air drag fluctuations are greater than ± 8% is sufficiently limited to make the success of the ARISTOTELES mission possible.

At this stage, the spectral analysis of fluctuations is crucial, including frequency and amplitude determination. Around one orbit the air drag acceleration can increase or decrease by a factor of two. As resulting from previous remarks, the precision of measurements at too great distances will be weak. Moreover, the temporal stability required for the "ε" coefficients is so high that it could not be guaranteed for periods of time greater than 200 s. Thus, the data will have to be processed in periods of time not greater than 200 s. Taking into account also the time response of the accelerometers, the useful interval of time could lie between 10 s and 200 s which corresponds to a frequency bandwidth of [5. 10^{-3} Hz, 0.1 Hz]or to a distance along the trajectory of the satellite of 70 km to 1400 km.

The fluctuations of air drag in such intervals as well as their amplitudes and their occurence are now analysed.

ANALYSIS OF AIR DRAG FLUCTUATIONS

In a classical form, the air drag acceleration modulus is written as:

$$\gamma = 1/2\ \rho\ C_D\ S/M\ V_S^2$$

where ρ stands for air density, S/M for the area to mass ratio, and V_S for the relative velocity between the satellite and the atmosphere. The fluctuations of the drag can be caused either by fluctuations of density or by fluctuations of winds. These kinds of fluctuations are presented in the next sections.

Fluctuations of Air Density due to Gravity Waves

It appears that fluctuations in the upper atmospheric parameters can often be related to the physics of gravity or acoustic waves. The observations of gravity waves travelling in the upper atmosphere have been extensively studied in the past. Some theoretical considerations can be found in Hines [1960], Testud and Vasseur [1969], Richmond [1978], Vidal-Madjar [1979], Klostermeyer [1989] and results, among many, in Georges [1968], Vasseur et al. [1972], Francis [1975], Reber et al. [1975], Potter et al. [1976], Prölss and von Zahn [1976], Bertin et al. [1978], Villain [1979, 1980], Villain et al. [1979], Mayr et al. [1985], Hedin and Mayr [1987], Mercier et al. [1989].

Generally speaking, these studies concern the propagation of gravity waves with speeds of several hundreds of meters per second, the periods of which range from 10 minutes to a few hours. These speeds, much smaller than the satellite velocity (7.8 kms^{-1}) allow us to observe on board a satellite a quasi-instantaneous cross section of the wave along the satellite trajectory. Let us consider gravity waves with a speed of 200 ms^{-1} and a period of 60 mn propagating in the direction of the velocity vector. They generate at the satellite level an apparent periodic signal with a wavelength of 720 km and a period of 92 s.

According to Georges [1968] and Francis [1975], two distinct categories of travelling disturbances relating to gravity waves have to be considered, the large-scale and medium-scale waves. Large-scale waves are characterized by horizontal speeds ranging between 400 and 1000 ms^{-1}, periods from 30 mn to 3 h and horizontal wavelengths greater than 1000 km. Medium-scale waves occur much more frequently than large-scale waves. They have horizontal speeds generally between 100 and 250 ms^{-1}, periods from 15 mn to more than 1 h and horizontal wavelengths of several hundred kilometers. As a result, the medium-scale gravity waves are the main source of perturbations for a gradiometer mission. In the considered bandwidth of the gradiometer [5. 10^{-3} Hz, 0.1 Hz], corresponding to a distance along the trajectory of 70 km to 1400 km, such gravity waves can be observed in the accelerometer and mass spectrometer signal (Figures 3, and 4).

The different parameters of the atmosphere (density, concentration, temperature, wind, ionospheric parameters) can be perturbed. In the present case, the parameters of interest are density and winds, but let us note that ion temperature, ion velocity, electron content are strongly coupled to the corresponding parameters of the neutral atmosphere. Consequently, it is also of interest to look at the behaviour of all these quantities (see for example Figure 5).

The occurence of fluctuations from the CACTUS accelerometer data have been studied as a function of their amplitude for day-time and night-time conditions at an altitude of about 300 km in the low latitude regions (Figure 6). During the night, the ion drag is weaker, making the propagation of gravity waves easier. As a result, they occur more frequently during the night than during the day. In the useful frequency bandwidth of the gradiometer, variations greater than 5% (full amplitude) can be considered, as a general rule, in 15% of cases by day and in 25% of cases by night [Villain, 1979]. A spectral analysis has shown that fluctuations with periods ranging from 100 s to 220 s are common features (Figure 3). These periods correspond to apparent wavelengths in the 700 km-1500 km range. It has also been shown from these data that a high geomagnetic activity level is generally well correlated with the greatest amplitudes observed but significant perturbations can also correspond to low or mean activity. In the equatorial and tropical regions, the perturbations appear generally later than the large increase in density occuring after a geomagnetic storm [Villain et al., 1979].

The same type of fluctuations have been observed with data obtained from mass spectrometers on board the Atmosphere Explorer-C satellite [Reber et al., 1975]. An example is given in Figure 7 for molecular nitrogen where variations up to 20 % (full amplitude) are evidenced at the altitude of the gradiometer mission. The scale sizes for the structure range from a few tens of kilometers to several hundred kilometers as seen on data from two orbits.

The occurence of fluctuations has also been statistically studied from the same data at an altitude of 250 km [Potter et al., 1976] as a function of magnetic latitude and local time and of their amplitudes (Figure 8). Wavelengths of interest for the gradiometer mission, from roughly 150 to 1500 km are readily discernible. Waves are most often found near local midnight and the occurence of strong fluctuations is significantly greater at high latitudes than at low ones. Large amplitude waves seem associated with large-scale gravity waves while smaller amplitude waves seem associated with medium-scale gravity waves.

A similar study of wavelike structure in Dynamics Explorer-2 neutral composition data was made, at all latitudes, by Hedin and Mayr [1987] for short (below 400 km) and long wavelengths (up to 4000 km). Touboul et al. [1991] performed a new study for wavelengths in the 70 km-1400 km range, especially dedicated to the ARISTOTELES

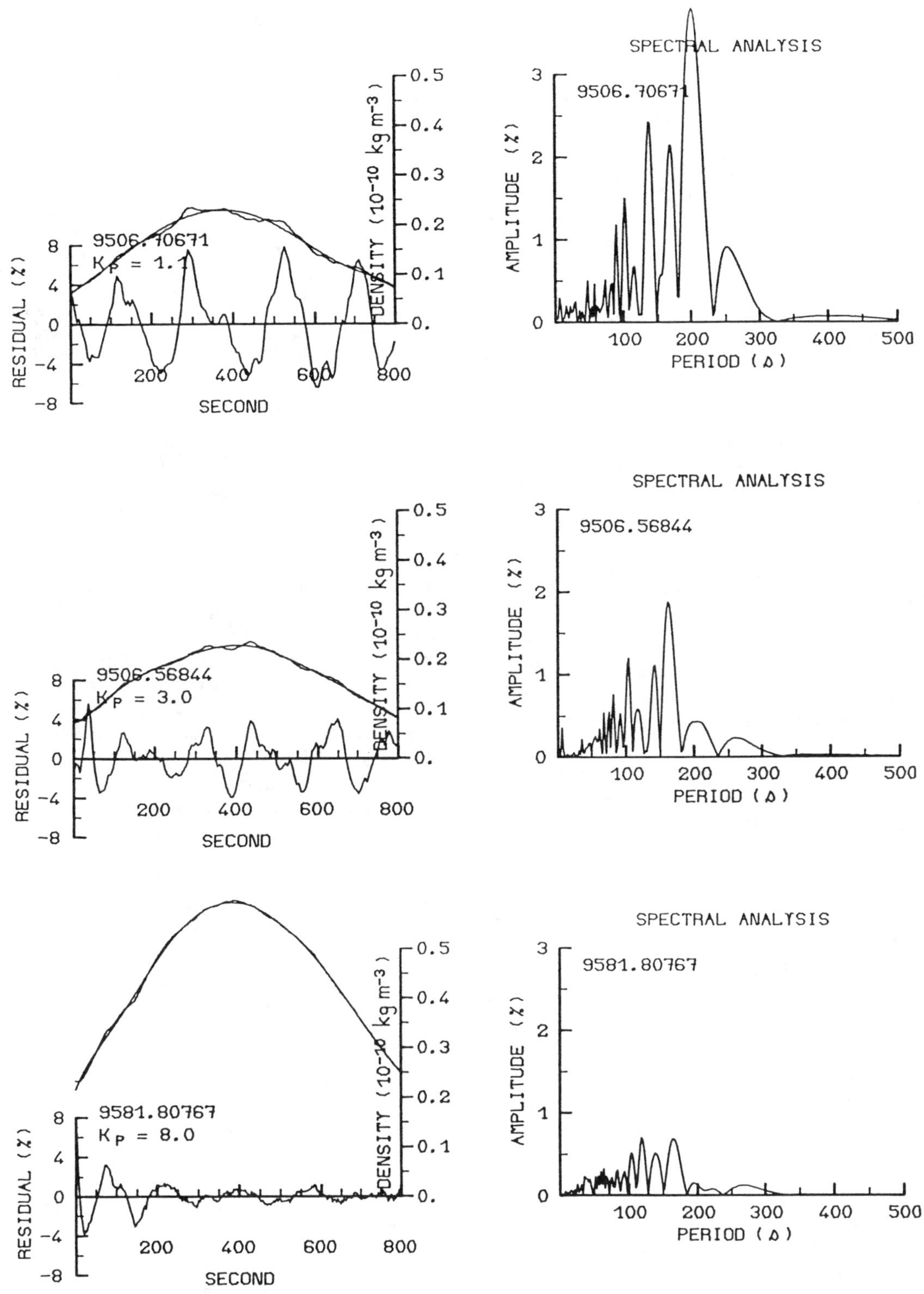

Fig. 3. Fluctuations of total density deduced from CACTUS accelerometer measurements near perigee (270 km) and spectral analysis of the residuals.
Examples of waves of great amplitude associated with a strong geomagnetic event (date: 9506 ≡ 01/11/76, 9581 ≡ 03/26/76) (Reproduced with permission of *Manuscripta Geodætica* [from Touboul et al., 1991]).

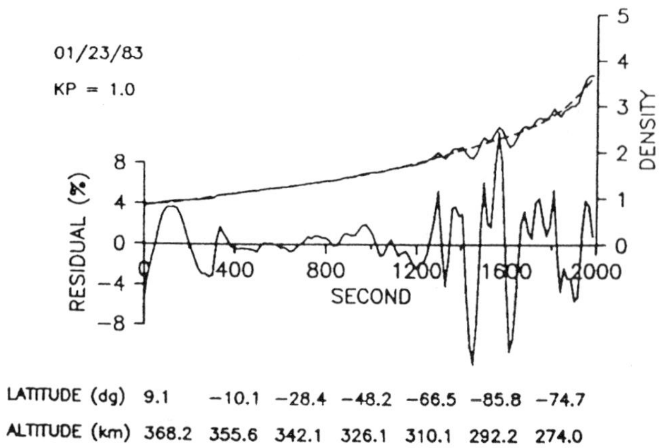

Fig. 4. Fluctuations of total density deduced from Dynamics Explorer–2 measurements (solid line) with respect to a polynomial fitting (dashed line). It corresponds to a low geomagnetic activity and to a mean solar flux of 140. 10^{-22} $\mathrm{Wm^{-1}}$ $\mathrm{Hz^{-1}}$. It exhibits different behaviour between low and high latitudes (Reproduced with permission of *Manuscripta Geodætica* [from Touboul et al., 1991]).

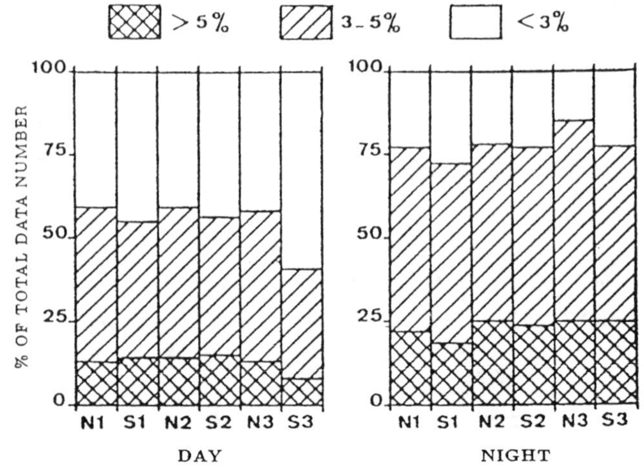

Fig. 6. Statistical analysis of the disturbances as a function of latitude, longitude and local time (Reproduced with permission of *Space Res.* [from Villain, 1979]).
N and S stand for latitude: N (0° ⟶30° N), S (30° S⟶0°)
1,2 and 3 stand for longitude: 1 (0° ⟶120°), 2 (120° ⟶240°), 3 (240° ⟶360°)

mission. It leads to an amplitude and occurence of fluctuations which decreases very significantly with latitude (Figure 4). Below 65° the amplitude of fluctuations is less than ±8% in 90% of cases. At higher latitudes for low geomagnetic activity 70% of cases are in the interval of ±8% but only 30% of cases are in this interval for high geomagnetic activity. Taking account of the low occurence of high geomagnetic conditions (about 15% of situations), it means that on average 35% of passes over the polar regions would be rejected or processed with a lower weight.

From these different studies, some conclusions can be drawn on the sources of these gravity waves for a better understanding of the physical process involved.

The fluctuations of the greatest amplitudes originate mainly in the auroral regions where particle precipitations occur generating gravity waves. They propagate towards the equator, being progressively attenuated. As noted by Hedin and Mayr [1987] wave activity maximizes near the magnetic poles and increases with magnetic activity for wavelengths both longer and shorter than 400 km indicating a significant auroral source for all wavelengths.

Many perturbations of medium–scale can be observed at any latitude for low geomagnetic conditions and can ,in some cases, be correlated with an intense meteorological disturbance. It follows that they can be formed from sources located not only in auroral regions but also at lower latitudes and in the very low atmosphere. It is possible to suggest

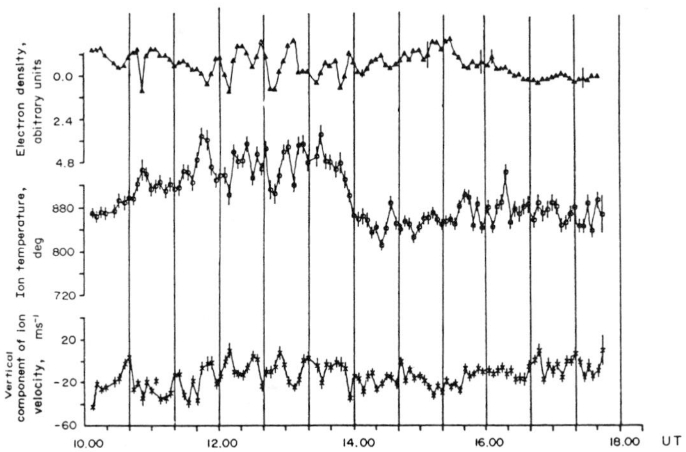

Fig. 5. The variations of electron density, ion temperature and vertical component of ion velocity observed at Mende (44° 31' N) on July 18, 1974 (Reproduced with permission of *J. Atmos. Terr. Phys.* [from Bertin et al., 1978]).

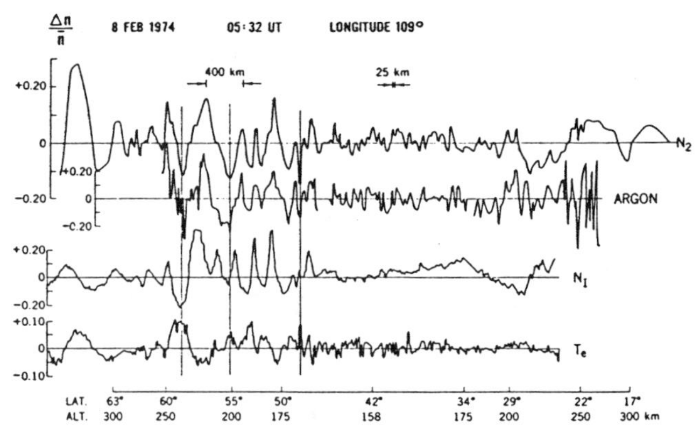

Fig. 7. The structural variations of data obtained from one orbit from the AE–C satellite. The entire data set was obtained in 14.7 mn at a satellite speed of 8.6 $\mathrm{kms^{-1}}$ [from Reber et al.,1975].

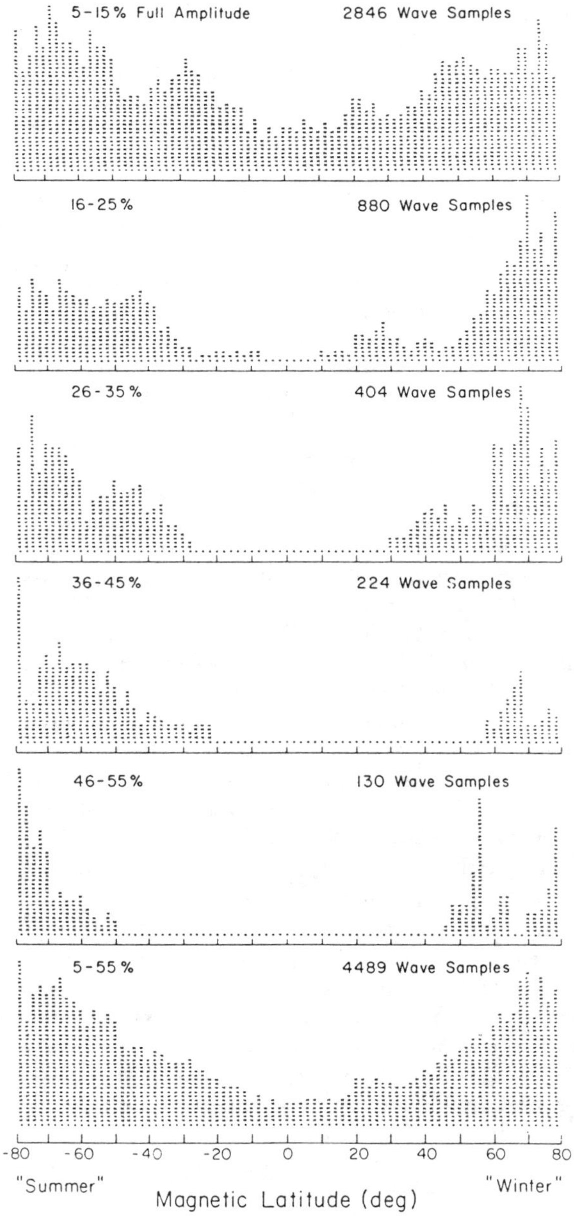

Fig. 8. Normalized wave occurence histograms are shown for a range of peak-to-peak wave amplitudes. The number of wave samples recorded in each range shows that the greatest number of waves are of relatively low amplitude and are found at all latitudes. Higher-amplitude waves were found only in high-latitude regions. The bottom histogram is a composite for all amplitudes measured [from Potter et al., 1976].

that a part of the medium-scale waves originate at the tropospheric level from meteorological phenomena [Bertin et al., 1978, Villain, 1980, Mercier et al., 1989]. The upwelling of heated neutrals can also be a secondary wave source, generating medium-scale waves at low latitude [Potter et al., 1976]. Indeed, such a source could explain the delay observed, in the CACTUS data, between the beginning of the increase in density and in fluctuations following a geomagnetic storm. Finally, the equatorial electrojet perturbed after a severe geomagnetic event could be also a source of gravity waves [Chimonas, 1970].

Another aspect of the physics of the gravity waves concerns their occurence with a specific type of periods. Thus, large-scale thermospheric waves which appear to have a period of about 2 h or more have been detected during large geomagnetic storms [Hernandez and Roble, 1978]. Large-scale neutral composition gravity waves with wavelengths of the order of 5000 km and periods of about 2.5 h were observed from Esro-4 data in the altitude region of 250 km [Trinks and Mayr, 1976]. Touboul et al. [1991] remarked that apparent periods were observed mainly beyond 100 s, probably corresponding to gravity waves with periods of about 1 h. Nygrén et al. [1990], from a spectral analysis of ion velocity data in the F-region, also observed the importance of an oscillation with a period of about 55 mn (Figure 9).

At the same time, the amplitude of the gravity waves is a function of the altitude when propagating upward. A theoretical example is given by Mercier et al. [1989] (Figure 10), the altitude of maximum lying between 230 km and 400 km. Potter et al. [1976], noted that waves of large amplitude are found near 250 km and are often seen above 350 km from the circular and elliptical orbit study of the AE-C satellite, respectively. More recently, Nygrén et al. [1990], indicated that the amplitude of the 55 mn-period fluctuation increases sharply above 170 km to reach a maximum at 250 km while the 25 mn-period oscillation is observed only near 150 km (Figure 9). Another example is given in Figure 11 which exhibits ion velocities parallel to the magnetic field line as measured by EISCAT for altitudes 190, 212, 256 km [Thuillier et al., 1990].

In conclusion, only a limited number of gravity waves with specific periods can be of significance. Perturbations corre-

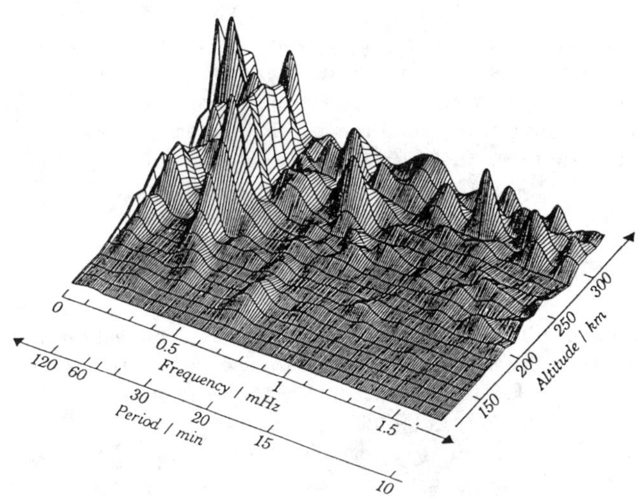

Fig. 9. Power spectra of ion velocities from the EISCAT UHF radar at Tromsø (Reproduced with permission of *J. Atmos. Terr. Phys.* [from Nygrén et al., 1990]).

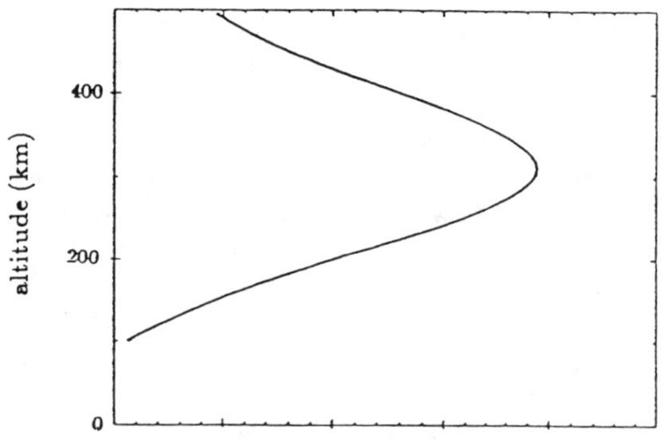

Fig. 10. Calculated amplitude versus height for an ascending gravity wave (Reproduced with permission of *Ann. Geophys.* [from Mercier et al., 1989]).

sponding to wavelengths ranging from 70 km to 5000 km or more are possible, the amplitude increasing with wavelength. Moreover, the mean amplitude of fluctuations could increase as a function of altitude above 200 km.

Fluctuations due to Acoustic Waves

Gravity waves are governed by basic equations in which two opposite forces act on the motion of a small box of atmosphere to maintain fluctuations around a relative equilibrium. One is due to the gravitational acceleration and the other to the atmospheric pressure. It can be shown that for frequencies greater than a given frequency (the Brunt–Väisälä frequency), no gravity waves can exist [Testud and Vasseur, 1969]. Accordingly, it is impossible to observe gravity waves with periods smaller than 10–15 minutes. Most of the observed gravity waves entail significant perturbations along the trajectory over distances of interest in the gradiometer experiment. The problem is now to see whether another type of fluctuations with higher frequencies could also perturb the gradiometer measurements. High frequency acoustic waves could propagate thanks to the compressibility of the upper atmosphere for some frequencies. The period of these waves could range from less than one second to a few minutes. Such waves have been observed in the vicinity of 250 km [Georges, 1968] from the Doppler sounding technique and in the F–region of the ionosphere [Shrestha, 1967, Rao et al., 1969]. According to the authors, it can be considered that the wave fronts are propagating in a near vertical direction. Their localized nature and their occasional persistence for several days suggests a connection with tropospheric weather activity [Georges, 1968]. But as a result of this localized nature and of the vertical propagation, the perturbations induced in the gradiometer measurement should correspond to a very short pulse along the vertical at the satellite level. Taking into account the very small effect of the air drag along this axis, such fluctuations should not affect the mission. However, the limited number of data does not allow us to be completely certain.

Mean Level and Fluctuations of Winds

In a first approximation, it can be assumed that the atmosphere is driven by the Earth rotation. Along the y–axis in the orbital reference frame, the rotation of the Earth leads to a positive acceleration at the ascending node of about $5.\,10^{-6}$ ms^{-2} (the relative air velocity is about 500 ms^{-1}). The acceleration is negative at the descending node. The physical reality is, of course, much more complex and winds are to be superimposed on this too simple schema [King–Hele and Walker, 1983]. A mean model of winds has been recently proposed [Hedin et al., 1991]. For example, zonal and meridional winds are predicted versus solar local time for six latitudes and three levels of the solar activity (Figures 12 a, b). Mean winds with amplitudes of 100 ms^{-1} can be observed.

Significant fluctuations can exist with respect to a mean model, in particular at high latitude regions, generally related to the propagation of thermospheric gravity waves. Examples are given in Figure 13 at about 250 km [Thuillier et al., 1990] from a ground–based interferometer, in Figure 14 [Mayr et al., 1985] from Dynamics Explorer 2 spectrometer data, and in Figure 15 [Forbes et al., 1987] from a satellite electrostatic triaxial accelerometer system between 170 and 210 km. Fluctuations of winds greater than 500 ms^{-1} can be observed. In such a case, gradiometer measurements will have to be processed carefully and probably rejected. Other results, among many, can also be found from ground–based measurements in Hernandez and Roble [1978], Babcock and Evans [1979], Biondi and Sipler [1985], Rees et al. [1985] and from space data in Marcos and Forbes [1985], Rees et al. [1985], Killeen and Roble [1988].

Wind fluctuations increase significantly with latitude (Figures 14 and 15). Strong and rapid variations can be observed in the gradiometer frequency bandwidth in such a way that the components of the gravity tensor corresponding to the y–axis could not be used in many cases at high latitudes. However, a statistical analysis has not been performed. For strong geomagnetic activity, it would

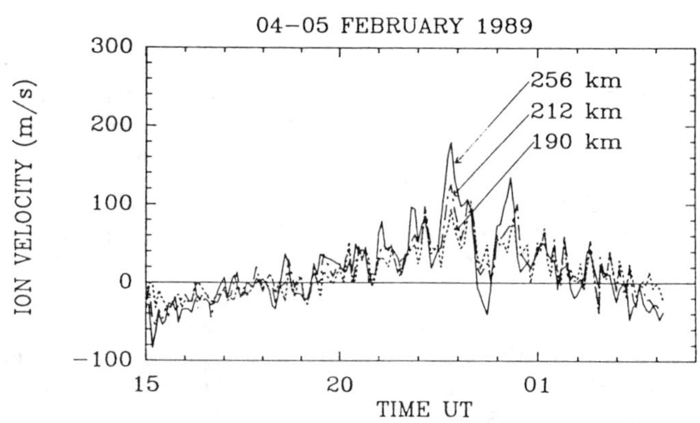

Fig. 11. Ion velocity parallel to the magnetic field as measured by EISCAT at Tromsø for three altitudes (Reproduced with permission of *J. Atmos. Terr. Phys.* [from Thuillier et al., 1990]).

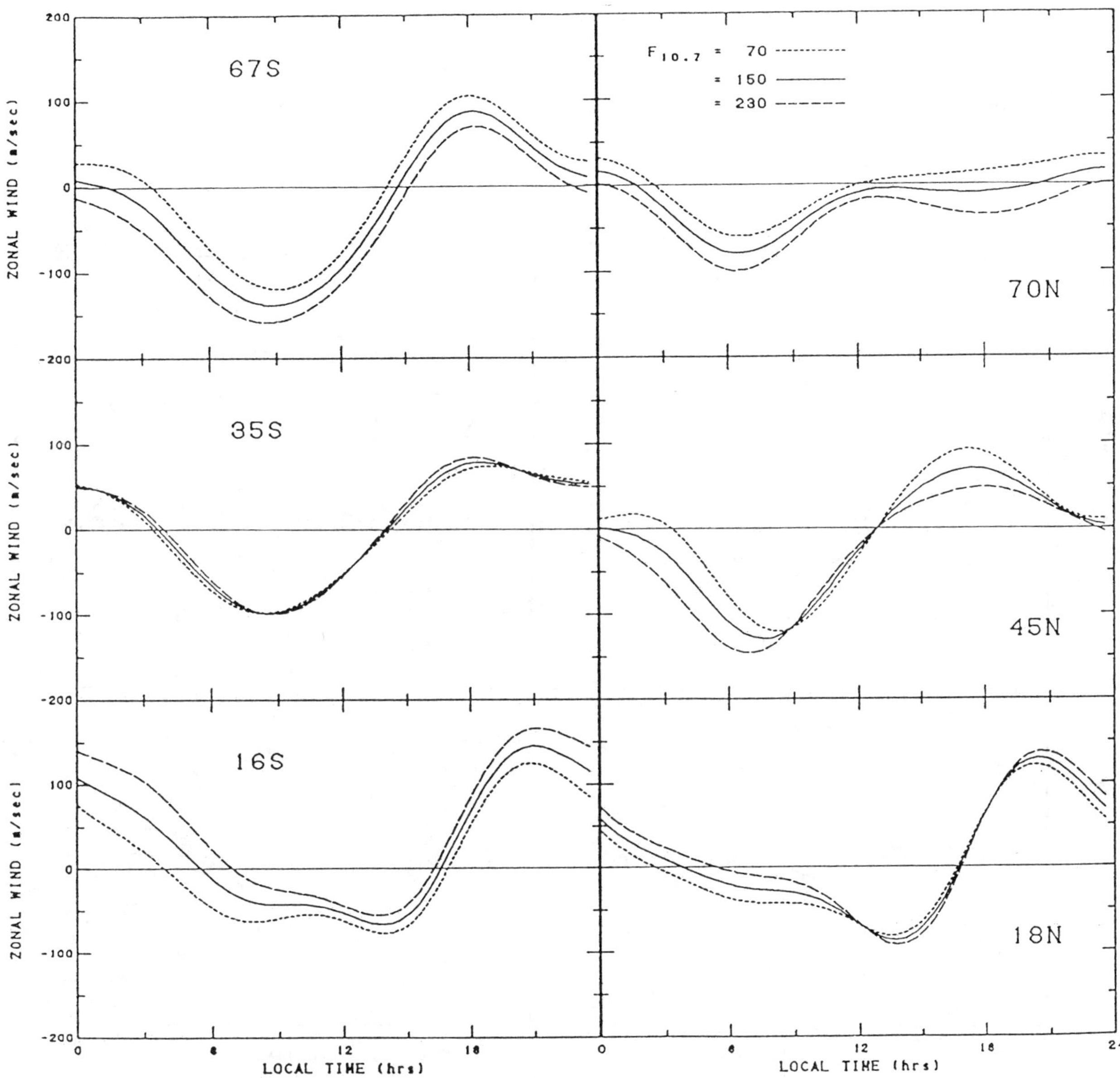

Fig. 12*a*. Zonal and seasonal average of zonal winds at 250 km from the Horizontal Wind Model 90 versus local time for six latitudes and three solar activities [from Hedin et al., 1991].

probably be difficult to use the y–axis components at high and medium latitudes when the fluctuations induced by the cross–winds are greater than 5. 10^{-7} ms^{-2}, as could generally be the case.

It can be noted that fluctuations in the vertical winds can be also observed but, generally speaking, their amplitudes are smaller, by nearly a factor of ten, than the amplitudes of horizontal winds [Herrero et al., 1984, Mayr et al., 1985, Wardill and Jacka, 1986]. Thus, the impact on the gradiometer mission should be negligible, making the vertical component $\delta\Gamma_{zz}$ not much perturbed and enabling the gravity field mission to be carried out.

In conclusion, horizontal winds and their fluctuations are able to generate significant air drag perturbations in a gradiometer mission along the x–axis and the y–axis. Frequent fluctuations greater than $\pm 8\%$ are possible making it mandatory to limit them. Along the x–axis, a flap device which will limit both the fluctuations of density and winds by varying the area to mass ratio, will be installed on board. Along the y–axis, the simplest way to minimize the impact should be to decrease the mean value of the air drag. To accomplish this, it could be sufficient to maintain the gradiometer plane as well as possible normal to the air drag vector but not normal to the velocity vector, making the

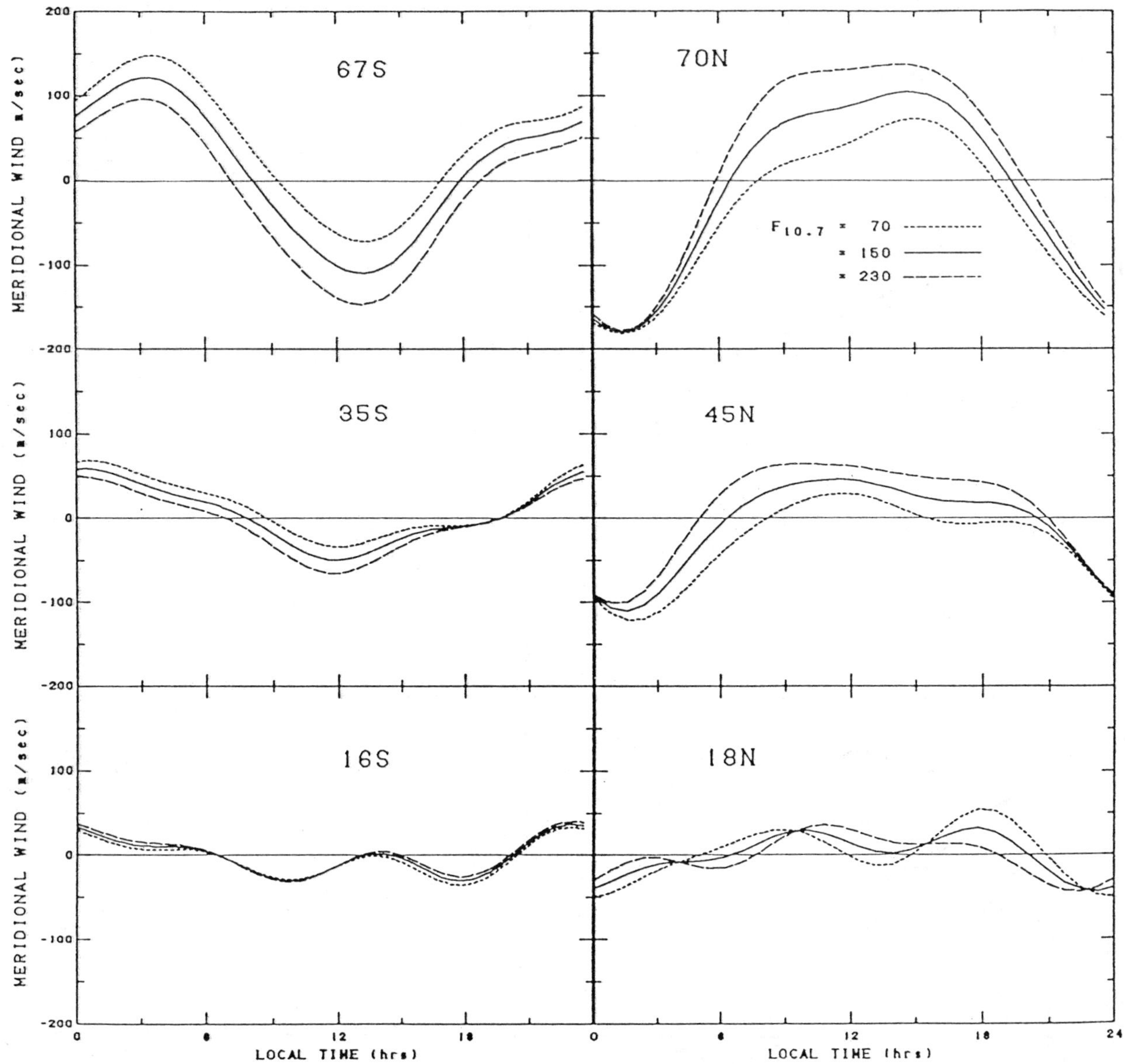

Fig. 12*b*. Same as Figure 12*a* but for meridional winds [from Hedin et al., 1991]).

term Γ_{iD} as small as possible. Thus, a yaw steering control system is required.

CONCLUSION

The impact of air drag on the gradiometer mission like ARISTOTELES is extremely important. The characteristics of a gradiometer cannot be determined as ideally as would be necessary. As a result, errors and biases induced by the air drag cannot be perfectly removed from the gradiometric signal. They are proportional to the mean level of the air density and to the air velocity square, to the fluctuations of these quantities and to their products.

Some precise points can be emphasized:

- the errors will be greater at high latitudes than at low ones due to the gravity waves generated in the auroral regions.
- the errors will be minimized for low solar activity level.
- the occurence of waves is greater at the poles than at the equator by a factor of two or three.
- the fluctuations are more important for night–time conditions than for day–time ones.
- the fluctuations increase as the wave frequency decreases.
- the amplitude of gravity waves propagating upward can increase with altitude above 200 km.

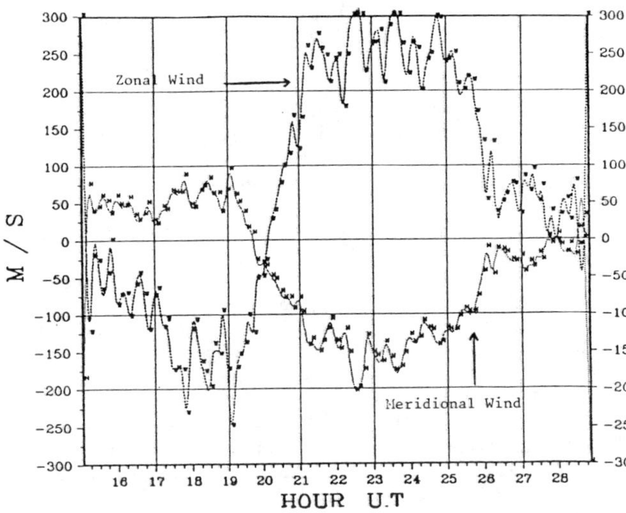

Fig. 13. Zonal and meridional winds, for the night 04–05 February 1989, measured by MICADO interferometer using the $O^{-1}D$ emission line (Reproduced with permission of *J. Atmos. Terr. Phys.* [from Thuillier et al., 1990]).

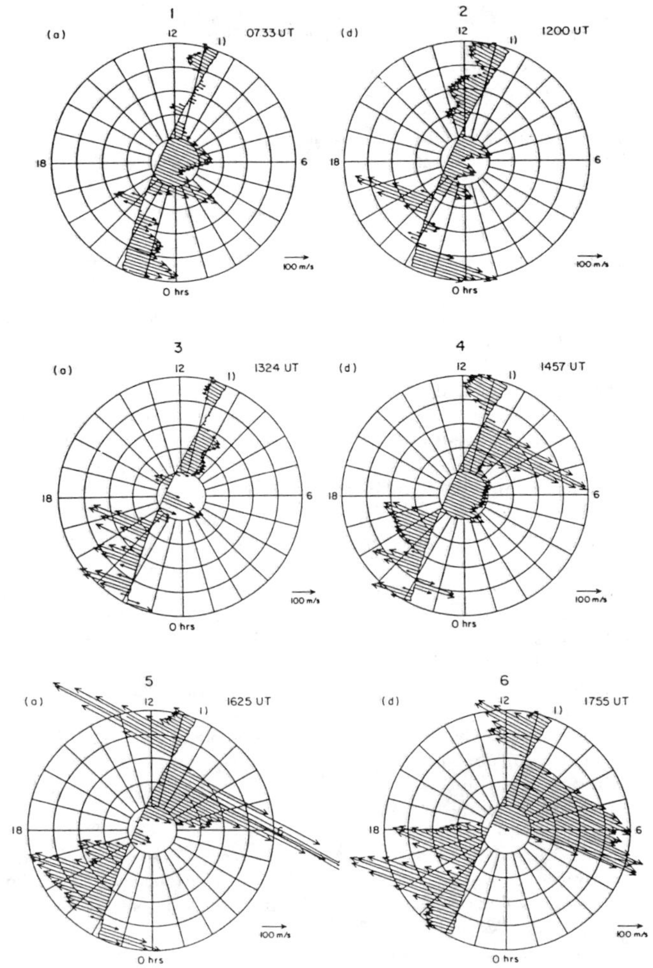

Fig. 15. Northern hemisphere winds along the satellite path as a function of magnetic latitude ($40^{\circ}-90^{\circ}$) and magnetic local time. Winds are extracted from the accelerations on the axis normal to the orbital plane from a satellite electrostatic triaxial accelerometer system (SETA) between 170 km and 240 km. (1) corresponds to quiet geomagnetic period and the others to the development of a storm [Adapted from Forbes et al., 1987].

In order to limit all these sources of errors in the useful frequency bandwidth and to make the mission successful, a flap device for compensating drag variations within 8% and a yaw steering system to maintain the gradiometer plane normal to the air drag vector must be installed on board. The ideal system would be a drag–free system.

The perturbations along the x–axis are too large to make it possible to obtain useful gradiometer measurements for this axis.

The perturbations along the y–axis can be controlled at an acceptable level and the perturbations along the z–axis should be, in many cases, acceptable.

Information would be usefully yield in the air drag field from the ARISTOTELES mission at 200 km where existing data are presently scarce.

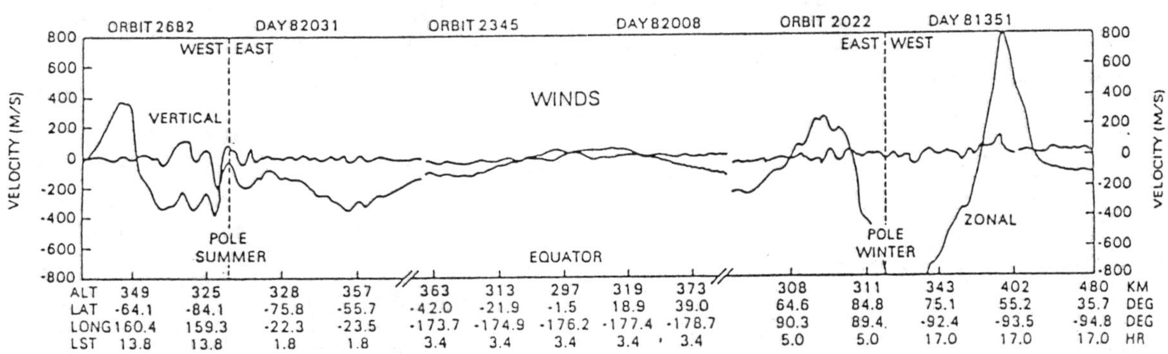

Fig. 14. Pole to pole measurements from WATS (Wind And Temperature Spectrometer) illustrating the wind structure observed from DE 2 (Reproduced with permission of *Adv. Space Res.* [from Mayr et al., 1985]).

Acknowledgements. A significant part of this review has been carried out in the frame of ALENIA SPAZIO contract. The authors are very grateful to this firm for its support as well as previously to the support of ESA (contract N° 7521/87/F/FL) and DORNIER. They are indebted to Dr. Hedin (GSFC) for providing Dynamics Explorer 2 total density data set, to Dr. Klinkrad (ESOC) for very useful information and to A. Bernard and P. Touboul ONERA for usefull discussions. The solar and geomagnetic data from "National Geophysical Data Center, N.O.A.A., Boulder, Colorado, USA" have been kindly provided by the "Centre de Prévision d'Activité Solaire" of Meudon Observatory.

REFERENCES

"ARISTOTELES" Baseline Definition Review. Additional phase (ESA), April, 1990.

Balmino, G., D. Letoquart, F. Barlier, M. Ducasse, A. Bernard, B. Sacleux, C. Bouzat, J.J. Runavot, X. Le Pichon, and M. Souriau, Le projet GRADIO et la détermination à haute résolution du géopotentiel, *Bull. Géod., 58*, 151–179, 1984.

Babcock R.R., and J.V. Evans, Effects of geomagnetic disturbances on neutral winds and temperatures in the thermosphere observed over Millstone Hill, *J. Geophys. Res., 84*, 5349–5354, 1979.

Bernard, A., and P. Touboul, A spaceborne gravity gradiometer for the nineties, *IAG, General Meeting*, Edinburgh, 1989.

Bertin F., J. Testud, L. Kersley, and P.R. Rees, The meteorological jet stream as a source of medium scale gravity waves in the thermosphere: an experimental study, *J. Atmos. Terr. Phys., 40*, 1161–1183, 1978.

Biondi M.A., and D.P. Sipler, Horizontal and vertical winds and temperatures in the equatorial thermosphere: measurements from Natal, Brazil during August–September 1982, *Planet. Space Sci., 33*, 817–823, 1985.

Chimonas G., The equatorial electrojet as a source of long long period travelling ionospheric disturbances, *Planet. Space Sci., 18*, 583–589, 1970.

Forbes J.M., R.G. Roble, and F.A. Marcos, Thermospheric dynamics during the March 22, 1979, magnetic storm. 2. Comparisons of model predictions with observations, *J. Geophys. Res., 92*, 6069–6081, 1987.

Francis S.H., Global propagation of atmospheric gravity waves: A review, *J. Atmos. Terr. Phys., 37*, 1011–1054, 1975.

Georges T.M., HF Doppler studies of traveling ionospheric disturbances, *J. Atmos. Terr. Phys., 30*, 735–746, 1968.

Hedin A.E., and H.G. Mayr, Characteristics of wavelike fluctuations in Dynamics Explorer neutral composition data, *J. Geophys. Res., 92*, 11,159–11,172, 1987.

Hedin A.E., M.A. Biondi, R.G. Burnside, G. Hernandez, R.M. Johnson, T.L. Killeen, C. Mazaudier, J.W. Meriwether, J.E. Salah, R.J. Sica, R.W. Smith, N.W. Spencer, V.B. Wickwar, and T.S. Virdi, Revised global model of thermosphere winds using satellite and ground–based observations, *J. Geophys. Res., 96*, 7657–7688, 1991.

Hernandez G., R.G. Roble, Observations of large–scale thermospheric waves during geomagnetic storms, *J. Geophys. Res., 83*, 5531–5538, 1978.

Herrero F.A., H.G. Mayr, I. Harris, and F. Varosi, Thermospheric gravity waves near the source: Comparison of variations in neutral temperature and vertical velocity at Sondre Stromfjord, *Geophys. Res. Let., 1*, 939–942, 1984.

Hines C.O., Internal atmospheric gravity waves at ionospheric heights, *Can. J. Phys., 38*, 1441–1480, 1960.

Jacchia L.G., Thermospheric temperature, density, and composition: new models, *Special Report 375*, Smithson. Astrophys. Observ., Cambridge, Massachusetts, 1977.

Killeen T.L., and R.G. Roble, Thermosphere dynamics: Contribution from the first 5 years of the Dynamics Explorer program, *Rev. Geophys., 26*, 329–367, 1988.

King–Hele D.G., and D.M.C. Walker, Upper–atmosphere zonal winds from satellite orbit analysis, *Planet. Space Sci., 31*, 509–535, 1983.

Klostermeyer J., Gravity waves and instabilities in the lower and middle atmosphere, in *Middle Atmosphere Program*, ed. S.Fukao, SCOSTEP, University of Illinois, Urbana, 269–298, 1989.

Lambeck K., ARISTOTELES: An ESA mission to study the Earth's gravity field. Impact of a high–resolution gravity–field mission for planet Earth, ESA *journal, 14*, 1–21, 1990.

Marcos F.A., and J.M. Forbes, Thermospheric winds from the satellite electrostatic triaxial accelerometer system, *J. Geophys. Res., 90*, 6543–6552, 1985.

Mayr H.G., I. Harris, F. Varosi, F.A. Herrero, H. Volland, N.W. Spencer, A.E. Hedin, R.E. Hartle, H.A. Taylor, L.E. Wharton, and C.R. Carignan, On the structure and dynamics of the thermosphere, *Adv. Space Res., 5*, 283–288, 1985.

Mercier C., F. Genova, and M.G. Aubier, Radio observations of atmospheric gravity waves, *Ann. Geophys., 7*, 195–202, 1989.

Nygrén T., B.S. Lanchester, A. Huuskonen, L. Jalonen, T. Turunen, H. Rishbeth, and A.P. van Eyken, Interference of tidal and gravity waves in the ionosphere and an associated sporadic E–layer, *J. Atmos. Terr. Phys., 52*, 609–623, 1990.

Potter W.E., D.C. Kayser, and K. Mauersberger K, Direct measurements of neutral wave characteristics in the thermosphere, *J. Geophys. Res., 81*, 5002–5012, 1976.

Prölß G.W., and U. von Zahn, Large and small scale changes in the disturbed upper atmosphere, *J. Atmos. Terr. Phys., 38*, 655–659, 1976.

Rao N.N., G.F. Lyon, and J.A. Klobuchar, Acoustic waves in the ionosphere, *J. Atmos. Terr. Phys., 31*, 539–545, 1969.

Reber C.A., A.E. Hedin, D.T. Pelz, W.E. Potter, and L.H. Brace, Phase and amplitude relationships of wave structure observed in the lower thermosphere, *J. Geophys. Res., 80*, 4576–4580, 1975.

Rees D., T.J. Fuller–Rowell, M.F. Smith, R. Gordon, T.L. Killeen, P.B. Hays, N.W. Spencer, L. Wharton, and N.C. Maynard, The westward thermospheric jet–stream of the evening auroral oval, *Planet. Space Sci., 33*, 425–456, 1985.

Richmond A.D., Gravity wave generation, propagation and dissipation in the thermosphere, *J. Geophys. Res., 83*, 4131–4145, 1978.

Shrestha K.L., Observation of infrasonic waves in the F–region of the ionosphere, *J. Atmos. Terr. Phys., 29*, 107–112, 1967.

Testud J., and G. Vasseur G, Ondes de gravité dans la thermosphère, *Ann. Géophys., 25*, 525–546, 1969.

Thuillier G., C. Lathuillere, M. Herse, C. Senior C, W. Kofman, M.L. Duboin, D. Alcaydé, F. Barlier, and J. Fontanari, Coordinated EISCAT– MICADO interferometer measurements of neutral winds and temperatures in the E– and F– regions, *J. Atmos. Terr. Phys., 52*, 625–636, 1990.

Touboul P., A. Bernard, F. Barlier, C. Berger, Air drag effect on gradiometer measurements, *Manuscripta Geodætica, 16*, 73–91, 1991.

Trinks H., and H.G. Mayr, Large–scale neutral composition gravity waves in the thermosphere observed by Esro–4, *J. Geophys. Res, 81*, 4023–4026, 1976.

Vasseur G., C.A. Reddy, and J. Testud, Observations of waves and travelling disturbances, *Space Res., 19*, 1109–1131, 1972.

Vidal–Madjar D., Medium scale gravity waves and their non linear interaction with the means flow: a numerical study, *J. Atmos. Terr. Phys. 41*, 279–292, 1979.

Villain J.P., Analysis of perturbations of the total density determined by the low–g accelerometer CACTUS, *Space Res., 19*, 231–235, 1979.

Villain J.P., Traitement des données brutes de l'accéléromètre CACTUS. Etude des perturbations de moyenne échelle de la densité thermosphérique, *Ann. Geophys., 36*, 41–49, 1980.

Villain J.P., C. Berger, and F. Barlier, Analyse de l'augmentation de la densité thermosphérique et de l'observation de perturbations en zone équatoriale durant un orage magnétique, *C.R. Acad. Sci., Paris, série B*, 25–28, 1979.

Wardill P., and F. Jacka, Vertical motions in the thermosphere over Mawson, Antartica, *J. Atmos. Terr. Phys., 48*, 289–292, 1986.

C. Berger and F. Barlier, OCA/CERGA, av. Copernic, 06130 Grasse, FRANCE.

Upper Atmosphere Model for the Prediction of the Position of Satellites

M. N. Vlasov

Institute of Applied Geophysics, Moscow, Russia

Our Upper Atmosphere Forecasts are aimed at the prediction of short-term and long-term variations of the parameters affecting satellite orbits. A thermospheric model is the foundation of this forecast. For highly accurate predictions it is necessary that the model can be corrected by current data from upper atmospheric sounding. A time-dependent model that is based on the numerical solution of the continuity, momentum, and energy equations has been developed. Empirical model data have been used to define the lower and upper boundary conditions. The photochemistry and heating forced by EUV and UV solar radiation, heating by currents and eddy diffusion, cooling by infrared radiation of atomic oxygen, carbon dioxide, and nitric oxide have all been taken into account, together with mass-averaged transport, molecular multicomponent diffusion, and eddy diffusion. The model calculates the total density, composition, and temperature of the upper atmosphere. The major features of the model are departures from barometric law of the height profiles of the main constituents and from the Bates' formula for the temperature height profiles. This model is used as the Basic Model of the Upper Atmosphere Forecast System (UAFS). Corrections to the Basic Model using current data for disturbed conditions may be introduced by changes in the upper and lower boundary conditions.

Introduction

The effects of the upper atmosphere on the evolution of satellite orbits are very significant. It is well known that disturbances in the upper atmosphere cause the significant changes in the trajectories and lifetimes of satellites. The major problem in the prediction of precise trajectories is in forecasting atmospheric parameters. Also, such forecasts are very important for numerous other applications such as the determination of upper atmosphere effects on various measurements performed in space, electro-magnetic wave propagation, etc.

A model of temperature, density, and composition is necessary for forecasting the effects of the upper atmosphere. At the present time, valuable progress has been achieved in developing empirical models of the thermosphere. As a result several such models are available, for example [CIRA, 1972; Kohnlein et al., 1979; Hedin et al., 1977; Barlier et al., 1978; Hedin et al., 1987]. These models are based on temperature, density, and composition data from a comprehensive summary of satellite, rocket, and incoherent scatter radar data. However, most experimental data were obtained by satellites. The main advantage of these empirical models is their simplicity. They were presented in such a way that any potential user could easily compute the main parameters. However all empirical models are based on a limited set of data and on the spherical harmonic expansions introduced by Hedin et al. [1977]. This implies that such models lead to extrapolated results when they are applied for geophysical conditions never encountered during the observation periods. It is, therefore, not too surprising that comparisons between various models may lead to very significant differences. At the present time, it is clear that discrepancies between the recent models are stable. Comparisons between model results and observations not involved in the construction of the model, always indicate significant discrepancies.

Environmental Effects on Spacecraft Positioning and Trajectories
Geophysical Monograph 73, IUGG Volume 13

The principal limitation of these empirical models is the assumption of a barometric distribution of the main constituents (O, O_2, N_2) and the Bates' formula for the temperature distribution with height.

It is well known that neither assumption agrees with experimental data [Vlasov et al., 1981]. It has been shown [Vlasov and Davydov, 1982; 1983] that the height distribution caused by the multicomponent diffusion of the mixture of O, O_2, N_2 differs significantly from the barometric distribution of these components.

Theoretical models of the thermal balance indicate that temperature height profiles do not correspond to the Bates' formula. Our analysis of the thermal balance [Vlasov and Korobeynikova, 1991] indicates that this discrepancy is primarily caused by cooling by the infrared radiation of nitric oxide. Therefore empirical models below 300 km are very approximate.

However, the major processes which control the distribution of the main constituents take place in the lower thermosphere and mesosphere. Thus the atomic oxygen distribution above 200 km is controlled by eddy diffusion and photochemistry which occur below 150 km. During geomagnetic storms, dissipation of energy takes place in the lower thermosphere. The energetics, dynamics, and chemistry of the Earth's mesosphere-lower thermosphere system are poorly understood owing to their inherent complexity. This region is too high for direct probing by balloons and, for most part, too low for extended direct measurements by satellite. At the same time rocket measurements of the vertical profiles of the main constituents indicate that large discrepancies appear between all model values and the measurements [Vlasov et al., 1981; Kockarts, 1980].

In this situation a combination of theoretical and empirical models may be very useful. The main aim of this work is to present a new type of upper atmosphere model based on the numerical solution of the hydrodynamic equations system with the upper and lower boundary conditions based on empirical models. This approach may improve the reliability of thermospheric models by including theoretical modelling, where there are deficiencies in the experimental data. At the same time, the theoretical description may be simplified by using experimental data. Such an approach is very important for the control and forecast of the upper atmosphere. Some aspects of the Upper Atmosphere Forecasting System are now discussed.

BASIC EQUATIONS OF MODEL

The continuity, momentum, and energy equations are considered. The continuity and momentum equations are given by

$$\frac{\partial n_i}{\partial t} = q_i - l_i n_i - \frac{\partial \phi_i}{\partial z} \tag{1}$$

$$\phi_i = n_i w - K_e \left[\frac{\partial n_i}{\partial z} + \frac{n_i}{\overline{H}} + \frac{n_i}{T}\frac{\partial T}{\partial z}\right] - \sum_{j=1}^{3} D_{ij}\left[\frac{\partial n_j}{\partial z} + \frac{n_j}{H_j} + \frac{n_j}{T}\frac{\partial T}{\partial z}\right] \tag{2}$$

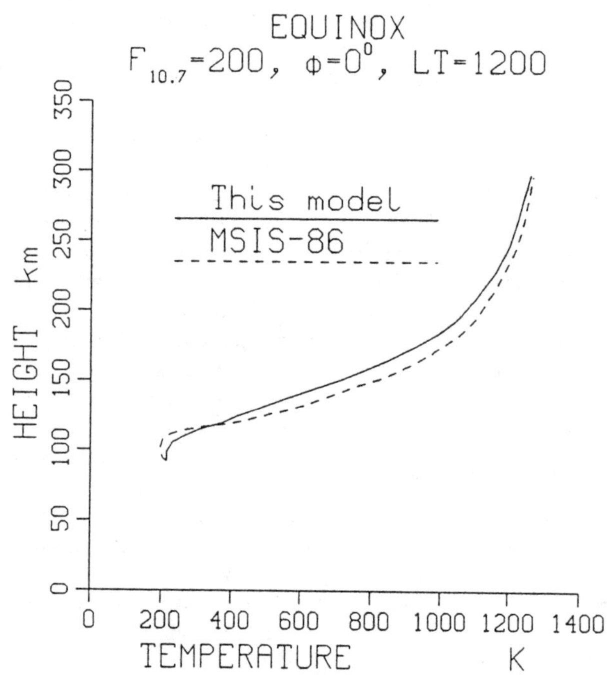

Fig. 1. Comparison of the temperature height profiles corresponding to this model and MSIS-86 at equator and daytime.

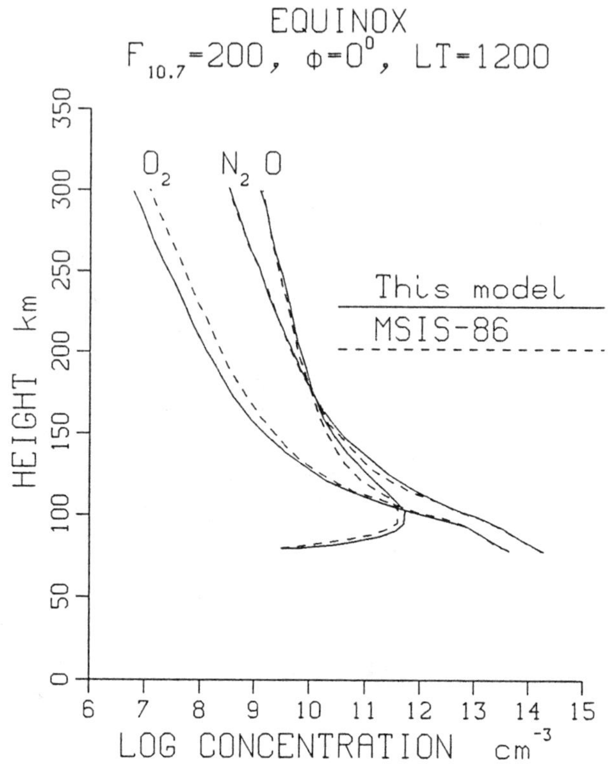

Fig. 2. Comparison of the main composition distributions corresponding to this model and MSIS-86 at equator and daytime.

where n_i is concentration of the ith component, (i = 1, 2, and 3 for the components O, O_2, N_2 respectively), ϕ_i is the flux due to molecular and eddy diffusion as well as mass-averaged transport, K_e is the eddy diffusion coefficient, D_{ij} is the coefficient of molecular diffusion for the ith component in the three-component mixture [Vlasov and Davydov, 1982], w is the velocity of mass-averaged transport, H_j is the scale height of the jth component, $\overline{H}$ is the scale height of the mixed atmosphere, T is the temperature, q_i is the production rate and l_i is loss rate due to photochemical processes. Lists of the photochemical reactions used in this model are given by [Vlasov and Davydov, 1982; 1983]. The photodissociation of molecular oxygen, the three-body recombination and radiative recombination of atomic oxygen are the most important of these reactions. The O_2 dissociation rate is calculated by using appropriate solar fluxes [Torr et al., 1980; Nusinov et al., 1989] and absorption cross sections [Ackerman, 1971].

The molecular diffusion term in equation 2 is characterized by the fact that the influence of the gradients $\partial n_j/\partial z$ on the diffusive velocity of the ith component is taken into account. The expression for coefficients D_{ij} were given by Vlasov and Davydov [1982]. An empirical profile of the eddy diffusion coefficient can be given in the form [Shimazaki, 1971]

$$\begin{aligned} K_e &= K_m \exp\left[-S_1 (z - z_m)^2\right] && z \geqslant z_m \\ K_e &= (K_m - K_0) \exp\left[-S_2 (z - z_m)^2\right] + K_0 \exp\left[S_3 (z - z_m)\right] && z < z_m \end{aligned} \tag{3}$$

where K_m is the maximum value of the coefficient K_e, z_m is the altitude where the maximum value of K_e is attained and K_0, S_1, S_2 and S_3 are empirical parameters according to Shimazaki [1971]. The velocity of mass averaged transport is determined by $w = A/\rho$, where ρ is total density, A is constant. This formula for the vertical velocity is based on Vlasov and Davydov [1983].

The heat balance equation may be presented by

$$c_V\rho\left(\frac{\partial T}{\partial t} + w\frac{\partial T}{\partial z}\right) + p\frac{\partial w}{\partial z} = \frac{\partial}{\partial z}\left(\lambda\frac{\partial T}{\partial z}\right) + Q_s + Q_e - L_{NO} - L_O - L_{CO_2} \qquad (4)$$

where c_V is specific heat per molecule at constant volume, p is the pressure, λ is the thermal conductivity, Q_s is the heating rate by solar ultraviolet radiation, precipitating electrons and currents, Q_e is the heating rate and heat transport by eddy diffusion, L_{NO}, L_O and L_{CO_2} are the cooling by the infrared radiation of NO, O and CO_2, respectively.

The eddy conduction rate and the heating rates of turbulent energy dissipation owing to the action of viscous and buoyancy forces are introduced in the form given by [Izakov, 1978]. Infrared emission in the NO 5.3μm band is one of the major cooling processes in the thermosphere [Kockarts, 1980]. The nitric oxide distribution is determined according to [Gérard et al., 1986]. Cooling by infrared radiation of atomic oxygen and carbon dioxide is calculated according to [Kockarts and Peetermans, 1970; Gordiets et al., 1982]. The ultraviolet solar flux and the heating efficiency are determined by [Torr et al., 1980; Nusinov et al., 1989].

The method of the numerical solution of the model equations system was described by [Vlasov and Davydov, 1982; 1983]. The implicit finite-difference scheme [Gershengorn, 1972; Samarsky, 1971] is set up for the whole system of equations. This scheme is solved by the method of factorization [Gershengorn, 1972]. The difference scheme is set up with a variable height interval Δz_i, and a constant time interval $\Delta\tau = 20$ min. The value Δz_i decreases to 1 km in the region of intensive turbulence and increases to 5 km in the thermosphere in accordance with physical considerations and accuracy requirements. At the lower boundary the concentrations of O, O_2, and N_2 together with the temperature, T, are taken from the empirical model, while at the upper boundary it is assumed that there are vertical fluxes of densities and heat corresponding to the same model.

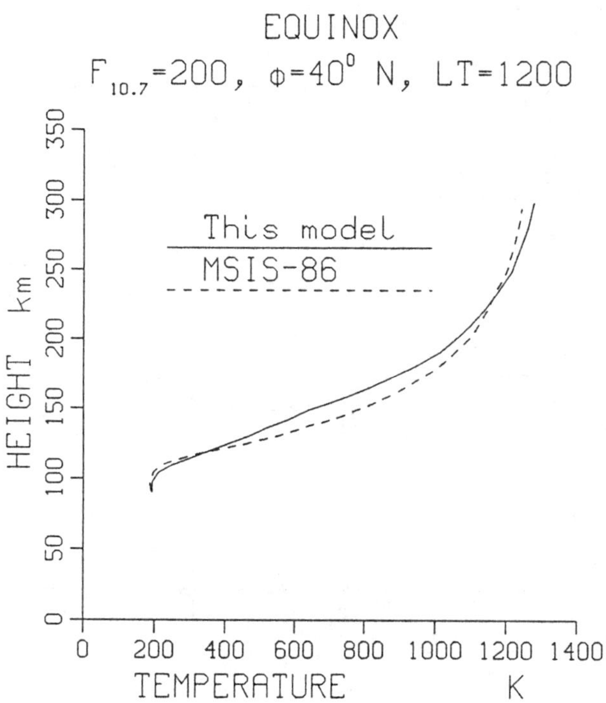

Fig. 3. Same as for Fig. 1 at middle latitude.

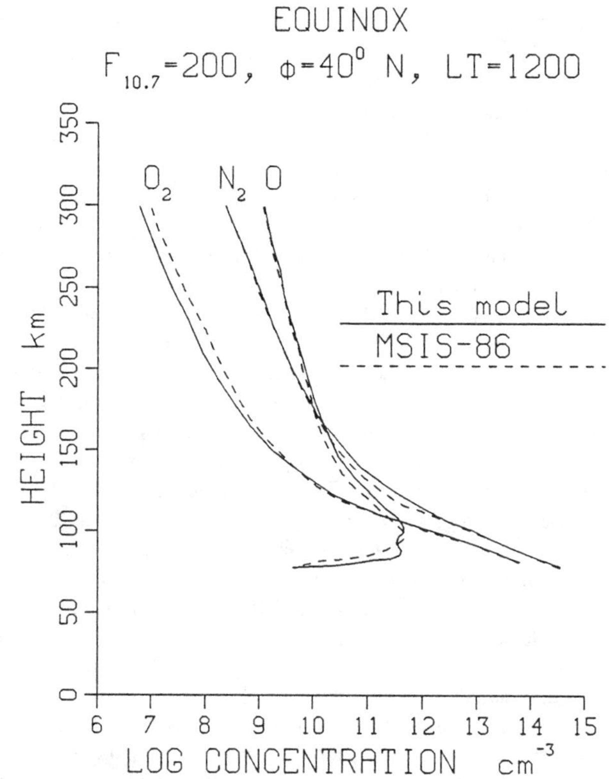

Fig. 4. Same as for Fig. 2 at middle latitude.

MODEL RESULTS AND COMPARISON WITH THE EMPIRICAL MODEL

Fig. 1 shows the temperature height distributions at the equator calculated by the theoretical model and by MSIS-86 [Hedin, 1987]. Significant discrepancies between the temperature profiles take place above 120 km and up to 250 km. The low thermospheric temperature derived by our model is caused by cooling due to the NO infrared radiation. In this case the theoretical height profiles of temperature cannot be described by the Bates' formula. Analysis of our results indicates that this is true for all conditions except during solar minimum, when cooling by the NO infrared radiation is negligible.

Comparison of the theoretical height profiles of the main constituents with those calculated by MSIS-86 indicates significant discrepancies (Fig. 2). The major cause of these discrepancies is multicomponent diffusion, which leads to results different from the barometric distribution used by empirical models. In this case the value of the theoretical concentrations of the main constituents is higher than those in MISI-86 in spite of the lower theoretical temperatures. The empirical barometric concentrations decrease faster than the theoretical concentrations. The same results were obtained by the comparison of the empirical model with the rocket measurements of the atomic oxygen concentration [Kockarts, 1980]. In [Vlasov et al., 1981] this fact was explained by multicomponent diffusion and mass-averaged transport. The departure from the barometric law increases with the rise of the solar activity. Similar results are obtained for the temperature and composition distributions at middle latitudes (Fig. 3 and Fig. 4). One of the important parameters in the theoretical model is the height distribution of the eddy diffusion coefficient. Our analysis indicates that the K_m value cannot exceed 5 $10^6 cm^2 s^{-1}$ for the maximum altitude above 90 km. If this value is exceeded theoretical temperature profiles in the 100 km region become unreliable because a temperature maximum is produced about 90 km as a result of heating by eddy diffusion. However, this limitation of the eddy diffusion coefficient contradicts the K_m values ordinarily used to describe composition in the time-dependent models (for example [Vlasov and Davydov, 1983, 1986]).

This contradiction indicates that the eddy diffusion and eddy heat conductivity coefficients may be different. However, there is another way to remove this contradiction. Analysis of the model indicates that the gradient of the eddy diffusion coefficient above z_m influences

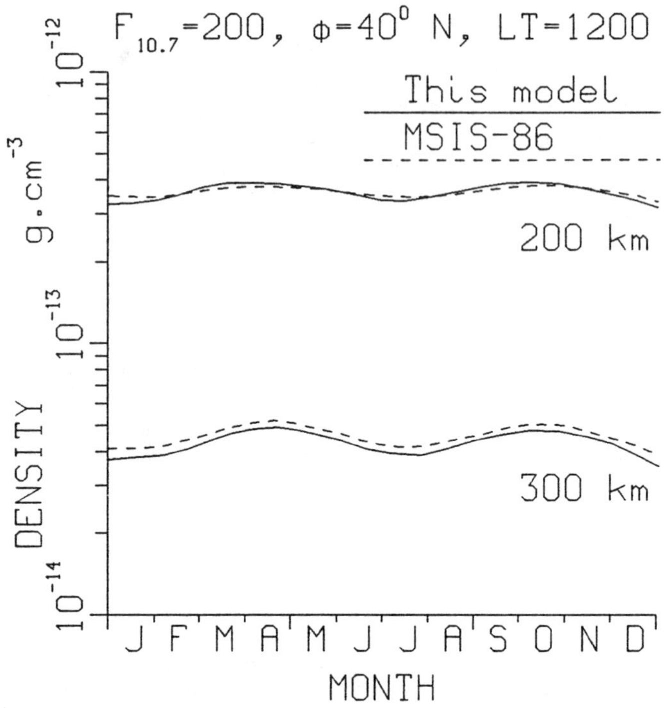

Fig. 6. Annual variations of the total density according to this model and MSIS-86 (equator).

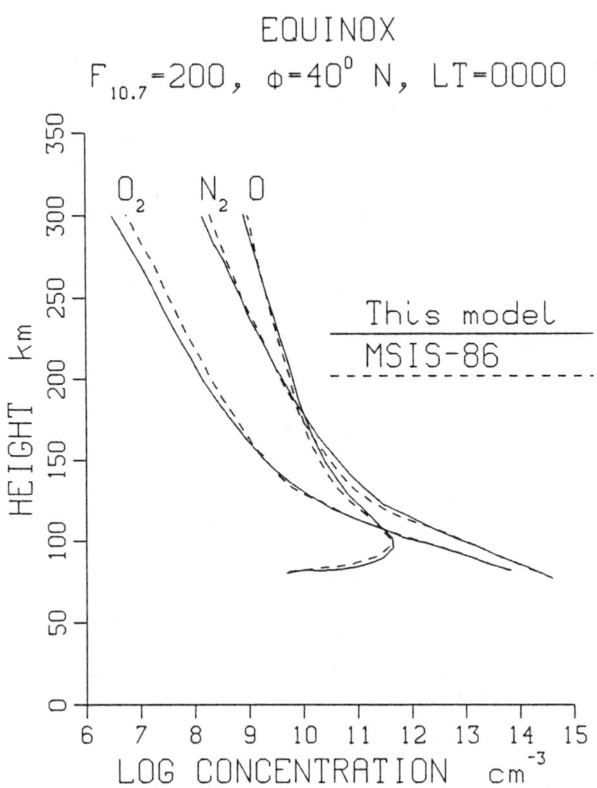

Fig. 5. Same as for Fig. 4 at night time.

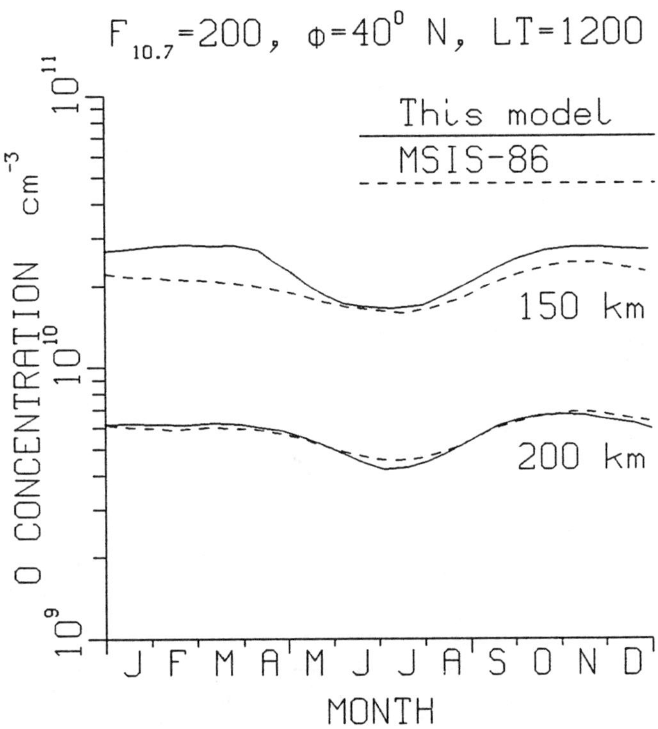

Fig. 7. Annual variations of the atomic oxygen concentrations according to this model and MSIS-86 at middle latitude.

the atomic oxygen distribution as well as the K_m value. This effect is very important and a set of features of atomic oxygen distribution may be explained by the way. For example, a splitting of the atomic oxygen height profiles may be produced, as it is shown on Fig. 4. At nighttime, the main features of the discrepancies between our model and MSIS-86 below 300 km are conserved (Fig. 5), however, the main constituent concentrations do not coincide around 300 km.

The annual variation of the total density at 200 km and 300 km is shown in Fig. 6 for solar maximum conditions at midlatitudes. The theoretical variation is qualitatively similar to the variation from the empirical model at all altitudes. However, the theoretical annual and semiannual variations of atomic oxygen differ quantitatively from the empirical model variation below 200 km (Fig. 7). This discrepancy reflects the fact that the empirical model may not necessarily represent real physical conditions, particularly for atomic oxygen which is influenced by photochemistry and by transport processes.

MONITORING AND FORECAST

The model presented above is the Basic Model of Upper Atmosphere Control and Forecast System. A schematic diagram of the system is presented in Fig. 8. The system takes account of both the model and the use of satellite, rocket, and ground based techniques for the sounding of upper atmosphere parameters. The sounding data are necessary for improvement in the model during disturbed conditions. In this case, the results of measurements are used as well as empirical model data.

This approach makes possible to develop the hydrodynamical method of forecast. The main features of this method were considered in [Vlasov, 1989], which indicates that statistical forecasts based on limited data cannot satisfy modern practical demands. The variability of the upper atmosphere is so large that it is impossible to provide reliable statistical data. At the same time the hydrodynamical approach may compensate for this deficiency of data. The goal of the hydrodynamical forecast is the prediction of the upper atmosphere parameters which determine the evolution of satellite orbits. The main external parameters are the levels of the solar and geomagnetic activity, which characterize the absorption of the solar energy by the upper atmosphere.

However, energy penetrating from the lower atmosphere may be important too. This energy influences eddy diffusion. The parametrization of this effect is a very important problem. An index of the wave activity of the lower atmosphere is necessary. The advantage of the hydrodynamical approach is that it can describe the interaction between the upper and lower atmosphere.

One of the main demands of the Basic model is for rapid numerical calculation of the upper atmosphere parameters. It is very desirable that this time does not exceed half the period of the satellite

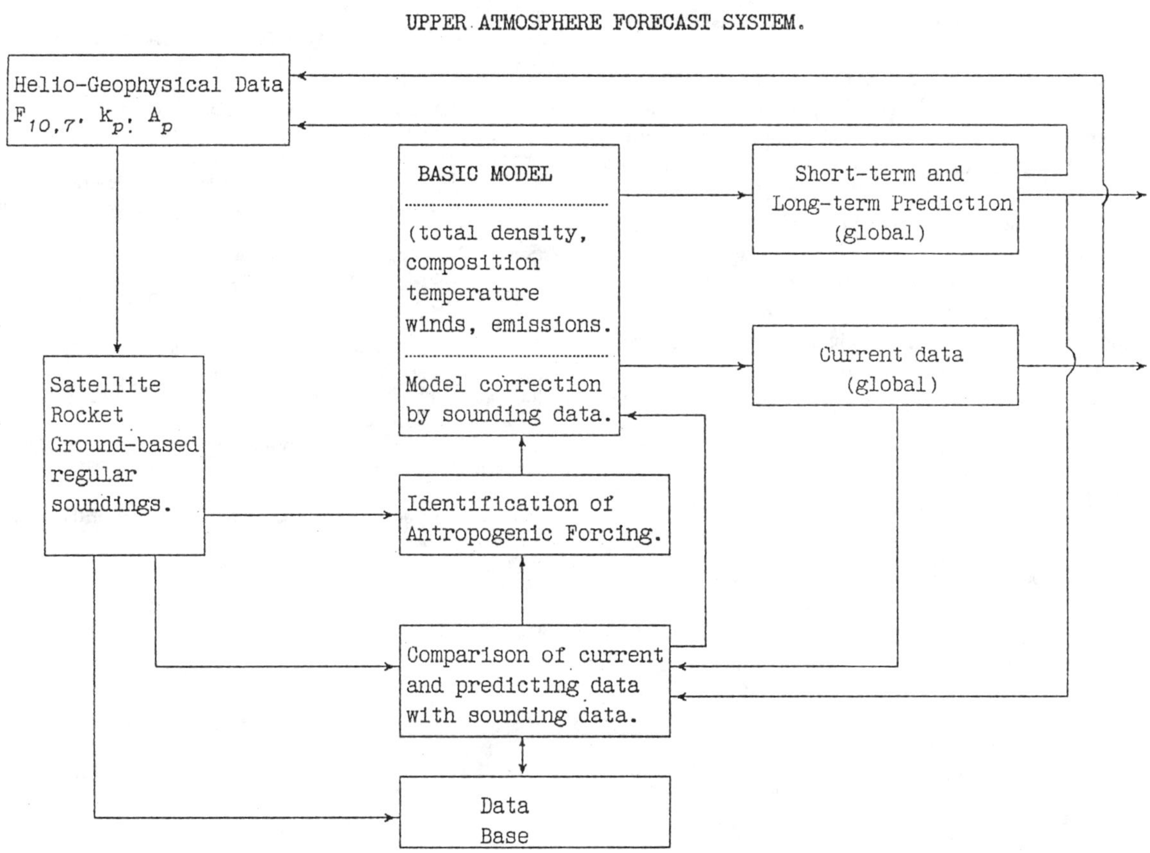

Fig. 8. The principal scheme of Upper Atmosphere Control and Forecast System.

orbit. In this case data for each orbit may be taken into account. This is very necessary for disturbed geomagnetic conditions. The establishment of analytical approximations for the numerical solution decreases the calculation time. Currently, the calculation time for temperature, total density, and composition is about twenty minutes for our model.

It is necessary to emphasize that current information on the upper atmosphere state is based on the hydrodynamical model as well as forecast, because the measurements cover a very small part of the upper atmosphere.

SUMMARY AND CONCLUSIONS

An upper atmosphere model based on the numerical solution of the hydrodynamical time-dependent equations and empirical model data has been constructed. Variations of the temperature and main constituent concentrations are described by this model.

A comparison of this model with the empirical model indicates significant discrepancies below 300 km. The temperature height profiles calculated by this model differ from the empirical model and do not correspond to the Bates' temperature profile. This discrepancy is caused by the cooling of the nitric oxide infrared radiation. The theoretical height profiles of the main constituents do not follow the barometric law. This is the result of multicomponent diffusion. These main differences between our model and empirical models occur for all conditions except at solar minimum.

The model analysis indicates that the gradient of the eddy diffusion coefficient above the altitude of the K_e maximum influences the atomic oxygen concentration as well as the K_m value. The main features of the atomic oxygen height distribution and variations may be described by this effect. This effect may explain the contradiction between the high K_e values necessary for the composition description and the low K_e value corresponding to the thermal balance. The splitting of the atomic oxygen height profiles about the maximum may arise because of a gradient in the eddy diffusion and photochemistry. The short time of the numerical calculations and possibility of the correction by measurements make it possible to use the model as the Basic model for the Upper Atmosphere Control and Forecasting System. A hydrodynamical forecast may be realized in this system. The short-term and long-term variations of the upper atmosphere parameters and the effects on the evolution of satellite orbits may be predicted.

REFERENCES

Ackerman, Mesospheric Models and Related Experiments, D. Reidel, Dordrecht, Holland, 140, 1971.

Barlier F., C. Berger, J. L. Falin, G. Kockarts, and G. Thuillier, A thermospheric model based on satellite drag data, Ann. Geophys. 34, 9-24, 1978.

CIRA 1972, COSPAR International Reference Atmosphere, Verlag, Berlin. 1972.

Gérard J. C., and C. E. Noel, AE-D measurement of the NO geomagnetic latitudal distribution and contaminations for N+(5S) emissions, J. Geophys Res., 91, 10136, 1986.

Gershengorn G. I. On the methods of the numerical solutions of the differential equations for the ionospheric electrons, Issled. Geomagn. Aeron. Fiz. Solntsa, 21, 283-287, 1972.

Gordietz, B. F., Yu. N. Kulikov, M. N. Markov, and M. Ya. Marov, Numerical modeling of the thermospheric heating budget, J. Geophys. Res., 87, 4504-4512, 1982.

Hedin A. E., MSIS-86 Thermospheric model, J. Geophys. Res., 92, 4649-4662. 1987.

Hedin A. E., C. A. Rober, G. P. Newton, N. W. Spencer, H. C. Brinton, H. G. Mayar, and W. E. Potter, A global thermospheric model based on mass spectrometer and incoherent scatter data MSIS, J. Geophys. Res., 82, 2139-2156, 1977.

Izakov, M. N., Influence of the eddy diffusion on the thermal balance of the thermosphere. Cosmic Res., 16, 324, 1978.

Kockarts G., Nitric oxide cooling in the terrestrial thermosphere, Geophys. Res. Lett., 7, 137-140, 1980.

Kockarts G., Some recent advances in the thermospheric models, Report presented at COSPAR, Budapest, 1980.

Kockarts G., and W. Peetemans, Atomic oxygen infrared emission in the earth's upper atmosphere, Planet Space Sci, 18, 271-277, 1970.

Kohnlein W., D. Krankowski, P. Lammerzahl, W. Joes, and H. Volland, A thermospheric model of the annual variations of the He, N, O, N2, and Ar from the AEROS NIMS data, J. Geophys. Res., 84, 4355-4632. 1979.

Nusinov A. A., T. V. Kazachevskaya, and V. V. Katyushina, Prediction and control of EUV and X- ray solar radiation by 10.7 cm radio emission, Solar-Terrestrial Predictions: Proceedings of a Workshop at Leura, Australia, October 16-20, 1, 546, 1989.

Samarsky A. A., Vvedenie v Teoriyu Raznostnykh Skhem, Nauka, Moscow, 1971.

Shimazaki T., Effective eddy diffusion coefficient and atmospheric composition in the lower thermosphere, J. Atmosph. Terr. Phys., 33, 1383-1401, 1971.

Torr M. R., D. G. Torr, and H. E. Hinteregger, Solar flux variability in the Schumann-Runge continuum as a function of solar cycle 21, J. Atmosph. Geophys. Res., 85, 6063, 1980.

Vlasov, M. N., and V. E. Davydov, Theoretical description of the main neutral constituents in the earth's upper atmosphere, J. Atmosph. Terr. Phys., 44, 641-647, 1982.

Vlasov, M. N., Fundamentals of the physical forecast of ionospheric plasma, Solar Terrestrial Predictions Proceedings, 1989.

Vlasov M. N., V. E. Davydov, and A. A. Lifshic, Analysis of the semiempirical models by using the theoretical calculations and the rocket measurements of the atomic oxygen, Kosmich Issled., 19, 638-640, 1981.

Vlasov, M. N., and V. E. Davydov, Influence of the vertical mass-averaged transport on the upper atmosphere composition during the geomagnetic disturbance, Cosmic Res., 21, 569, 1983.

Vlasov, M. N., and V. E. Davydov, On the role of the mass-averaged transport for the seasonal variations of the thermospheric composition, Cosmic Res., 24, 691, 1986.

Vlasov, M. N., and T. V. Korobeynikova, Influence of the eddy diffusion, the infrared radiation, and the mass-averaged transport on the height of the temperature in the middle and high latitude atmosphere, Cosmic Res., 29, 544-550, 1991.

M. N. Vlasov, Institute of Applied Geophysics, Moscow, Russia.

Progress In Geomagnetic Storm Prediction

RITA C. SAGALYN

Phillips Laboratory, Geophysics Directorate
Hanscom AFB, MA 01731-5000

SIDNEY A. BOWHILL

Aeronomy Laboratory, Dept of Electrical Engineering
University of Massachusetts Lowell
Lowell, MA 01854

Significant advances have been made in recent years in our understanding of the origins of major solar-terrestrial disturbances which impact spacecraft operation and tracking, as well as ground-based power and communication systems. The finding that coronal mass ejections are a major source of enhanced geomagnetic activity has led to increased efforts to measure their transport through the interplanetary medium. Two to three day forecasts of near-Earth geomagnetic disturbances could be obtained from a space-based solar mass ejection imager. Real-time libration-point measurements of the solar wind and interplanetary magnetic field on the WIND and ACE spacecraft scheduled for launch by NASA in 1993 and 1996, respectively, can provide approximately one-hour alerts. In the past few years, the first deterministic global models of space weather processes have been developed for space environment specification and prediction. The most successful prediction techniques rely on a combination of empirical and numerical techniques driven by observation. Because of the complex coupling between the solar, interplanetary, magnetosphere, ionosphere and thermosphere regions, progress in magnetic storm prediction is expected to depend increasingly on the utilization of new techniques such as expert systems.

INTRODUCTION

It has been known for many decades that solar activity is the source of geomagnetic storms. Yet, little was known about the role of currents, fields, plasma pressure, and particle entry in the development of magnetic storms until data was returned from instruments on board NASA, DoD, ESA and Soviet Union satellites. These include ISEE, Helios, Solar Maximum Mission (SMM), GOES, DMSP, Prognoz, Skylab and P78-1.

The first observational support of the importance of the interplanetary magnetic field (IMF) in the development of geomagnetic storms came from the work of *Fairfield and Cahill* [1966]. They showed that a change in the IMF from northward to southward resulted in geomagnetic disturbances at the ground. They suggested that this effect is due to "field connection", or merging of the IMF with the Earth's magnetic field at the day-side magnetopause.

Until a few years ago, it was generally assumed that solar flares were the primary source of non-recurrent geomagnetic storms. Coronagraph data from P78-1, Skylab and SMM showed the existence of coronal mass ejections (CMEs). In-situ observations from ISEE-3 showed that CMEs rather than flares produce interplanetary conditions which can result in large geomagnetic storms. The recent work of *Gosling et al.* [1991], *Tsurutani et al.* [1992] and others has shown that

Environmental Effects on Spacecraft Positioning and Trajectories
Geophysical Monograph 73, IUGG Volume 13

fast coronal mass ejections, magnetized clouds in the solar wind, drive fast shocks, which, when characterized by a strong southward IMF component, can produce major magnetic storms. *Burlaga* [1988] and *Lepping et al.* [1990] have reported on modeling and observations of magnetized clouds in the heliosphere. There are extensive data [*Harrison*, 1986; *Kahler*, 1987; *Sheeley et al.*, 1985; *Hundhausen*, 1987; and *Harrison et al.*, 1990] which show that coronal mass ejections (CMEs) usually precede solar flares. There are also spectacular data from the Dynamics Explorer satellites that show the intensification of the many space environmental features that characterize a geomagnetic storm, that is, increased current, particle precipitation, auroral expansions and their dependence on the direction of the IMF [*Craven and Frank*, 1987; *Heppner and Maynard*, 1987; *Hoffman et al.*, 1985; *Craven and Frank*, 1989; and *Hairston and Heelis*, 1990].

We do not discuss geomagnetic storms at times of low solar activity, which often show a 27-day recurrence. These tend to be smaller, and not associated with CMEs.

The rapid expansion, during the past decade, in the use of satellites for commercial and defense purposes has led to a growing awareness that the space environment is not benign and that knowledge of great space weather disturbances such as geomagnetic storms is crucial to spacecraft design and operation.

This report reviews (a) the effects of magnetic storms on space and ground based systems; (b) advances in our understanding of observations of CMEs as they relate to geomagnetic storms; (c) prediction techniques and space environment model development; and, (d) space measurements necessary for the establishment of reliable geomagnetic storm forecasts.

EFFECTS OF STORMS

It has long been known that variations of the space environment can alter or degrade satellite operations. Intense perturbations usually originate at the sun and propagate through the interplanetary medium into near-earth space where most satellites are located. The operation of space systems, commercial communications satellites, meteorological spacecraft or military missions, for example, are affected by space radiation, hot plasmas, fluxes of solar and galactic cosmic rays and high-speed coronal mass ejections. Satellite charging, orbit changes caused by increasing density of the upper atmosphere, and degradation of microelectronics and satellite attitude sensors can result [*Gorney*, 1989]. There is a growing need to forecast space environment disturbances in order to minimize or mitigate their effect on costly, complex space systems.

Significant variations can occur over a broad temporal range, from seconds to days to years. The solar flares and coronal disturbances of March 22 and 23, 1991, illustrate short term (seconds to days) effects on several space and communications systems. Space-based data show that during this period there were enhancements in the energetic proton populations. Four energy channels at the GOES-7 spacecraft over the period 22 through 25 March are shown in Figure 1. The dotted line shows that the peak fluxes for all four energy channels occur at 03:50 UT on March 24. This follows the observation at earth of a sudden commencement at 03:42 UT which signals the beginning of a geomagnetic storm in the magnetosphere-ionosphere system. It is important to note that the near-earth disturbances of the geomagnetic field and the peak of the energetic particle populations occurred approximately 29 hours after the onset of the solar disturbance. A flare which began at 22.43 UT on March 22 was classified as Importance 3 in area, though there were no CME observations at this time. Some of the effects of this March 91 magnetic storm on satellite systems and communications are listed in Tables 1 and 2.

It is of interest to discuss results from the Combined Release Radiation Effects Satellite (CRRES) during the March 1991 storm. Within a minute of a shock's arrival at the magnetosphere boundary, the CRRES magnetic field measurements increased by a factor of 30 over the period of a minute. The peak in the magnetic field intensity coincides with a peak in all the CRRES energetic proton detector channels, shown in Figure 2. Within thirty seconds of the SSC at 3:42 UT, the 20 to 80 MeV channels showed a significant flux increase. Totally unexpected was the formation of a second peak within the inner proton radiation belt located at approximately 2.55 Earth radii for protons with energy > 29 MeV, Figure 3. Subsequent orbits showed that this

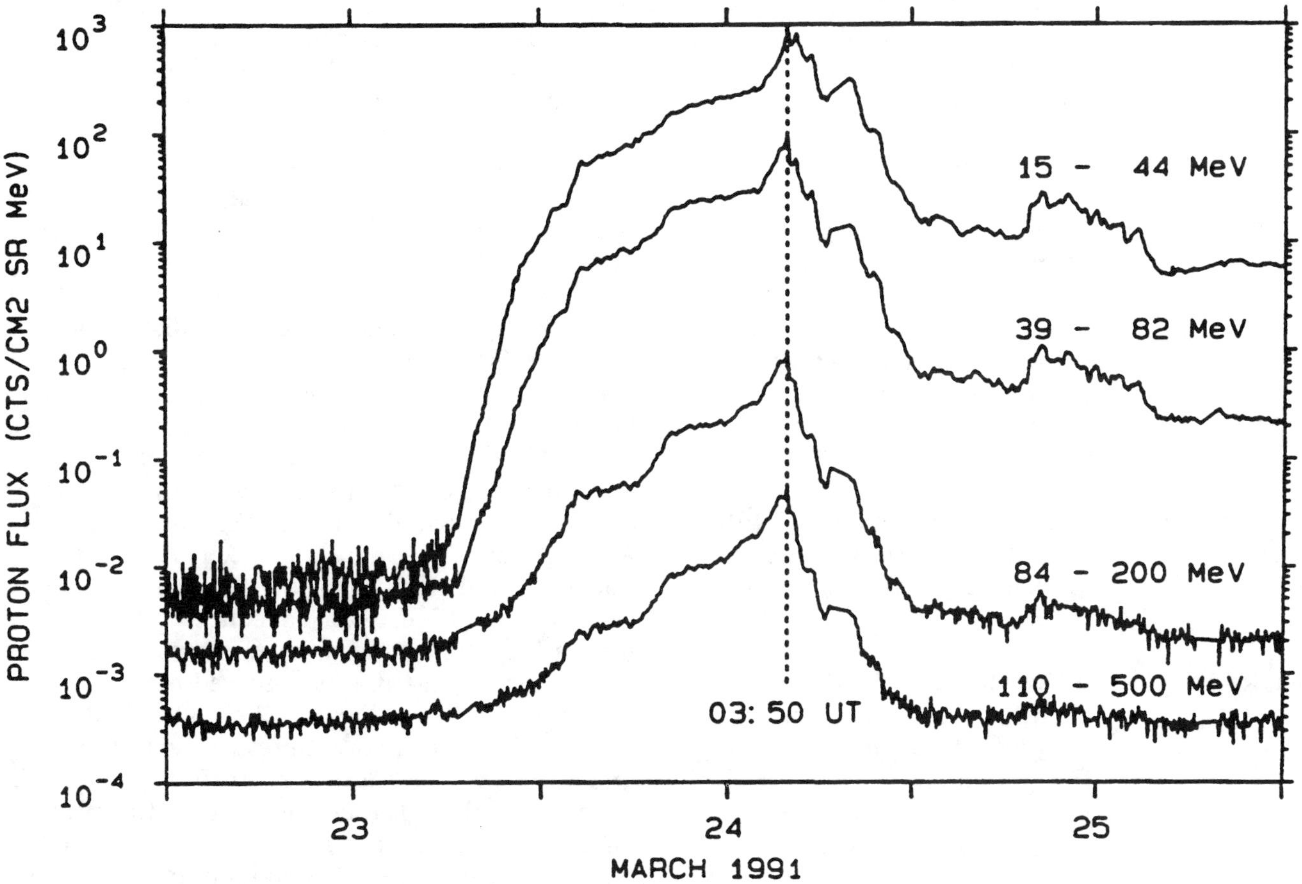

Fig. 1. Five-minute-averaged Goes-7 proton flux in 4 energy channels for the 23-25 March 1991 solar event. The dotted line shows the peak fluxes for all 4 channels at 03:50 UT on 24 March, following the Sudden Storm Commencement at 03:42 UT [*Mullen and Gussenhoven*, 1991].

second peak in the high energy proton belt persisted for over 40 days with only a slight decrease in its location to 2.34 Earth radii for 54 MeV particles. These second inner-belt proton peaks persisted through October 1991 when the operation of the CRRES spacecraft was terminated. Studies are underway to determine the mechanism by which the high energy protons in the second inner-belt remain stably trapped for extensive periods of time [*Mullen and Gussenhoven*, 1991].

SOLAR AND INTERPLANETARY PHENOMENA

One of the most important findings of the last decade is the crucial role of CMEs in the generation of large geomagnetic storms. A CME observed by SMM is illustrated in Figure 4. They originate from global magnetic changes on the sun [*Wagner*, 1984; *Hundhausen*, 1987]. The outward motion of low density plasma containing intense magnetic field compresses the background plasma ahead of it, giving the bright frontal loop seen in Figure 4. Measurements with the white-light photometer on Helios-2 [*Webb and Jackson* 1990] and the white-light coronagraph on P78-1 [*Howard et al.*, 1985] have shown that CMEs occur once or twice per day during times of high sunspot activity [*Webb*, 1991]. Nearly all of the shocks observed, out to 1AU, are associated with CMEs, and these tend to be large and fast moving, and to originate at low latitudes on the sun [*Sheeley et al.*, 1985], though not usually associated with solar flares [*Cane*, 1988]. *Gonzalez and Tsurutani* [1987] and *Tsurutani et al.* [1988], using ISEE-3 data,

Table 1

SATELLITE EFFECTS
March 91 Storm

MARECS-A	Satellite failure
INTELSAT 602 GOES-7	Permanent power degradation (decrease of satellite lifetime by 2-3 years)
TDRSS INTELSAT CRRES	Microelectronic effects: single effects upsets, computer soft errors increase from 1 per day to 10-20 per day.
DOD NOAA-11	High spacecraft potential with degradation of sensors and cable discharges. Loss of automatic attitude control
NORAD	Atmospheric density enhancement increased. Satellite drag loss of 200 vehicles from catalogue

Table 2

TERRESTRIAL EFFECTS
March 91 Storm

HF Links	Loss of polar and auroral zone point-to-point communications
HF Links	Midlatitudes Point-to-point HF com degraded
Eastern US	Power line loss (Service interrupted & equipment damage)

investigated the interplanetary source of the strong southward fields that caused ten major storms. They found that various combinations of magnetic field configurations were associated with ten major storms during the 1978-1979 solar maximum; e.g., shock effects on the heliospheric current sheet or on pre-existing fields, driver gas fields, and disturbances following the shock. *Gosling et al.* [1990, 1991] show that major storms originate in transient disturbances in the solar wind driven by fast CMEs characterized by strong southward directed interplanetary magnetic fields. In the majority of the largest geomagnetic storms examined, both a shock and a CME driving the shock encountered the Earth.

A sketch of a meridional cut through a hypothetical interplanetary shock disturbance driven by a fast CME, and the corresponding radial variation of solar wind speed and field strength along the centerline of the disturbance, is shown in Figure 5.

According to *Gosling et al.* [1991], shocks may be observed by a spacecraft at 1 AU even when the CME driving the shock is not. Observations suggest this is due to the relatively broader extent of shocks compared to the CMEs that drive them. Recent work by *Crooker et al.* [1992] indicates that, in addition to the meridional draping depicted in Figure 5, in-ecliptic draping of magnetic fields is also important, specifically for the frequent great storms observed near the times of the equinoxes at Earth. The reason is that the Earth's dipole is tilted with respect to the ecliptic, and enhances the southward IMF component in geocentric solar-magnetospheric coordinates. The ambient magnetic field drapes about the CME because CMEs generally originate in closed field regions in the corona not previously participating in the solar wind expansion. The strong fields and the high flow speeds commonly associated with interplanetary disturbances driven by fast CMEs are what make such disturbances effective in stimulating geomagnetic activity [*Gosling et al.*, 1991].

Actual observations of solar wind plasmas and magnetic field parameters measured at ISEE-3 during a geomagnetic storm on August 28-30, 1979, are shown in Figure 6 together with the geomagnetic index Kp. From top to bottom, the plasma and field parameters plotted are the bulk flow speed, the log of the combined ion, electron, and magnetic field pressure, the magnetic field strength, the out-of-the ecliptic component of the magnetic field and Kp. More complete presentations of the plasma and field data for this event can be found elsewhere [*Gosling and McComas*, 1987; *Gosling et al.*, 1987]. The dashed vertical line marks passage of a transient interplanetary shock, and the solid vertical lines bracket the CME driving the shock.

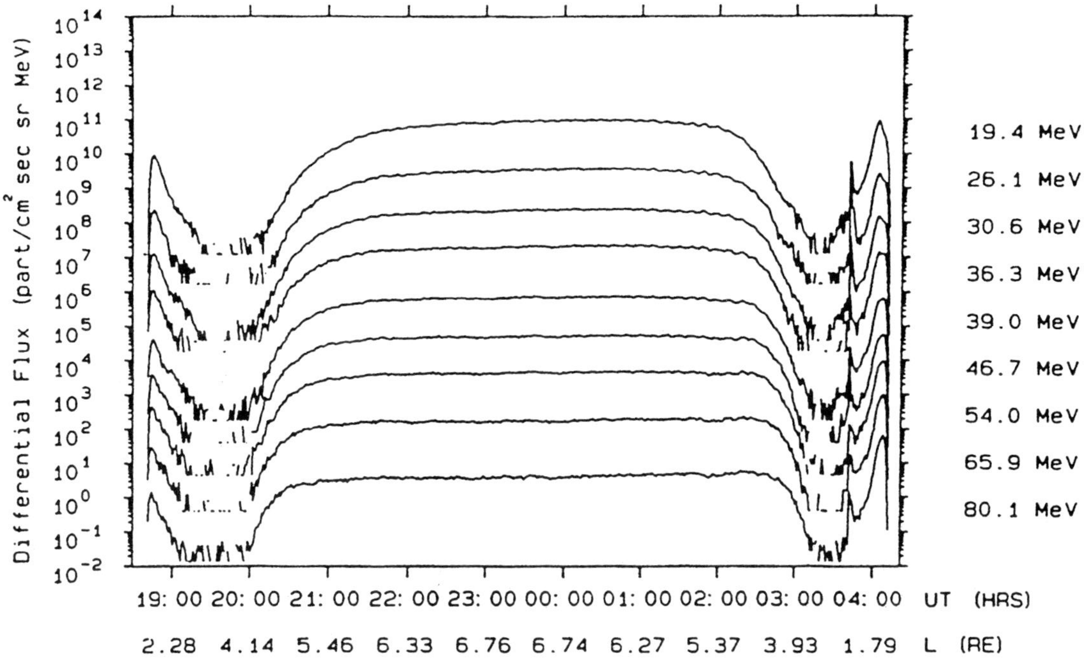

Fig. 2. One-minute average differential proton flux measurements for 9 energy channels between 19.4 and 80.1 MeV versus UT for CRRES Orbit 587 on 23-24 March 91. The sharp peak in fluxes at 03:42 UT (L = 2.55 R_E) coincides with the time of the Sudden Storm Commencement [*Mullen and Gussenhoven*, 1991].

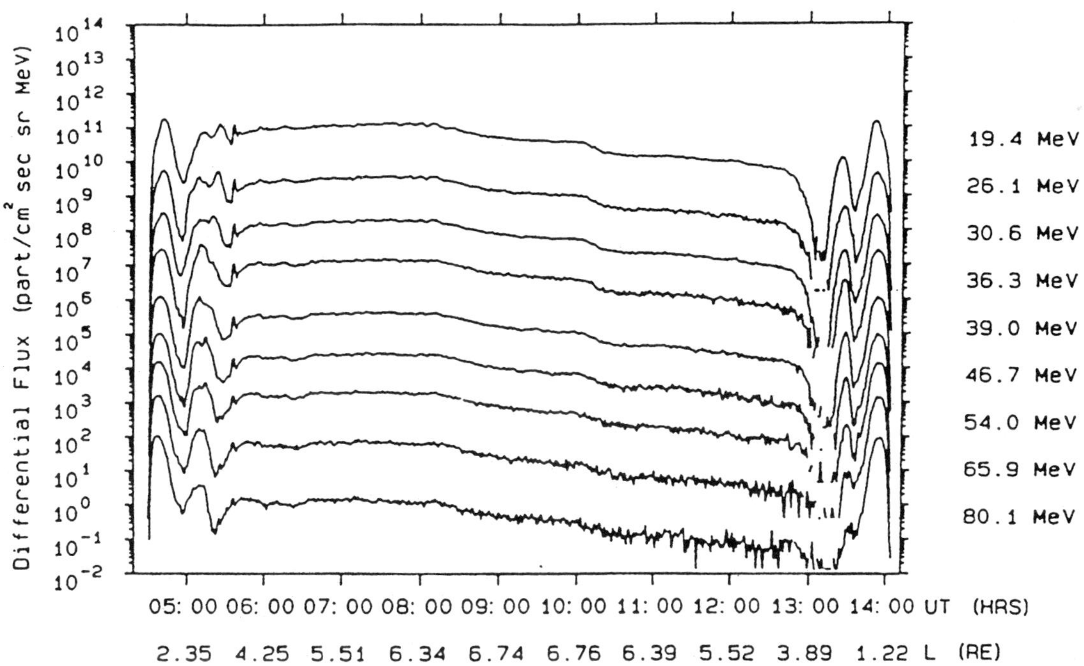

Fig. 3. Same as Figure 2, but for CRRES Orbit 588 on 24 March 1991 showing the formation of the double-peaked inner proton radiation belt during the early portion of the pass and its stable condition at the end of the pass. The outer peak of the 54 MeV protons is at a L = 2.6 R_E [*Mullen and Gussenhoven*, 1991].

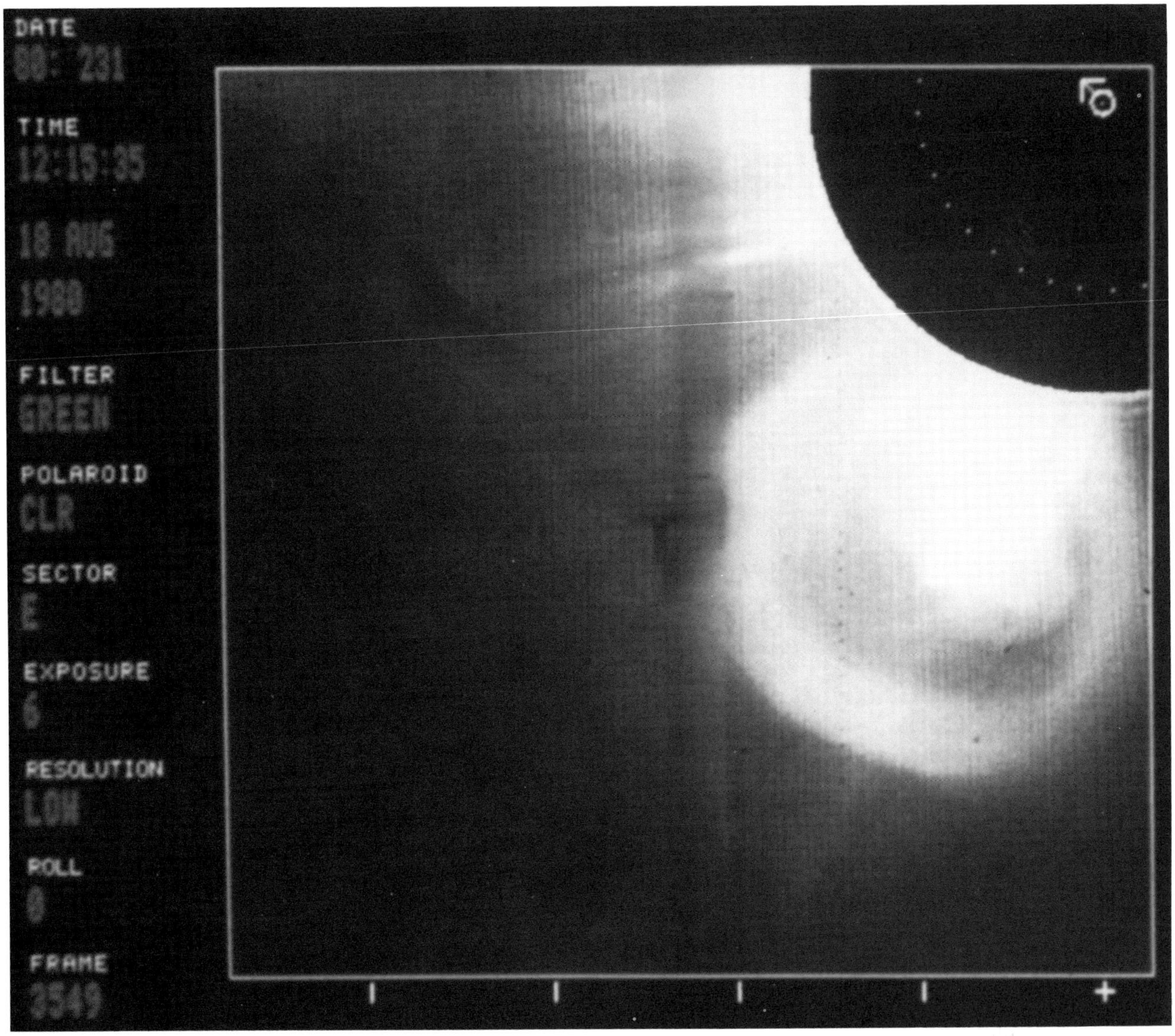

Fig. 4. Coronal mass ejection observed with SMM Coronagraph in 1980 [*Hundhausen*, 1987].

In the pie chart (Figure 7), *Gosling et al.* [1991] summarized the association of magnetic storms with Earth-passage of shock disturbances and CME events observed near the last solar maximum. The CMEs were identified by observations of counter-streaming solar-wind halo electrons at energies above about 80 eV. Only one large geomagnetic storm was unassociated with Earth-passage of either a shock or a CME event. Most storms (25 out of 34) were associated with passage of both a shock and a CME. A strong southward interplanetary magnetic field (IMF) was also found at the onset time of each geomagnetic storm. These results demonstrate the importance of real time measurements of CMEs and the interplanetary solar wind and magnetic field for alerts of hazardous space environments and for global models of the solar space environment operating in real time, as well as for long-term (2-3 day) predictions of geomagnetic activity.

MODELS AND FORECASTS

In recent years, many scientists have worked to establish analytical and empirical relationships between the highly variable transport of energy in

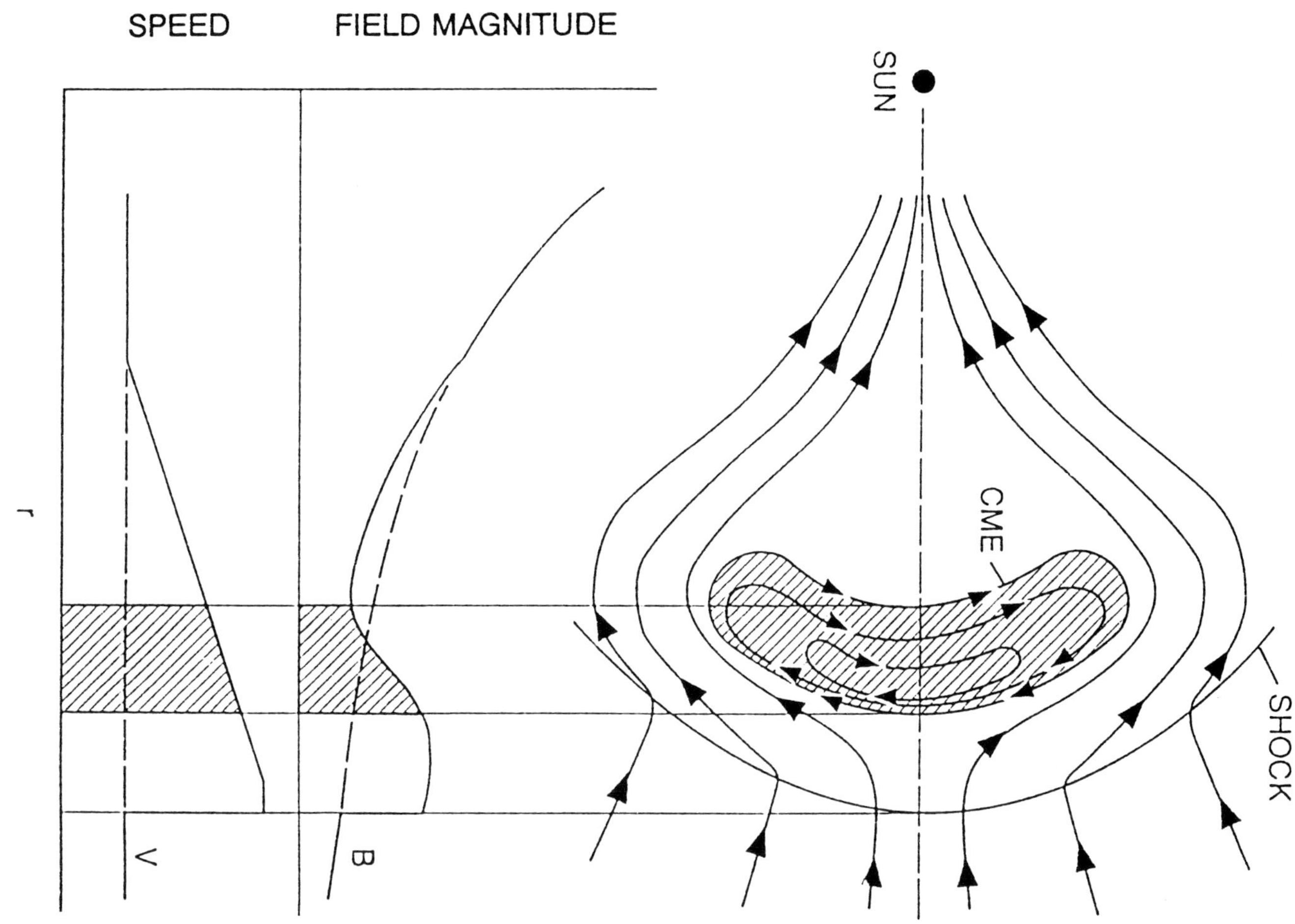

Fig. 5. Interplanetary shock wave disturbance driven by a fast CME [*Gosling et al.*, 1991].

the solar wind and disturbances in near Earth space. For example, *Snyder et al.* [1963] found a relationship between the Earth's 24 hour geomagnetic index ΣKp and solar wind velocity. *Crooker et al.* [1977] established a relationship between Ap (the planetary magnetic activity index), the solar wind velocity and interplanetary magnetic field component, B_z. The importance of the interplanetary magnetic field (IMF) to magnetic storm activity has become increasingly clear, as suggested originally by *Alfven* [1950], *Dungey* [1961], *Fairfield and Cahill* [1966] and others. Since then, much has been written on the correlation between various magnetospheric phenomena and the IMF southward component. The importance of the IMF southward component on magnetic substorm activity was later confirmed by several studies [cf. *Rostoker and Falthammar*, 1967; *Arnoldy*, 1971; *Meng et al.*, 1973; *Akasofu and Lepping*, 1977; *Akasofu*, 1980a,b]. Further, the response of the magnetosphere to a "sharp southward turning" of the IMF vector has been well documented [*Burton et al.*, 1975; *Akasofu and Lepping*, 1977; *Akasofu and Kamide*, 1980]. It includes an Earthward shift of the dayside magnetopause, and a change in the size and shape of the Earth's magnetotail; equatorward shift of the cusp; an equatorward expansion of the auroral oval and increase of the electric field and convection across the polar cap. The collective changes in the magnetosphere result in a geomagnetic storm or substorm.

Akasofu [1981], *Akasofu and Fry* [1986] and *Akasofu et al.* [1988] developed a storm prediction scheme. It has as input the characteristics of the initial solar disturbance, which include the date, time, duration, solar latitude and longitude and area of the disturbance, as well as its initial speed.

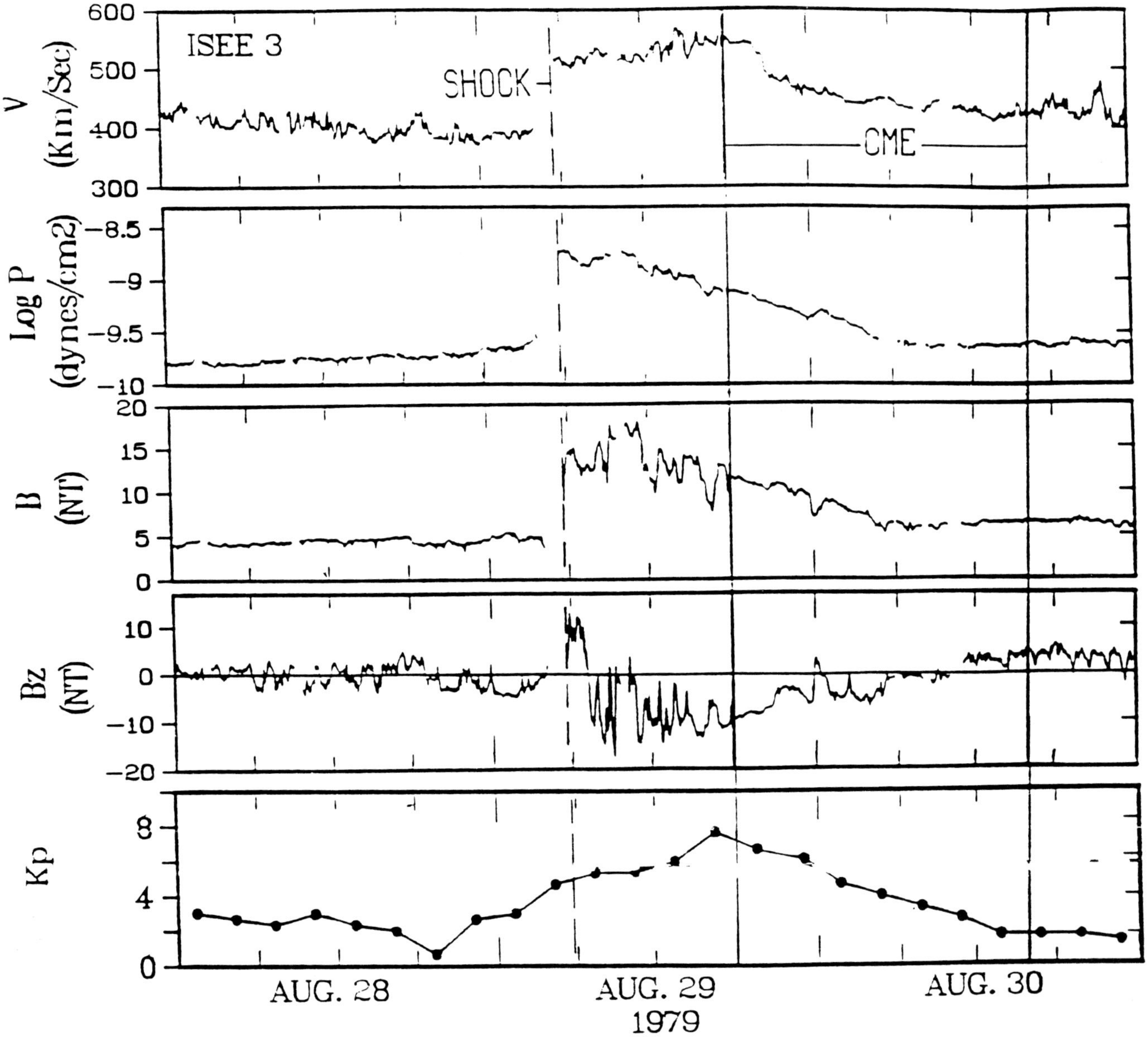

Fig. 6. Relation between interplanetary parameters and magnetic index [*Gosling et al.*, 1990].

From these data the solar wind speed and IMF magnitude and direction are derived. The interaction between these quantities in the magnetosphere constitutes a generator. Increasing solar wind increases the power output of this generator. An increased power causes geomagnetic-ionospheric storms. The power can be expressed as:

Power (Megawatts) = $20VB^2\sin^2(\Theta/2)$,

where

V = Solar wind speed in km/s;
B = Solar wind magnetic field magnitude in nT;
Θ = Solar wind magnetic field polar angle.

This expression requires knowledge of V, B, and Θ, for each event. The following characteristics of the storm in the near-Earth environment are derived from the input power as a function of time: AE, Dst, Kp and auroral oval size.

Given the importance of magnetic storms and substorms in the magnetosphere environment, many suggestions for predicting their occurrence

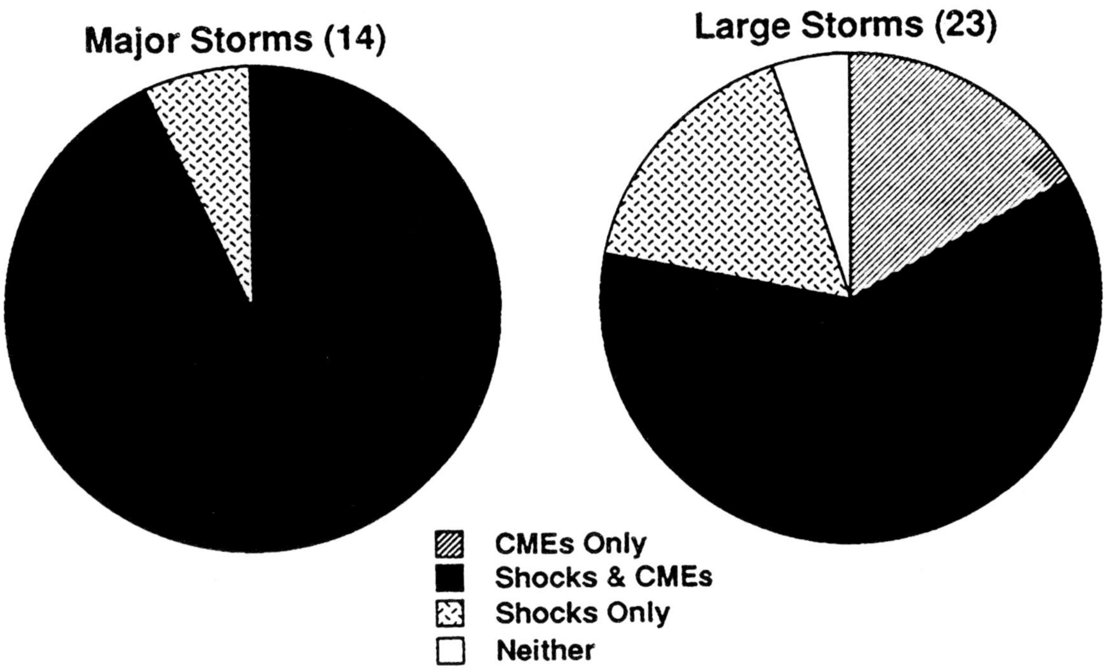

Fig. 7. Pie charts illustrating the association of major and large geomagnetic storms with Earth passage of shock disturbances and coronal mass ejection events (CMEs) during the interval August 15, 1978 - October 17, 1982, spanning the last solar maximum. The numbers in parentheses indicate the number of storms in each category. [*Gosling et al.*, 1990].

have been considered [*Baker et al.*, 1979; *Baker*, 1990; *Baker et al.*, 1990; *Crooker et al.*, 1992; *Holzer and Slavin*, 1982, 1983; *Kamide*, 1983; *Klimas et al.*, 1991; *Lam*, 1987; *and Marubashi*, 1989]. This has included the study of systematic changes in the upstream solar wind conditions [*Rodger*, 1990] and changes in the internal magnetotail configuration [*Baker et al.*, 1979]. The Solar Terrestrial Predictions Workshop in Leura, Australia, summarized requirements for improved predictions [*Rodger et al.*, 1990].

Bargatze et al. [1985] studied the temporal relationship between geomagnetic activity and the solar wind by examining results from many different time intervals. These authors studied the temporal response of the magnetosphere to changes in the solar wind energy input. They employed the linear prediction filtering (LPF) method in which the response of the Earth's magnetosphere to variations in the solar wind input is embedded in an impulse response filter function [e.g., *Wiener*, 1949]. *Bargatze et al.* [1985] used the AL index as a measure of substorm activity and used the product VB_z as the measure of solar wind input. Here V is the solar wind speed and B_z is the southward IMF component.

The LPF method assumes that the convolution of the input parameter over a range of lag times with the impulse response function produces the output parameter.

This essentially linear analysis cannot be applied directly to the relation between AL and VB_z, because the relationship is different at high and low values of geomagnetic activity. Accordingly, *Bargatze et al.* ranked about 30 intervals (each of several-day duration) by level of geomagnetic activity, and then computed the filters relating AL and VB_z. *Bargatze et al.* [1985] concluded that there were two dominant peaks in their computed filters at low and medium geomagnetic activity, the first at a time lag of approximately 20 min and the second at approximately 1 hr (See Figure 8). They ascribed the first peak to the action of the driven component of magnetospheric response [e.g., *Perreault and Akasofu*, 1978] and the second to the energy loading-unloading component [*Baker et al.*, 1984]. At high values of geomagnetic activity the 20-min peak predominates.

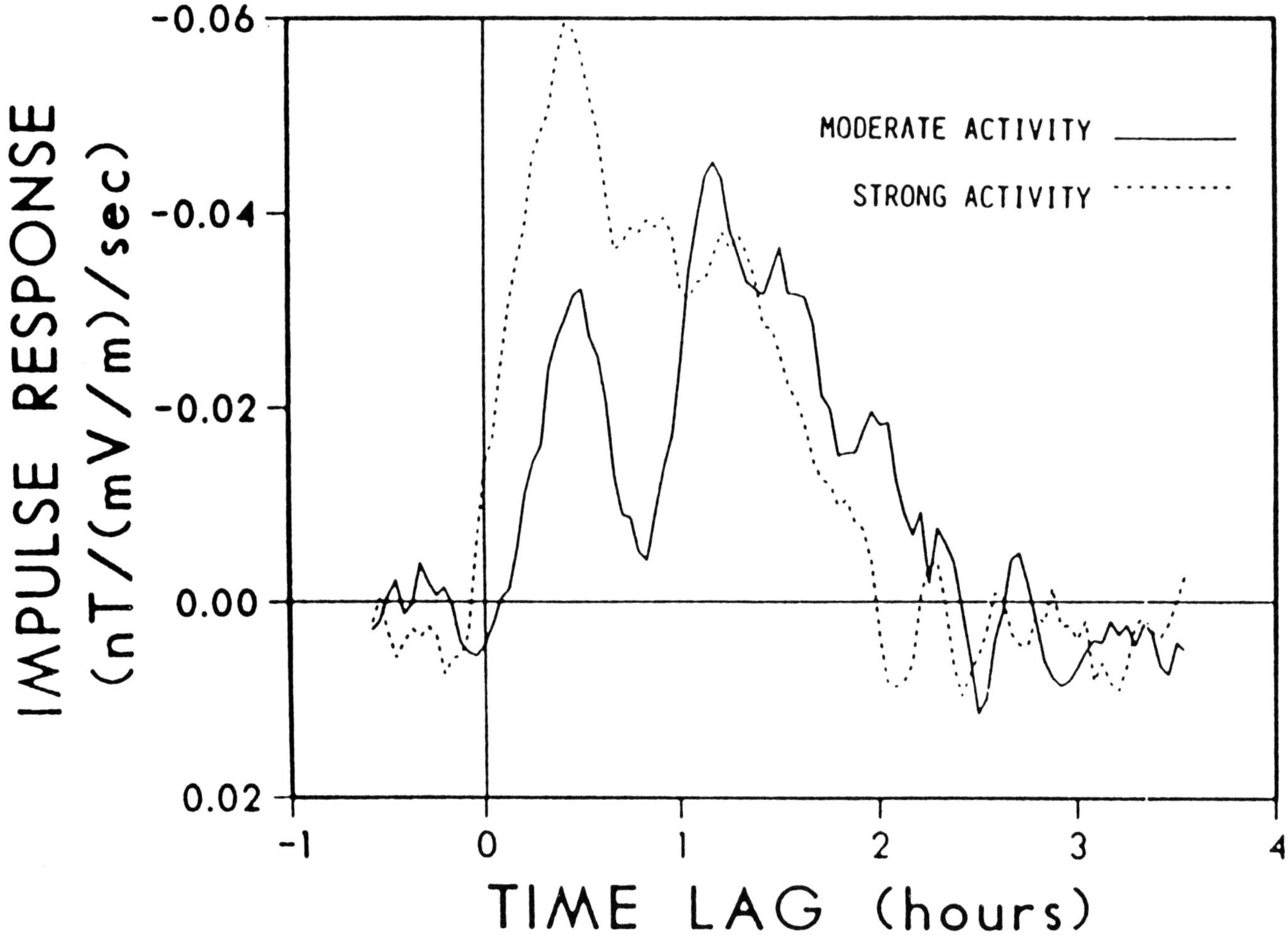

Fig. 8. Impulse response filters relating VB_z and the AL index for moderate and strong geomagnetic activity [*Bargatze et al.*, 1985].

The interpretation of these results is that the shorter 20-min time lag is a measure of the time scale over which the magnetosphere adjusts so that its direct dissipation rate is appropriate to the new solar-wind energy input rate: it is a measure of the dynamic response of the magnetosphere and it is associated with direct dissipation. The longer (1 hr) lag is a measure of the time necessary for the loading of energy in the magnetotail plus the time necessary for its unloading at expansion phase onset: it is the period of the loading-unloading cycle [*Baker*, 1990].

It has been increasingly recognized that empirical, statistical or morphological models do not lead to a realistic representation of the highly variable space environment. Physical models are needed which address all the processes operating in each region as well as their coupling to the solar wind.

DETERMINISTIC MODELS/FORECASTS

For over a decade, the Geophysics Directorate (GP) of the Phillips Laboratory (PL) has been engaged in the development of physics based models and forecasting of the space environment. Its scientists work closely with the Air Weather Service, the Air Force agency responsible for operational forecasts. The total effort involves the development of deterministic models and codes of the magnetosphere, ionosphere, thermosphere and the solar wind regions which are subsequently converted to operational versions. The physical coupling among the models is being built in each stage of development. The coupled models, taken together, represent a major step in our efforts to build a near-earth space weather specification and prediction system. In the operational version the models are interconnected through a smart executive system which can be readily accessed by Space Forecast Center operators at Falcon AF Station, CO.

The magnetospheric specification and forecast models (MSFM) will be discussed first to illustrate each model development process. The MSFM is based on the Rice convection model [*Harel et al.*, 1981a,b] and is designed to provide physically based deterministic predictions of geosynchronous particle fluxes, auroral precipitation, ionospheric current and Joule heating in the ionosphere on a time scale of a few hours. The model is based on the physical processes governing the motion of magnetospheric plasma and its coupling to the ionosphere in the inner and/or middle magnetosphere. The models are data driven and require a suite of input data that can include Kp, Dst, DMSP particle and field data, geosynchronous particle and field data, as well as solar wind plasma and interplanetary magnetic field (IMF) data, when available. The models are deterministic and predict the short term future according to the laws determining the temporal evolution of the system and the appropriate initial and boundary conditions. A flow diagram for the Rice convection model is given in Figure 9 [*Erickson et al.*, 1991].

There are two uses to which the models can be put: specification and forecasting. Specification implies that the models cannot be run in real time but can be used only in a post-analysis mode of operation. This mode is useful for analysis, for example, of spacecraft failures that are suspected to be of environmental origin. Forecasting means that the model can be used to predict events that can cause such failures [*Heinemann*, 1990].

An illustration of results from the operation of the MSFM is given in Figure 10. Inputs to the model are shown in the bottom panels. They include the ground-based geomagnetic indices Kp and Dst, the location of the equatorward edge of the auroral zone (Eq E) and the polar cap potential drop (PCP) obtained from DMSP satellite data, and the energy density of magnetospheric plasma measured at geosynchronous orbit. Given these inputs, the model computes electron particle fluxes

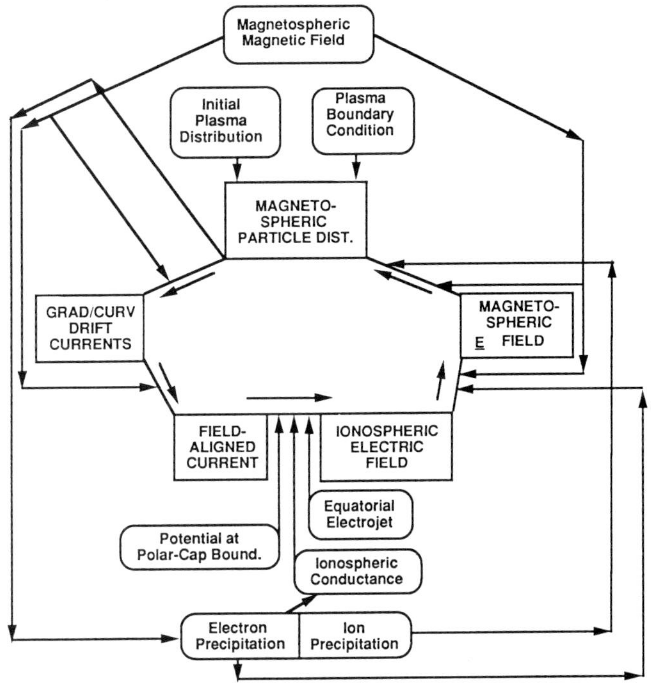

Fig. 9. The logical configuration of the Rice Convection Model (RCM) with the addition of Harang discontinuity physics [*Erickson et al.*, 1991].

in the near-Earth magnetosphere. The outputs for this input data set are shown in the top panels as contours of particle fluxes in the equatorial plane. A simple mapping along magnetic field lines (not shown) then gives the fluxes in all of the three-dimensional space in the inner magnetosphere. Separate application routines then compute from the global simulation output the fluxes along particular spacecraft trajectories. This kind of output can inform the satellite operator of periods along the orbit when satellite hazards can be expected. A similar model is being developed for the ionosphere based on the models of *Schunk* [1988], *Roble et al.* [1988] and *Schunk and Sojka* [1988].

The neutral atmosphere or thermosphere model referred to as VSH, is based on the model of *Killeen et al.* [1987]. The structure of the VSH model is given in Figure 11.

The interconnections between these models, the input data to operate them, and their utilization at the Space Forecast Center are illustrated in Figure 12. In the lower left the magnetosphere, ionosphere, thermosphere and solar wind models

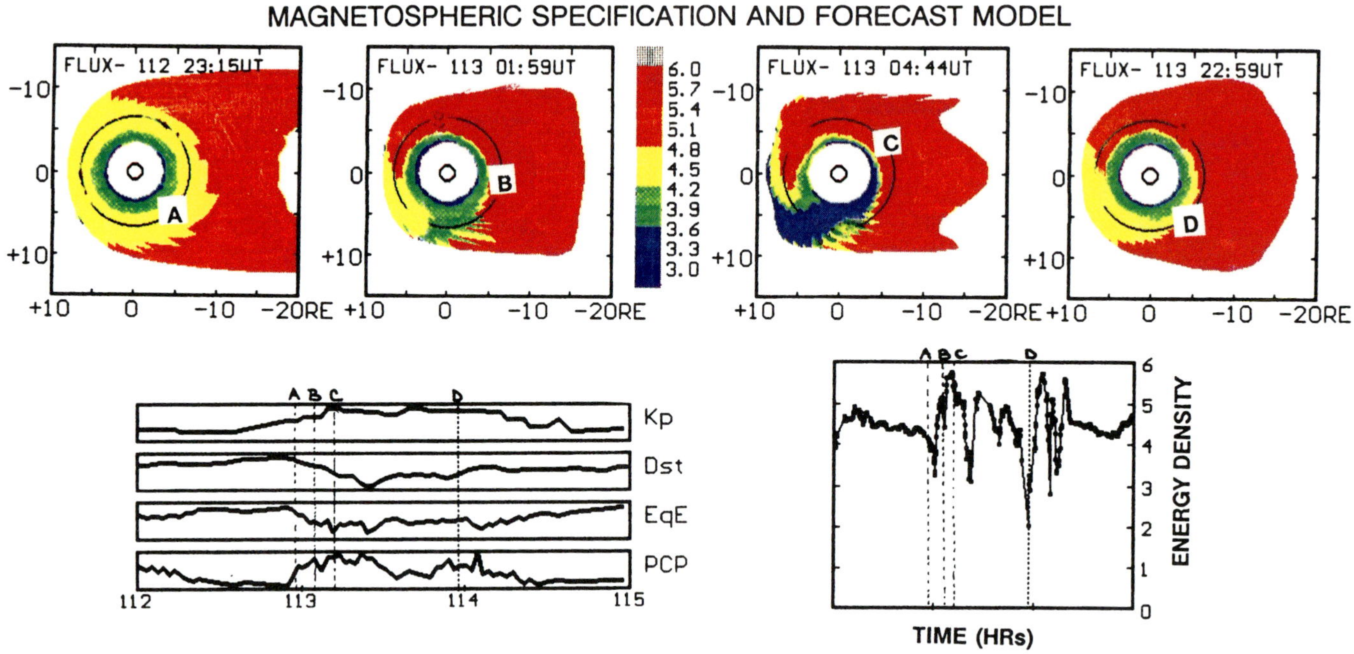

Fig. 10. Illustration of results from the magneto- spheric specification and forecast model.

are shown as inputs to a smart executive program, the Integrated Space Environment Model (ISEM). The acronyms of the models are defined below in Table 3.

In the upper left, data sources, such as DMSP, WIND, geosynchronous satellites, etc., are transmitted through the Smart Executive System. The operator at the Space Forecast Center can request information about a certain region of space at a specific time. Once the operator receives a response to his query, he is in a position to provide the satellite operator with information on the probability of radiation hazards, spacecraft charging, microelectronic upsets, satellite drag predictions, communication outages, etc., and to provide information for the analysis of specific anomalies.

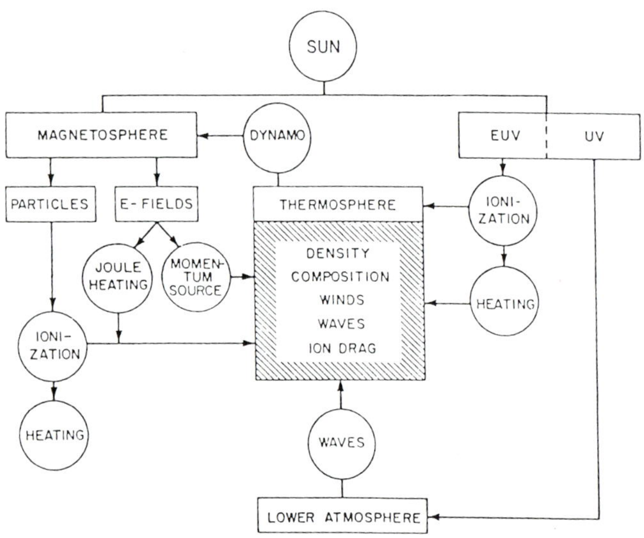

Fig. 11. Structure of the thermosphere model.

DATA REQUIREMENTS

Measurements of the space environment parameters obtained from many U.S. and European Satellites such as Skylab, Solar Maximum Mission, ISEE, DMSP, GOES, Helios, etc., have provided the basis for the development of realistic physical models. Plans to continue space measurements from DMSP, GOES, the NASA WIND and POLAR spacecraft, etc. will ensure that the models will continue to advance. It must be pointed out, however, that the data from these satellites with the exception of GOES and DMSP will not be provided in real time. Further, the recent findings that large geomagnetic storms originate with CMEs characterized by a southward IMF point to the need for real time IMF data and measurements of the size and speeds of earthward-moving CMEs. The only planned satellites from which the

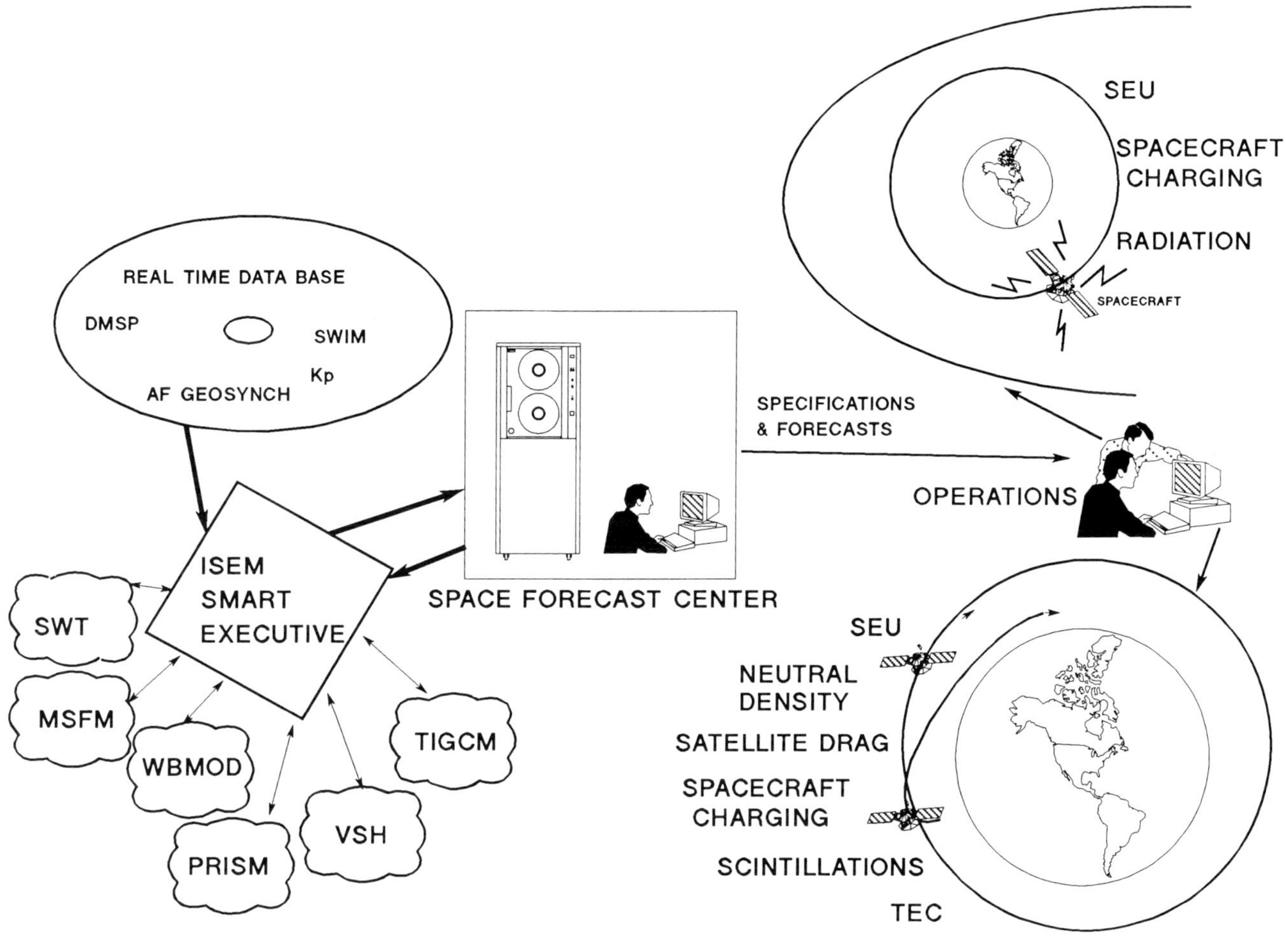

Fig. 12. Interconnection between models, the input data to operate them, and their utilization at the Space Forecast Center are illustrated in Figure 12. In the lower left the magnetosphere, ionosphere, thermosphere and solar wind models are shown as inputs to a smart executive program, the Integrated Space Environment Model (ISEM).

required interplanetary data are planned are the NASA/WIND spacecraft scheduled for launch in 1993, and the ACE spacecraft planned for 1996. The Air Force has negotiated with NASA to obtain real time interplanetary data from the WIND spacecraft. This will only partially satisfy the requirement for interplanetary real time data. Thirty- to sixty-min alerts can be obtained of hazardous conditions (such as satellite charging and high-energy-particle fluxes) which impact spacecraft operations. These data will also be used as inputs to global operational models of the space environment.

For many applications, such as satellite launch decisions, longer term, 2-3 day, forecasts are required. An instrument, the Solar Mass Ejection Imager (SMEI), is being developed by the Air Force Phillips Laboratory based on results from a photometer flown on the Helios spacecraft to satisfy this need. *Webb and Jackson* [1990] have shown that the Helios photometer, designed to observe zodiacal light, can also identify and characterize CMEs. SMEI will measure the Thompson scattering from high speed electrons in CMEs. The flight of a coronagraph on a low-altitude sun-synchronous satellite, planned for the late 1990s, would provide 2-3 day alerts which would also be used as an input for the physical models to predict particle and field disturbances throughout the space environment.

TABLE 3
MODELS AND THEIR ACRONYMS

ACRONYMS	MODEL
SWT	Solar Wind Transport Model
MSFM	Magnetosphere Specification and Forecast Model
WBMOD	DNA Wide-Band Satellite Global Scintillation Model
PRISM	Parameterized Real Time Ionosphere Specification Model
VSH	Vector Spherical Harmonic Model of Thermosphere
TIGCM	Thermosphere Ionosphere Global Circulation Model

TABLE 4
PREDICTION OF GEOMAGNETIC DISTURBANCES

REQTS	SHORT TERM ~ 1 HOUR	LONG TERM > 1 DAY
GLOBAL MODELS	X	X
SPACE DATA		
L_1(Solar Wind, IMF)	X	X
PLASMA	X	X
ELECTRIC/ MAGNETIC FIELDS	X	X
CORONAL MASS EJECTIONS		X
GROUND MEASUREMENTS		
SOLAR MAGNETIC FIELD		X

SUMMARY

Significant progress has been made in space science and technology during the past four decades. This has come about as a result of extensive measurement, modeling and simulation programs. Recognition of the important role of coronal mass ejections, associated shocks and the southward turning of the interplanetary magnetic field B_z component in the development of geomagnetic storms is encouraging the solar interplanetary scientist to focus on new aspects of solar activity and its consequences in near earth-space. Global model development/prediction schemes for the magnetosphere, ionosphere, thermosphere, and solar wind are progressing rapidly as a result of focused efforts within the space science community. There continues to be a urgent need for space based measurements of the solar wind and interplanetary magnetic field for the forecasting of geomagnetic storms. A summary of the models and data required for short and longer term prediction is given in Table 4.

Acknowledgments. The authors are grateful for helpful discussions with E. Cliver, S. Kahler, M. Heinemann, W.J. Burke and E.G. Mullen during preparation of this report. We also thank M. Outwater for her assistance in the preparation of the manuscript.

REFERENCES

Akasofu, S.-I., and R. P. Lepping, Interplanetary magnetic field and magnetospheric substorms, *Planet. Space Sci., 25,* 895-897, 1977.

Akasofu, S.-I., Study of individual geomagnetic storms in terms of the solar wind, *Planet. Space Sci., 28,* 933-944, 1980a.

Akasofu, S.-I., The solar wind-magnetosphere energy coupling and magnetospheric disturbances, *Planet. Space Sci., 28,* 495-509, 1980b.

Akasofu, S.-I. and Y. Kamide, Recent progress in studies of magnetospheric storms and substorms, *J. Geophys. Res., 32*, 585-615, 1980.

Akasofu, S.-I., Prediction of development of geomagnetic storms using the solar wind-magnetosphere energy coupling function-epsilon, *Planet. Space Sci., 29,* 1151- 1158, 1981.

Akasofu, S.-I., and C.F. Fry, A 1st generation

numerical geomagnetic storm prediction scheme, *Planet. Space Sci., 34*, 77-92, 1986.

Akasofu, S.-I., T. Oki, C. Olmsted and T. Saito, Quantitative forecasting of the 27-day recurrent magnetic activity, *Planet. Space Sci., 36*, 1133-1147, 1988.

Alfven, H., *Cosmical Electrodynamics*, Clarendon Press, Oxford, England 1950.

Arnoldy, R.L., Signature in the interplanetary medium for substorms, *J. Geophys. Res., 76*, 5189-5201, 1971.

Baker, D.N., P.R. Higbie, E.W. Hones, Jr., and R.D. Belian, The use of > 30 keV electron anisotropies of 6.6 r_E to predict magnetosphere substorms in *Solar-Terrestrial Predictions Proceedings, Vol. 4* (R.F. Donnelly, Ed.) p. B- 12, NOAA, Boulder, CO, 1979.

Baker, D.N., S.-I. Akasofu, W. Baumjohann, J.W. Bieber, D.H. Fairfield, E.W. Hones, Jr., B. Mauk, R.L. McPherron and T.E. Moore, Substorms in the magnetosphere, *Solar Terrestrial Physics-Present and Future, NASA Pub. 1120, Chapter 8*, Wash D.C. 1984.

Baker, D.N., The magnetosphere environment, *Solar-Terrestrial Predictions: Proceedings of a Workshop at Leura, Australia, October 16-20, 1989*, Vol 1, 26-45, 1990.

Baker, D.N., R.L. McPherron, T.E. Cayton and R.W. Klebesadel, Linear prediction filter analysis of relativistic electron-properties at 6.6 Re, *J. Geophys. Res., 95*, 5133-5140, 1990.

Bargatze, L.F., D.N. Baker, R.L. McPherron, and E.W. Hones, Magnetospheric impulse-response for many levels of geomagnetic activity, *J. Geophys. Res., 90*, 6387-6394, 1985.

Burlaga, L.F., Magnetic clouds and force-free fields with constant alpha, *J. Geophys. Res. 93*, 7217-7224, 1988.

Burton, R.K., R.L. McPherron, and C.T. Russell, An empirical relationship between interplanetary conditions and Dst, *J. Geophys. Res. 80*, 4204 -4214, 1975.

Cane, H.V., The large-scale structure of flare-associated interplanetary shocks, *J. Geophys. Res., 93*, 1-6, 1988.

Craven, J.D. and L.A. Frank, Latitudinal motions of the aurora during substorms, *J. Geophys. Res. 92*, 4565-4573, 1987.

Craven, J.D. and L.A. Frank, Diagnosis of auroral dynamics using global auroral imaging with emphasis on large-scale evolution, *Auroral Physics*, Cambridge University Press, Cambridge (England), 1989.

Crooker, N.U., E.W. Cliver and B.T. Tsurutani, The semiannual variation of great geomagnetic storms and the postshock Russell-McPherron effect preceding coronal mass ejecta, *Geophys. Res. Lett., 19*, 429-432, 1992.

Dungey, J.W., Interplanetary magnetic field and the auroral zones, *Phys. Rev. Lett., 6*, 47, 1961.

Erickson, G.M., R.W. Spiro, and R.A. Wolf, The physics of the Harang discontinuity, *J. Geophys. Res. 96*, 1633-1645, 1991.

Fairfield, D.H. and L.J. Cahill, Jr., Transition region magnetic field and polar magnetic disturbances, *J. Geophys. Res, 71*, 155-169, (1966).

Gonzalez, W.D. and B.T. Tsurutani, Criteria of interplanetary parameters causing intense magnetic storms, (D_{st}< -100 nT), *Planet. Space Sci., 35*, 1101-1109, 1987.

Gorney, D.J., Solar cycle effects on near-earth plasmas and space systems, *J. Spacecraft, 26*, 428-438, 1989.

Gosling, J.T. and D.J. McComas, Field line draping about fast coronal mass ejecta - a source of strong out-of-the-ecliptic interplanetary magnetic fields, *Geophys. Res. Lett., 14*, 355-358, 1987.

Gosling, J.T., D.N. Baker, S.J. Bame, W.C. Feldman, E.J Smith, and R.D. Zwickl, Bidirectional solar wind electron heat flux events, *J. Geophys. Res.,92*, 8519-8535, 1987.

Gosling, J.T., S.J. Bame, D.J. McComas, and J.L. Phillips, Coronal mass ejections and large geomagnetic storms, *Geophys. Res. Lett., 17*, 901-904, 1990.

Gosling, J.T., D.J. McComas, J.L. Phillips, and S.J. Bame, Geomagnetic activity associated with earth-passage of interplanetary shock disturbances and coronal mass ejections, *J. Geophys. Res. 96*, 7831- 7839, 1991.

Harel, M., R.A. Wolf, P.H. Reiff, and R.W. Spiro, Quantitative simulation of a magnetospheric substorm (1. Model logic and overview), *J. Geophys. Res., 86*, 2217-2241, 1981a.

Harel, M., R.A. Wolf, R.W. Spiro, P.H. Reiff, and C.-K. Chen, Quantitative simulation of a magnetospheric substorm (2. Comparison with observations), *J. Geophys. Res., 86*, 2242-2260, 1981b.

Hairston, M.R. and R.A. Heelis, Model of the

high-latitude ionosphere convection pattern during southward interplanetary magnetic field using DE 2 data, *J. Geophys. Res.*, *95*, 2333-2344, 1990.

Harrison, R.A., Solar coronal mass ejections and flares, *Astron. Astrophys. 162*, 283-291, 1986.

Harrison, R.A., E. Hildner, A.J. Hundhausen, D.G. Sime, and G.M. Simnett, The launch of solar coronal mass ejections: Results from the Coronal Mass Ejection Onset Program, *J. Geophys. Res. 95*, 917-937, 1990.

Heinemann, M., Magnetospheric specification and forecasting models, *Solar-Terrestrial Predictions: Proceedings of a Workshop at Leura, Australia, October 16-20, 1989*, Vol. 2, 62-64, NOAA, Boulder CO, 1990.

Heppner, J.P. and N.C. Maynard, Empirical high-latitude electric field models, *J. Geophys. Res, 92*, 4467-4489, 1987.

Hoffman, R.A., M. Sugiura, and N.C. Maynard, Current carriers for the field-aligned current system, *Adv. Space Res.*, *5*, 109-126, 1985.

Holzer, R.E., and J.A. Slavin, An evaluation of 3 predictors of geomagnetic activity, *J. Geophys. Res.*, *87*, 2558-2562, 1982.

Holzer, R.E., and J.A. Slavin, An evaluation of 3 predictors of geomagnetic activity, - Reply, *J. Geophys. Res.*, *88*, 4955-4958, 1983.

Howard R.A., N.R. Sheeley Jr., M.J. Koomen and D.J. Michels, Coronal mass ejections: 1979-1981, *J. Geophys. Res. 90*, 8173-8191, 1985.

Hundhausen, A.J., The origin and propagation of coronal mass ejections, *Proceedings of the Sixth International Solar Wind Conference*, Sponsored by HAO of NCAR, 181-214, Aug. 1987.

Kahler, S.W., Solar flares and coronal mass ejections, *Ann. Rev. Astron. Astrophys. 30*, 113-141, 1992.

Kahler, S.W., Coronal mass ejections, *Rev. Geophys. 25*, 663-675, 1987.

Kamide, Y., An evaluation of 3 predictors of geomagnetic activity, Comment, *J. Geophys. Res.*, *88*, 4953-4954, 1983.

Killeen, T.L., R.G. Roble and N.W. Spencer, A computer model of global thermospheric winds and temperatures, *Adv. Space Res.*, *7*, 207-215, 1987.

Klimas, A.J., D.N. Baker, and D.A. Roberts, Linear prediction filters for linear and nonlinear modeled geomagnetic activity, *Geophys. Res. Lett.*, *18*, 1635-1638, 1991.

Lam, H.L., Forecasts of geomagnetic activity in Canada by linear prediction filtering, *J. Geomagn. Geoelectr. 39*, 535-542, 1987.

Lepping, R.P., J.A. Jones, and L.F. Burlaga, Magnetic field structure of interplanetary magnetic clouds at 1 AU, *J. Geophys. Res. 95*, 11957-11965, 1990.

Marubashi, K., The space weather forecast program, *Space Sci. Rev.*, *51*, 197-214, 1989.

Meng, C.-I., B. Tsurutani, K. Kawaski and S.-I. Akasofu, Cross-correlation analysis of the AE index and the interplanetary magnetic field B_z component, *J. Geophys. Res.*, *78*, 617-629, 1973.

Mullen, E.G. and S. Gussenhoven, Results of space experiments: CRRES, (to be published in the *Proceedings of the Behavior of Systems in the Space Environment*, NATO Adv. Study Inst. Pitlochry, Scotland, 1991).

Perreault, P., and S.-I. Akasofu, A study of geomagnetic storms, *Geophys. J. RAS, 54*, 547-573, 1978.

Roble, R.G., E.C. Ridley, A.D. Richmond, and R.E. Dickinson, A coupled thermosphere/ ionosphere general circulation model, *Geophys. Res. Lett.*, *15*, 1325-1328, 1988.

Rodger, A.S., J. Hruska, J.A. Joselyn and K. Marabashi, Report of the geomagnetic and space environment working group, *Solar Terrestrial Predictions, Proceedings of a Workshop at Leura, Australia, 16-20 Oct 89*, Vol. 1, 7-17, NOAA, Boulder, CO 1990.

Rodger, A.S., Recent progress in understanding the magnetospheric environment, *Solar Terrestrial Predictions, Proceedings of a Workshop at Leura, Australia Workshop, 16-20 Oct 89*, Vol. 1, 183-196, NOAA, Boulder, CO 1990.

Rostoker, G. and C.G. Falthammar, Relationship between changes in the interplanetary magnetic field and variations in the magnetic field at the earth's surface, *J.Geophys.Res.*, *23*, 5853-5863, 1967.

Schunk, R.W., A mathematical model of the middle and high latitude ionosphere, *PAGEOPH , 127*, 255-303, 1988.

Schunk, R.W. and J.J. Sojka, Ionospheric climate and weather modeling, *EOS, 69, 11*, 153-168, 1988.

Sheeley, N.R., Jr., R.A. Howard, M.J. Koomen, D.J. Michels, R. Schwenn, K.H. Muhlhauser, and H. Rosenbauer, Coronal mass ejections and interplanetary shocks, *J. Geophys. Res.*, *90*, 163-175, 1985.

Snyder, C.W., M. Neugebauer, and U.R. Rao, The solar wind velocity and its correlation with cosmic-ray variations and with solar and geomagnetic activity, *J. Geophys. Res., 68,* 6361-6370, 1963.

Tsurutani, B.T., W.D. Gonzalez, F. Tang, and Y.T. Lee, Great magnetic storms, *J. Geophys. Res., 19*, 73-77, 1992.

Tsurutani, B.T., F. Tang, E.J. Smith, S.-I. Akasofu, and W.D. Gonzales, Origin of interplanetary southward magnetic-fields responsible for major magnetic storms near solar maximum (1978-1979), *J. Geophys. Res., 93*, 8519-8531, 1988.

Wagner, W.W., Coronal mass ejections, *Ann. Rev. Astron. Astrophys., 22*, 267-289, 1984.

Webb, D.F., and B.V. Jackson, The identification and characteristics of solar mass ejections observed in the heliosphere by the Helios 2 photometers, *J. Geophys. Res., 95*, 20641-20661, 1990.

Webb, D.F., The solar cycle variation of the rates of CMEs and related activity, *Adv. Space Res., 11,* 37-40, 1991.

Wiener, N. (Ed.), *Extrapolation, interpolation and smoothing of stationary time series with engineering applications*, John Wiley, New York, 1949.